FUNDAMENTALS OF WIRELESS SENSOR NETWORKS

Wiley Series on Wireless Communications and Mobile Computing

Series Editors: Dr Xuemin (Sherman) Shen, *University of Waterloo, Canada*
                Dr Yi Pan, *Georgia State University, USA*

The "Wiley Series on Wireless Communications and Mobile Computing" is a series of comprehensive, practical and timely books on wireless communication and network systems. The series focuses on topics ranging from wireless communication and coding theory to wireless applications and pervasive computing. The books provide engineers and other technical professionals, researchers, educators, and advanced students in these fields with invaluable insight into the latest developments and cutting-edge research.

Other titles in the series:

FUNDAMENTALS OF WIRELESS SENSOR NETWORKS

THEORY AND PRACTICE

Waltenegus Dargie
Technical University of Dresden, Germany

Christian Poellabauer
University of Notre Dame, USA

A John Wiley and Sons, Ltd., Publication

This edition first published 2010
© 2010 John Wiley & Sons Ltd.

Registered office
John Wiley & Sons Ltd, The Atrium, Southern Gate, Chichester, West Sussex, PO19 8SQ, United Kingdom

For details of our global editorial offices, for customer services and for information about how to apply for permission to reuse the copyright material in this book please see our website at www.wiley.com.

Library of Congress Cataloging-in-Publication Data:

Dargie, Waltenegus.
  Fundamentals of wireless sensor networks : theory and practice / Waltenegus Dargie, Christian Poellabauer.
    p. cm.
  Includes index.
  ISBN 978-0-470-99765-9 (cloth)
  1. Wireless sensor networks. I. Poellabauer, Christian. II. Title.
  TK7872.D48D37 2010
  681'.2 – dc22

                                                                2010003984

A catalogue record for this book is available from the British Library.

ISBN 978-0-470-99765-9 (H/B)

Typeset in 10/12 Times by Laserwords Private Limited, Chennai, India.

To my wife, Kathy,
and my children, Joshua and Pheben

Waltenegus Dargie

To my wife, Rumana,
and my children, Adam and Maya

Christian Poellabauer

Contents

–

About the Series Editors

Xuemin (Sherman) Shen (M'97-SM'02) received the B.Sc degree in electrical engineering from Dalian Maritime University, China in 1982, and the M.Sc. and Ph.D. degrees (both in electrical engineering) from Rutgers University, New Jersey, USA, in 1987 and 1990 respectively. He is a Professor and University Research Chair, and the Associate Chair for Graduate Studies, Department of Electrical and Computer Engineering, University of Waterloo, Canada. His research focuses on mobility and resource management in interconnected wireless/wired networks, UWB wireless communications systems, wireless security, and ad hoc and sensor networks. He is a co-author of three books, and has published more than 300 papers and book chapters in wireless communications and networks, control and filtering. Dr. Shen serves as a Founding Area Editor for IEEE Transactions on Wireless Communications; Editor-in-Chief for Peer-to-Peer Networking and Application; Associate Editor for IEEE Transactions on Vehicular Technology; KICS/IEEE Journal of Communications and Networks, *Computer Networks*; ACM/Wireless Networks; and Wireless Communications and Mobile Computing (Wiley), etc. He has also served as Guest Editor for IEEE JSAC, IEEE Wireless Communications, and IEEE Communications Magazine. Dr. Shen received the Excellent Graduate Supervision Award in 2006, and the Outstanding Performance Award in 2004 from the University of Waterloo, the Premier's Research Excellence Award (PREA) in 2003 from the Province of Ontario, Canada, and the Distinguished Performance Award in 2002 from the Faculty of Engineering, University of Waterloo. Dr. Shen is a registered Professional Engineer of Ontario, Canada.

Dr. Yi Pan is the Chair and a Professor in the Department of Computer Science at Georgia State University, USA. Dr. Pan received his B.Eng. and M.Eng. degrees in computer engineering from Tsinghua University, China, in 1982 and 1984, respectively, and his Ph.D. degree in computer science from the University of Pittsburgh, USA, in 1991. Dr. Pan's research interests include parallel and distributed computing, optical networks, wireless networks, and bioinformatics. Dr. Pan has published more than 100 journal papers with over 30 papers published in various IEEE journals. In addition, he has published over 130 papers in refereed conferences (including IPDPS, ICPP, ICDCS, INFOCOM, and GLOBECOM). He has also co-edited over 30 books. Dr. Pan has served as an editor-in-chief or an editorial board member for 15 journals including five IEEE Transactions and has organized many international conferences and workshops. Dr. Pan has delivered over 10 keynote speeches at many international conferences. Dr. Pan is an IEEE Distinguished Speaker (2000–2002), a Yamacraw Distinguished Speaker (2002), and a Shell Oil Colloquium Speaker (2002). He is listed in Men of Achievement, Who's Who in America, Who's Who in American Education, Who's Who in Computational Science and Engineering, and Who's Who of Asian Americans.

Preface

Rapid advances in the areas of sensor design, information technologies, and wireless networks have paved the way for the proliferation of wireless sensor networks. These networks have the potential to interface the physical world with the virtual (computing) world on an unprecedented scale and provide practical usefulness in developing a large number of applications, including the protection of civil infrastructures, habitat monitoring, precision agriculture, toxic gas detection, supply chain management, and health care. However, the design of wireless sensor networks introduces formidable challenges, since the required body of knowledge encompasses a whole range of topics in the field of electrical and computer engineering, as well as computer science.

Wireless sensor networks are currently being offered as a subject at advanced undergraduate and graduate levels at many universities around the world. Moreover, they are the focus of countless graduate theses and student projects. Therefore, this book is primarily written as a textbook aimed at students of engineering and computer science. It provides an introduction into the fundamental concepts and building blocks of wireless sensor network design. An attempt has been made to maintain a balance between theory and practice, as well as established practices and the latest developments. At the end of each chapter, a number of practical questions and exercises are given to help the students to assess their understanding of the main concepts and arguments presented in the chapter. Furthermore, the chapters and parts of the book are sufficiently modular to provide flexibility in course design.

The book will also be useful to the professional interested in this field. It is suitable for self-study and can serve as an essential reference. For such a reader, the material can be viewed as a tutorial in the basic concepts and surveys of recent research results and technological developments.

Structure of the Book

This book provides an introduction to the fundamental concepts and principles of wireless sensor networks (WSNs) and a survey of protocols, algorithms, and technologies at different layers of a sensor system, including the network protocol stack, middleware, and application level.

The text is broken into three parts. In Part One, **Introduction**, Chapter 1 provides an overview of WSN applications, sensor nodes, and basic system structure. Chapter 2 continues the introduction into the WSN domain by providing an overview of representative sensor network applications. Chapter 3 presents different node architectures and discusses in detail the sensing and processing subsystems as well as communication interfaces. Moreover, it provides several examples of representative prototype implementations. Chapter 4 describes functional and nonfunctional aspects of operating systems and provides a survey of state-of-the-art examples.

Part Two, **Basic Architectural Framework**, provides a detailed discussion of protocols and algorithms used at different network protocol layers in sensor systems. The design choices at these layers significantly impact the operation and resource efficiency of sensor nodes and networks. Chapter 5 begins this discussion with an introduction into physical layer architectures and concepts. Since the wireless medium is shared between many sensor nodes, MAC-layer protocols are required to arbitrate access to the wireless channels. MAC-layer solutions are discussed in Chapter 6. Chapter 7 discusses multi-hop communications in WSNs and the associated challenges. It also surveys existing and proposed routing protocols.

Part Three, **Node and Network Management**, discusses several additional techniques and presents solutions for a variety of challenges. Chapter 8 begins the discussion with an overview of power management techniques for wireless sensor networks. When multiple sensor nodes observe the same event in the physical world, it is important to correctly correlate these observations from the different sensors. This requires the clocks of the sensor nodes to be synchronized with each other. Synchronized clocks are also required by a variety of protocols and algorithms, e.g., many MAC protocols rely on accurate timing to ensure that no two nodes transmit packets at the same time. Therefore, Chapter 9 introduces the concept of time synchronization and provides an overview of several synchronization strategies. For many sensor network applications, it is essential that sensor nodes estimate their own position, either using absolute coordinates (e.g., using GPS) or relative to other nodes or landmarks in the environment. Chapter 10 presents a variety of localization strategies and compares their tradeoffs. Wireless sensor networks pose several security challenges due to the nature of many sensor applications (military, emergency response) and the unique characteristics of sensor networks (e.g., scale and unattended operation). Therefore, security challenges and defenses against attacks on sensor networks are discussed in Chapter 11. Finally, Chapter 12 concludes the book with a description of development environments and programming techniques for sensor networks, including an overview of frequently used sensor network simulators.

Part One

Introduction

1

Motivation for a Network of Wireless Sensor Nodes

Sensors link the physical with the digital world by capturing and revealing real-world phenomena and converting these into a form that can be processed, stored, and acted upon. Integrated into numerous devices, machines, and environments, sensors provide a tremendous societal benefit. They can help to avoid catastrophic infrastructure failures, conserve precious natural resources, increase productivity, enhance security, and enable new applications such as context-aware systems and smart home technologies. The phenomenal advances in technologies such as very large scale integration (VLSI), microelectromechanical systems (MEMS), and wireless communications further contribute to the widespread use of distributed sensor systems. For example, the impressive developments in semiconductor technologies continue to produce microprocessors with increasing processing capacities, while at the same time shrinking in size. The miniaturization of computing and sensing technologies enables the development of tiny, low-power, and inexpensive sensors, actuators, and controllers. Further, embedded computing systems (i.e., systems that typically interact closely with the physical world and are designed to perform only a limited number of dedicated functions) continue to find application in an increasing number of areas. While defense and aerospace systems still dominate the market, there is an increasing focus on systems to monitor and protect civil infrastructure (such as bridges and tunnels), the national power grid, and pipeline infrastructure. Networks of hundreds of sensor nodes are already being used to monitor large geographic areas for modeling and forecasting environmental pollution and flooding, collecting structural health information on bridges using vibration sensors, and controlling usage of water, fertilizers, and pesticides to improve crop health and quantity.

This book provides a thorough introduction to the fundamental aspects of *wireless sensor networks* (WSNs), covering both theoretical concepts and practical aspects of network technologies and protocols, operating systems, middleware, sensor programming, and security. The book is targeted at researchers, students, and practitioners alike, with the goal of helping them to gain an understanding of the challenges and promises of this exciting field. It has been written primarily as a textbook for graduate or advanced undergraduate courses in wireless sensor networks. Each chapter ends with a number of exercises and questions that will allow students to practice the described concepts and techniques. As the field of wireless sensor networks is based on numerous other domains, it is recommended that

Fundamentals of Wireless Sensor Networks: Theory and Practice   Waltenegus Dargie and Christian Poellabauer
© 2010 John Wiley & Sons, Ltd

students have taken courses such as networking and operating systems (or comparable courses) before they take a course on sensor networks. Also, some topics covered in this book (e.g., security) assume previous knowledge in other areas or require that an instructor provides an introduction into the basics of these areas before teaching these topics.

1.1 Definitions and Background

1.1.1 Sensing and Sensors

Sensing is a technique used to gather information about a physical object or process, including the occurrence of events (i.e., changes in state such as a drop in temperature or pressure). An object performing such a sensing task is called a *sensor*. For example, the human body is equipped with sensors that are able to capture optical information from the environment (eyes), acoustic information such as sounds (ears), and smells (nose). These are examples of *remote sensors*, that is, they do not need to touch the monitored object to gather information. From a technical perspective, a sensor is a device that translates parameters or events in the physical world into signals that can be measured and analyzed. Another commonly used term is *transducer*, which is often used to describe a device that converts energy from one form into another. A sensor, then, is a type of transducer that converts energy in the physical world into electrical energy that can be passed to a computing system or controller. An example of the steps performed in a sensing (or *data acquisition*) task is shown in Figure 1.1. Phenomena in the physical world (often referred to as *process*, *system*, or *plant*) are observed by a sensor device. The resulting electrical signals are often not ready for immediate processing, therefore they pass through a *signal conditioning* stage. Here, a variety of operations can be applied to the sensor signal to prepare it for further use. For example, signals often require amplification (or attenuation) to change the signal magnitude to better match the range of the following analog-to-digital conversion. Further, signal conditioning often applies *filters* to the signal to remove unwanted noise within certain frequency ranges (e.g., highpass filters can be used to remove 50 or 60 Hz noise picked up by surrounding power lines). After conditioning, the analog signal is transformed into a digital signal using an *analog-to-digital converter* (ADC). The signal is now available in a digital form and ready for further processing, storing, or visualization.

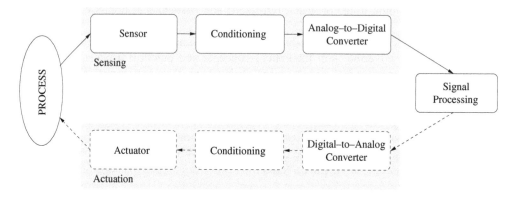

Figure 1.1   Data acquisition and actuation.

Many wireless sensor networks also include *actuators* which allow them to directly control the physical world. For example, an actuator can be a valve controlling the flow of hot water, a motor that opens or closes a door or window, or a pump that controls the amount of fuel injected into an engine. Such a *wireless sensor and actuator network* (WSAN) takes commands from the processing device (controller) and transforms these commands into input signals for the actuator, which then interacts with a physical process, thereby forming a closed control loop (also shown in Figure 1.1).

1.1.1.1 Sensor Classifications

Which sensors should be chosen for an application depends on the physical property to be monitored, for example, such properties include temperature, pressure, light, or humidity. Table 1.1 summarizes some common physical properties, including examples of sensing technologies that are used to capture them. Besides physical properties, the classification of sensors can be based on a variety of other methods, for example, whether they require an external power supply. If the sensors require external power, they are referred to as *active* sensors. That is, they must emit some kind of energy (e.g., microwaves, light, sound) to trigger a response or to detect a change in the energy of the transmitted signal. On the other hand, *passive* sensors detect energy in the environment and derive their power from this energy input – for example, passive infrared (PIR) sensors measure infrared light radiating from objects in the proximity.

The classification of sensors can also be based on the methods they apply and the electrical phenomena they utilize to convert physical properties into electrical signals. *Resistive* sensors rely on changes to a conductor's electrical resistivity, ρ, based on physical properties such as temperature. The resistance, R, of a conductor can be determined as:

$$R = \frac{l \times \rho}{A} \tag{1.1}$$

where l is the length of the conductor and A is the area of the cross-section. For example, the well-known Wheatstone bridge (Figure 1.2) is a simple circuit that can be used to convert a physical property into an observable electric effect. In this bridge, R_1, R_2, and R_3 are

Table 1.1 Classification and examples of sensors

Type	Examples
Temperature	Thermistors, thermocouples
Pressure	Pressure gauges, barometers, ionization gauges
Optical	Photodiodes, phototransistors, infrared sensors, CCD sensors
Acoustic	Piezoelectric resonators, microphones
Mechanical	Strain gauges, tactile sensors, capacitive diaphragms, piezoresistive cells
Motion, vibration	Accelerometers, gyroscopes, photo sensors
Flow	Anemometers, mass air flow sensors
Position	GPS, ultrasound-based sensors, infrared-based sensors, inclinometers
Electromagnetic	Hall-effect sensors, magnetometers
Chemical	pH sensors, electrochemical sensors, infrared gas sensors
Humidity	Capacitive and resistive sensors, hygrometers, MEMS-based humidity sensors
Radiation	Ionization detectors, Geiger–Mueller counters

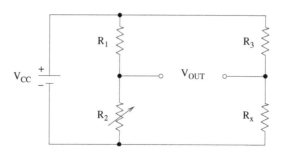

Figure 1.2 Wheatstone bridge circuit.

resistors of known resistance (where the resistance of R_2 is adjustable) and R_x is a resistor of unknown value. If the ratio R_2/R_1 is identical to the ratio R_x/R_3, the measured voltage V_{OUT} will be zero. However, if the resistance of R_x changes (e.g., due to changes in temperature), there will be an imbalance, which will be reflected by a change in voltage V_{OUT}. In general, the relationship between the measured voltage V_{OUT}, the resistors, and the supply voltage (V_{CC}) can be expressed as:

$$V_{OUT} = V_{CC} \times \left(\frac{R_x}{R_3 + R_x} - \frac{R_2}{R_1 + R_2} \right) \tag{1.2}$$

A similar principle can be applied to *capacitive* sensors, which can be used to measure motion, proximity, acceleration, pressure, electric fields, chemical compositions, and liquid depth. For example, in the parallel plate model, that is, a capacitor consisting of two parallel conductive plates separated by a dielectric with a certain permittivity ε, the capacitance is determined as:

$$C = \frac{\varepsilon \times A}{d} \tag{1.3}$$

where A is the plate area and d is the distance between the two plates. Similar to the resistive model, changes in any of these parameters will change the capacitance. For example, if pressure is applied to one of the two plates, the separation d can be reduced, thereby increasing the capacitance. Similarly, a change in the permittivity of the dielectric can be caused by an increase in temperature or humidity, thereby resulting in a change in capacitance.

Inductive sensors are based on the electrical principle of inductance, that is, where an electromagnetic force is induced by a fluctuating current. Inductance is determined by the dimensions of the sensor (cross-sectional area, length of coil), the number of turns of the coil, and the permeability of the core. Changes in any of these parameters (e.g., caused by movements of the core within the coil) change the inductance. Inductive sensors are often used to measure proximity, position, force, pressure, temperature, and acceleration.

Finally, *piezoelectric* sensors use the piezoelectric effect of some materials (e.g., crystals and certain ceramics) to measure pressure, force, strain, and acceleration. When a pressure is applied to such a material, it causes a mechanical deformation and a displacement of charges, proportional to the amount of pressure. The main advantage of piezoelectric devices over other approaches is that the piezoelectric effect is not sensitive to electromagnetic fields or radiation.

1.1.2   Wireless Sensor Networks

While many sensors connect to controllers and processing stations directly (e.g., using local area networks), an increasing number of sensors communicate the collected data wirelessly to a centralized processing station. This is important since many network applications require hundreds or thousands of sensor nodes, often deployed in remote and inaccessible areas. Therefore, a *wireless* sensor has not only a sensing component, but also on-board processing, communication, and storage capabilities. With these enhancements, a sensor node is often not only responsible for data collection, but also for in-network analysis, correlation, and fusion of its own sensor data and data from other sensor nodes. When many sensors cooperatively monitor large physical environments, they form a *wireless sensor network* (WSN). Sensor nodes communicate not only with each other but also with a *base station* (BS) using their wireless radios, allowing them to disseminate their sensor data to remote processing, visualization, analysis, and storage systems. For example, Figure 1.3 shows two *sensor fields* monitoring two different geographic regions and connecting to the Internet using their base stations.

The capabilities of sensor nodes in a WSN can vary widely, that is, simple sensor nodes may monitor a single physical phenomenon, while more complex devices may combine many different sensing techniques (e.g., acoustic, optical, magnetic). They can also differ in their communication capabilities, for example, using ultrasound, infrared, or radio frequency technologies with varying data rates and latencies. While simple sensors may only collect and communicate information about the observed environment, more powerful devices (i.e., devices with large processing, energy, and storage capacities) may also perform extensive processing and aggregation functions. Such devices often assume additional responsibilities in a WSN, for example, they may form communication backbones that can be used by other resource-constrained sensor devices to reach the

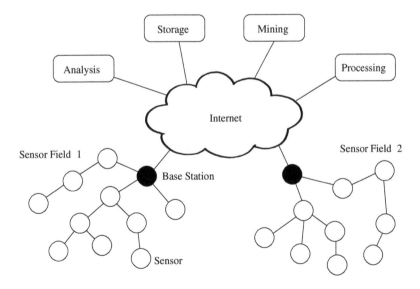

Figure 1.3   Wireless sensor networks.

base station. Finally, some devices may have access to additional supporting technologies, for example, Global Positioning System (GPS) receivers, allowing them to accurately determine their position. However, such systems often consume too much energy to be feasible for low-cost and low-power sensor nodes.

1.1.2.1 History of Wireless Sensor Networks

As with many other technologies, the military has been a driving force behind the development of wireless sensor networks. For example, in 1978, the Defense Advanced Research Projects Agency (DARPA) organized the Distributed Sensor Nets Workshop (DAR 1978), focusing on sensor network research challenges such as networking technologies, signal processing techniques, and distributed algorithms. DARPA also operated the Distributed Sensor Networks (DSN) program in the early 1980s, which was then followed by the Sensor Information Technology (SensIT) program.

In collaboration with the Rockwell Science Center, the University of California at Los Angeles proposed the concept of Wireless Integrated Network Sensors or WINS (Pottie 2001). One outcome of the WINS project was the Low Power Wireless Integrated Microsensor (LWIM), produced in 1996 (Bult *et al.* 1996). This smart sensing system was based on a CMOS chip, integrating multiple sensors, interface circuits, digital signal processing circuits, wireless radio, and microcontroller onto a single chip. The Smart Dust project (Kahn *et al.* 1999) at the University of California at Berkeley focused on the design of extremely small sensor nodes called *motes*. The goal of this project was to demonstrate that a complete sensor system can be integrated into tiny devices, possibly the size of a grain of sand or even a dust particle. The PicoRadio project (Rabaey *et al.* 2000) by the Berkeley Wireless Research Center (BWRC) focuses on the development of low-power sensor devices, whose power consumption is so small that they can power themselves from energy sources of the operating environment, such as solar or vibrational energy. The MIT μAMPS (micro-Adaptive Multidomain Power-aware Sensors) project also focuses on low-power hardware and software components for sensor nodes, including the use of microcontrollers capable of dynamic voltage scaling and techniques to restructure data processing algorithms to reduce power requirements at the software level (Calhoun *et al.* 2005).

While these previous efforts are mostly driven by academic institutions, over the last decade a number of commercial efforts have also appeared (many based on some of the academic efforts described above), including companies such as Crossbow (www.xbow.com), Sensoria (www.sensoria.com), Worldsens (http://worldsens.citi.insa-lyon.fr), Dust Networks (http://www.dustnetworks.com), and Ember Corporation (http://www.ember.com). These companies provide the opportunity to purchase sensor devices ready for deployment in a variety of application scenarios along with various management tools for programming, maintenance, and sensor data visualization.

1.1.2.2 Communication in a WSN

The well-known IEEE 802.11 family of standards was introduced in 1997 and is the most common wireless networking technology for mobile systems. It uses different frequency bands, for example, the 2.4-GHz band is used by IEEE 802.11b and IEEE 802.11g, while the IEEE 802.11a protocol uses the 5-GHz frequency band. IEEE 802.11 was frequently used in early wireless sensor networks and can still be found in current networks when bandwidth

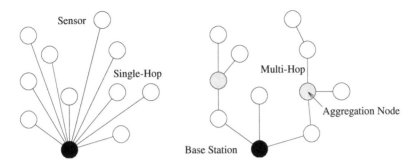

Figure 1.4 Single-hop versus multi-hop communication in sensor networks.

demands are high (e.g., for multimedia sensors). However, the high-energy overheads of IEEE 802.11-based networks makes this standard unsuitable for low-power sensor networks. Typical data rate requirements in sensor networks are comparable to the bandwidths provided by dial-up modems, therefore the data rates provided by IEEE 802.11 are typically much higher than needed. This has led to the development of a variety of protocols that better satisfy the networks' need for low power consumption and low data rates. For example, the IEEE 802.15.4 protocol (Gutierrez *et al.* 2001) has been designed specifically for short-range communications in low-power sensor networks and is supported by most academic and commercial sensor nodes.

When the transmission ranges of the radios of all sensor nodes are large enough and the sensors can transmit their data directly to the base station, they can form a star topology as shown on the left in Figure 1.4. In this topology, each sensor node communicates directly with the base station using a single hop. However, sensor networks often cover large geographic areas and radio transmission power should be kept at a minimum in order to conserve energy; consequently, *multi-hop communication* is the more common case for sensor networks (shown on the right in Figure 1.4). In this *mesh topology*, sensor nodes must not only capture and disseminate their own data, but also serve as *relays* for other sensor nodes, that is, they must collaborate to propagate sensor data towards the base station. This *routing* problem, that is, the task of finding a multi-hop path from a sensor node to the base station, is one of the most important challenges and has received immense attention from the research community. When a node serves as a relay for multiple routes, it often has the opportunity to analyze and pre-process sensor data in the network, which can lead to the elimination of redundant information or aggregation of data that may be smaller than the original data.

1.2 Challenges and Constraints

While sensor networks share many similarities with other distributed systems, they are subject to a variety of unique challenges and constraints. These constraints impact the design of a WSN, leading to protocols and algorithms that differ from their counterparts in other distributed systems. This section describes the most important design constraints of a WSN.

1.2.1 Energy

The constraint most often associated with sensor network design is that sensor nodes operate with limited energy budgets. Typically, they are powered through batteries, which must be either replaced or recharged (e.g., using solar power) when depleted. For some nodes, neither option is appropriate, that is, they will simply be discarded once their energy source is depleted. Whether the battery can be recharged or not significantly affects the strategy applied to energy consumption. For nonrechargeable batteries, a sensor node should be able to operate until either its *mission time* has passed or the battery can be replaced. The length of the mission time depends on the type of application, for example, scientists monitoring glacial movements may need sensors that can operate for several years while a sensor in a battlefield scenario may only be needed for a few hours or days.

As a consequence, the first and often most important design challenge for a WSN is energy efficiency. This requirement permeates every aspect of sensor node and network design. For example, the choices made at the physical layer of a sensor node affect the energy consumption of the entire device and the design of higher-level protocols (Shih *et al.* 2001). The energy consumption of CMOS-based processors is primarily due to switching energy and leakage energy (Sinha and Chandrakasan 2000):

$$E_{\text{CPU}} = E_{\text{switch}} + E_{\text{leakage}} = C_{\text{total}} V_{\text{dd}}^2 + V_{\text{dd}} I_{\text{leak}} \Delta t \qquad (1.4)$$

where C_{total} is the total capacitance switched by the computation, V_{dd} is the supply voltage, I_{leak} is the leakage current, and Δt is the duration of the computation. While the switching energy still dominates the energy consumption of processors, it is expected that in future processor designs, the leakage energy will be responsible for more than half the energy consumption (De and Borkar 1999). Some techniques to control leakage energy include progressive shutdown of idle components and software-based techniques such as Dynamic Voltage Scaling (DVS).

The medium access control (MAC) layer is responsible for providing sensor nodes with access to the wireless channel. Some MAC strategies for communication networks are *contention-based*, that is, nodes may attempt to access the medium at any time, potentially leading to collisions among multiple nodes, which must be addressed by the MAC layer to ensure that transmissions will eventually succeed. Downsides of these approaches include the energy overheads and delays incurred by the collisions and recovery mechanisms and that sensor nodes may have to listen to the medium at all times to ensure that no transmissions will be missed. Therefore, some MAC protocols for sensor networks are *contention-free*, that is, access to the medium is strictly regulated, eliminating collisions and allowing sensor nodes to shut down their radios when no communications are expected. The network layer is responsible for finding routes from a sensor node to the base station and route characteristics such as length (e.g., in terms of number of hops), required transmission power, and available energy on relay nodes determine the energy overheads of multi-hop communication.

Besides network protocols, the goal of energy efficiency impacts the design of the operating system (e.g., small memory footprint, efficient switching between tasks), middleware, security mechanisms, and even the applications themselves. For example, *in-network processing* is frequently used to eliminate redundant sensor data or to aggregate multiple sensor readings. This leads to a tradeoff between computation (processing the sensor data) and communication (transmitting the original versus the processed data), which can often be exploited to obtain energy savings (Pottie and Kaiser 2000; Sohrabi *et al.* 2000).

1.2.2   Self-Management

It is the nature of many sensor network applications that they must operate in remote areas and harsh environments, without infrastructure support or the possibility for maintenance and repair. Therefore, sensor nodes must be *self-managing* in that they configure themselves, operate and collaborate with other nodes, and adapt to failures, changes in the environment, and changes in the environmental stimuli without human intervention.

1.2.2.1   Ad Hoc Deployment

Many wireless sensor network applications do not require predetermined and engineered locations of individual sensor nodes. This is particularly important for networks being deployed in remote or inaccessible areas. For example, sensors serving the assessment of battlefield or disaster areas could be thrown from airplanes over the areas of interest, but many sensor nodes may not survive such a drop and may never be able to begin their sensing activities. However, the surviving nodes must autonomously perform a variety of setup and configuration steps, including the establishment of communications with neighboring sensor nodes, determining their positions, and the initiation of their sensing responsibilities. The mode of operation of sensor nodes can differ based on such information, for example, a node's location and the number or identities of its neighbors may determine the amount and type of information it will generate and forward on behalf of other nodes.

1.2.2.2   Unattended Operation

Many sensor networks, once deployed, must operate without human intervention, that is, configuration, adaptation, maintenance, and repair must be performed in an autonomous fashion. For example, sensor nodes are exposed to both system dynamics and environmental dynamics, which pose a significant challenge for building reliable sensor networks (Cerpa and Estrin 2004). A *self-managing* device will monitor its surroundings, adapt to changes in the environment, and cooperate with neighboring devices to form topologies or agree on sensing, processing, and communication strategies (Mills 2007). Self-management can take place in a variety of forms. Self-organization is the term frequently used to describe a network's ability to adapt configuration parameters based on system and environmental state. For example, a sensor device can choose its transmission power to maintain a certain degree of connectivity (i.e., with increasing transmission power it is more likely that a node will reach more neighbors). Self-optimization refers to a device's ability to monitor and optimize the use of its own system resources. Self-protection allows a device to recognize and protect itself from intrusions and attacks. Finally, the ability to self-heal allows sensor nodes to discover, identify, and react to network disruptions. In energy-constrained sensor networks, all these self-management features must be designed and implemented such that they do not incur excessive energy overheads.

1.2.3   Wireless Networking

The reliance on wireless networks and communications poses a number of challenges to a sensor network designer. For example, *attenuation* limits the range of radio signals, that is, a radio frequency (RF) signal fades (i.e., decreases in power) while it propagates through a medium and while it passes through obstacles. The relationship between the received power and transmitted power of an RF signal can be expressed using the *inverse-square law*:

$$P_r \propto \frac{P_t}{d^2} \qquad (1.5)$$

which states that the received power P_r is proportional to the inverse of the square of the distance d from the source of the signal. That is, if P_r^x is the power at distance x, doubling the distance to $y = 2x$ decreases the power at the new distance to $P_r^y = P_r^x/4$.

As a consequence, an increasing distance between a sensor node and a base station rapidly increases the required transmission power. Therefore, it is more energy-efficient to split a large distance into several shorter distances, leading to the challenge of supporting *multi-hop* communications and routing. Multi-hop communication requires that nodes in a network cooperate with each other to identify efficient routes and to serve as relays. This challenge is further exacerbated in networks that employ *duty cycles* to preserve energy. That is, many sensor nodes use a power conservation policy where radios are switched off when they are not in use. As a consequence, during these down-times, the sensor node cannot receive messages from its neighbors nor can it serve as a relay for other sensors. Therefore, some networks rely on *wakeup on demand* strategies (Shih *et al.* 2002) to ensure that nodes can be woken up whenever needed. Usually this involves devices with two radios, a low-power radio used to receive wakeup calls and a high-power radio that is activated in response to a wakeup call. Another strategy is *adaptive duty cycling* (Ye *et al.* 2004), when not all nodes are allowed to sleep at the same time. Instead, a subset of the nodes in a network remain active to form a network backbone.

1.2.4 Decentralized Management

The large scale and the energy constraints of many wireless sensor networks make it infeasible to rely on *centralized* algorithms (e.g., executed at the base station) to implement network management solutions such as topology management or routing. Instead, sensor nodes must collaborate with their neighbors to make localized decisions, that is, without global knowledge. As a consequence, the results of these *decentralized* (or *distributed*) algorithms will not be optimal, but they may be more energy-efficient than centralized solutions. Consider routing as an example for centralized and decentralized solutions. A base station can collect information from all sensor nodes, establish routes that are optimal (e.g., in terms of energy), and inform each node of its route. However, the overhead can be significant, particularly if the topology changes frequently. Instead, a decentralized approach allows each node to make routing decisions based on limited local information (e.g., a list of the node's neighbors, including their distances to the base station). While this decentralized approach may lead to nonoptimal routes, the management overheads can be reduced significantly.

1.2.5 Design Constraints

While the capabilities of traditional computing systems continue to increase rapidly, the primary goal of wireless sensor design is to create smaller, cheaper, and more efficient devices.

Driven by the need to execute dedicated applications with little energy consumption, typical sensor nodes have the processing speeds and storage capacities of computer systems from several decades ago. The need for small form factor and low energy consumption also prohibits the integration of many desirable components, such as GPS receivers. These constraints and requirements also impact the software design at various levels, for example, operating systems must have small memory footprints and must be efficient in their resource management tasks. However, the lack of advanced hardware features (e.g., support for parallel executions) facilitates the design of small and efficient operating systems. A sensor's hardware constraints also affect the design of many protocols and algorithms executed in a WSN. For example, routing tables that contain entries for each potential destination in a network may be too large to fit into a sensor's memory. Instead, only a small amount of data (such as a list of neighbors) can be stored in a sensor node's memory. Further, while in-network processing can be employed to eliminate redundant information, some sensor fusion and aggregation algorithms may require more computational power and storage capacities than can be provided by low-cost sensor nodes. Therefore, many software architectures and solutions (operating system, middleware, network protocols) must be designed to operate efficiently on very resource-constrained hardware.

1.2.6 Security

Many wireless sensor networks collect sensitive information. The remote and unattended operation of sensor nodes increases their exposure to malicious intrusions and attacks. Further, wireless communications make it easy for an adversary to eavesdrop on sensor transmissions. For example, one of the most challenging security threats is a *denial-of-service* attack, whose goal is to disrupt the correct operation of a sensor network. This can be achieved using a variety of attacks, including a *jamming attack*, where high-powered wireless signals are used to prevent successful sensor communications. The consequences can be severe and depend on the type of sensor network application. While there are numerous techniques and solutions for distributed systems that prevent attacks or contain the extent and damage of such attacks, many of these incur significant computational, communication, and storage requirements, which often cannot be satisfied by resource-constrained sensor nodes. As a consequence, sensor networks require new solutions for key establishment and distribution, node authentication, and secrecy.

1.2.7 Other Challenges

From the discussion so far, it becomes clear that many design choices in a WSN differ from the design choices of other systems and networks. Table 1.2 summarizes some of the key differences between traditional networks and wireless sensor networks. A variety of additional challenges can affect the design of sensor nodes and wireless sensor networks. For example, some sensors may be mounted onto moving objects, such as vehicles or robots, leading to continuously changing network topologies that require frequent adaptations at multiple layers of a system, including routing (e.g., changing neighbor lists), medium access control (e.g., changing density), and data aggregation (e.g., changing overlapping sensing regions). A heterogeneous sensor network consists of devices with varying hardware

Table 1.2 Comparison of traditional networks and wireless sensor networks

Traditional networks	Wireless sensor networks
General-purpose design; serving many applications	Single-purpose design; serving one specific application
Typical primary design concerns are network performance and latencies; energy is not a primary concern	Energy is the main constraint in the design of all node and network components
Networks are designed and engineered according to plans	Deployment, network structure, and resource use are often ad hoc (without planning)
Devices and networks operate in controlled and mild environments	Sensor networks often operate in environments with harsh conditions
Maintenance and repair are common and networks are typically easy to access	Physical access to sensor nodes is often difficult or even impossible
Component failure is addressed through maintenance and repair	Component failure is expected and addressed in the design of the network
Obtaining global network knowledge is typically feasible and centralized management is possible	Most decisions are made localized without the support of a central manager

capabilities, for example, sensor nodes may have more hardware resources if their sensing tasks require more computation and storage or if they are responsible for collecting and processing data from other sensors within the network. Also, some sensor applications may have specific performance and quality requirements, for example, low latencies for critical sensor events or high throughput for data collected by video sensors. Both heterogeneity and performance requirements affect the design of wireless sensors and their protocols. Finally, while traditional computer networks are based on established standards, many protocols and mechanisms in wireless sensor networks are proprietary solutions, while standards-based solutions emerge only slowly. Standards are important for interoperability and facilitate the design and deployment of WSN applications; therefore, a key challenge in WSN design remains the standardization of promising solutions and the harmonization of competing standards.

Exercises

1.1 What is the difference between passive sensors and active sensors and can you name a few examples for each category (e.g., using Table 1.1)?

1.2 Consider a Wheatstone bridge circuit using a resistive temperature sensor R_x as shown in Figure 1.2. Further assume that $R_1 = 10\,\Omega$ and $R_3 = 20\,\Omega$. Assume that the current temperature is $80\,°F$ and $R_x(80) = 10\,\Omega$. You wish to calibrate the sensor such that the output voltage V_{OUT} is zero whenever the temperature is $80\,°F$.

(a) What is the desired value of R_2?

(b) What is the output voltage (as a function of the supply voltage) at a temperature of $90\,°F$, when this increase in temperature leads to an increase in resistance of 20% for R_x?

1.3 As described in this chapter, using multiple communication hops instead of a single hop affects the overall energy consumption. Describe other advantages or disadvantages of multi-hop communications, for example, in terms of performance (latency, throughput), reliability, and security.

1.4 The relationship between the transmitted and the received power of an RF signal follows the inverse-square law shown in Equation (1.5), that is, power density and distance have a quadratic relationship. This can be used to justify multi-hop communication (instead of single-hop), that is, energy can be preserved by transmitting packets over multiple hops at lower transmission power. Assume that a packet p must be sent from a sender A to a receiver B. The energy necessary to directly transmit the packet can be expressed as the simplified formula $E_{AB} = d(A,\ B)^2 + c$, where $d(x,\ y)$ (or simply d in the remainder of this question) is the distance between two nodes x and y and c is a constant energy cost. Assume that you can turn this single-hop scenario into a multi-hop scenario by placing any number of equidistant relay nodes between A and B.

    (a) Derive a formula to compute the required energy as a function of d and n, where n is the number of relay nodes (that is, $n = 0$ for the single-hop case).

    (b) What is the optimal number of relay nodes to send p with the minimum amount of energy required and how much energy is consumed in this optimal case for a distance $d(A,\ B) = 10$ and (i) $c = 10$ and (ii) $c = 5$?

1.5 Name at least four techniques to reduce power consumption in wireless sensor networks.

References

Bult, K., Burstein, A., Chang, D., Dong, M., Fielding, M., Kruglick, E., Ho, J., Lin, F., Lin, T.H., Kaiser, W.J., Marcy, H., Mukai, R., Nelson, P., Newburg, F.L., Pister, K.S.J., Pottie, G., Sanchez, H., Sohrabi, K., Stafsudd, O.M., Tan, K.B., Yung, G., Xue, S., and Yao, J. (1996) Low power systems for wireless microsensors. *Proc. of the International Symposium on Low Power Electronics and Design*.

Calhoun, B.H., Daly, D.C., Verma, N., Finchelstein, D.F., Wentzloff, D.D., Wang, A., Cho, S.H., and Chandrakasan, A.P. (2005) Design considerations for ultralow energy wireless microsensor nodes. *IEEE Transactions on Computers* **54** (6), 727–749.

Cerpa, A., and Estrin, D. (2004) Ascent: Adaptive self-configuring sensor network topologies. *IEEE Transactions on Mobile Computing* **3** (3), 272–285.

DAR (1978) *Proceedings of the Distributed Sensor Nets Workshop*. Pittsburgh, PA, Department of Computer Science, Carnegie Mellon University.

De, V., and Borkar, S. (1999) Technology and design challenges for low power and high performance. *Proc. of the International Symposium on Low Power Electronics and Design (ISLPED)*.

Gutierrez, J.A., Naeve, M., Callaway, E., Bourgeois, M., Mitter, V., and Heile, B. (2001) IEEE 802.15.4: A developing standard for low-power low-cost wireless personal area networks. *IEEE Network* **15** (5), 12–19.

Kahn, J.M., Katz, R.H., and Pister, K.S.J. (1999) Mobile networking for smart dust. *Proc. of the ACM/IEEE International Conference on Mobile Computing and Networking (MobiCom)*.

Mills, K.L. (2007) A brief survey of self-organization in wireless sensor networks. *Wireless Communications and Mobile Computing* **7** (7), 823–834.

Pottie, G.J. (2001) Wireless integrated network sensors (WINS): The web gets physical. *National Academy of Engineering: The Bridge* **31** (4), 22–27.

Pottie, G.J., and Kaiser, W.J. (2000) Wireless integrated network sensors. *Communications of the ACM*.

Rabaey, J., Ammer, J., da Silva, Jr J.L., and Patel, D. (2000) Picoradio: Ad hoc wireless networking of ubiquitous low-energy sensor/monitor nodes. *Proc. of the IEEE Computer Society Annual Workshop on VLSI*.

Shih, E., Bahl, P., and Sinclair, M. (2002) Wake-on wireless: An event driven energy saving strategy for battery operated devices. *Proc. of the ACM/IEEE International Conference on Mobile Computing and Networking (MobiCom)*.

Shih, E., Cho, S.H., Ickes, N., Min, R., Sinha, A., Wang, A., and Chandrakasan, A. (2001) Physical layer driven protocol and algorithm design for energy-efficient wireless sensor networks. *Proc. of the 7th Annual International Conference on Mobile Computing and Networking*.

Sinha, A., and Chandrakasan, A.P. (2000) Energy aware software. *Proc. of the 13th International Conference on VLSI Design*.

Sohrabi, K., Gao, J., Ailawadhi V., and Pottie, G. (2000) Protocols for self-organization of a wireless sensor network. *IEEE Personal Communications Magazine* **7** (5), 16–27.

Ye, W., Heidemann, J., and Estrin, D. (2004) Medium access control with coordinated adaptive sleeping for wireless sensor networks. *IEEE/ACM Transactions on Networking* **12** (3), 493–506.

2

Applications

Wireless sensor networks have inspired many applications. Some of them are futuristic while a large number of them are practically useful. The diversity of applications in the latter category is remarkable – environment monitoring, target tracking, pipeline (water, oil, gas) monitoring, structural health monitoring, precision agriculture, health care, supply chain management, active volcano monitoring, transportation, human activity monitoring, and underground mining, to name a few. In this chapter some of these applications and the prototype implementations for these applications will be discussed in some detail.

2.1 Structural Health Monitoring

On 2 August 2007, a highway bridge unexpectedly collapsed in Minnesota into the fast-flowing Mississippi river. Nine people were killed in the event. The National Transportation Safety Board investigators were unable to determine the cause of the accident, but they short-listed three potential causes, namely, wear and tear, weather, and the weight of a nearby construction project which was taking place at the time. The construction project was closing half of the bridge's eight lanes when the accident happened. Two weeks later – on 14 August 2007 – another bridge collapsed at a popular Chinese tourist spot in Fenghuang county in Hunan province, killing 86 people on the spot. In fact, the BBC reported (14 August 2007) that China had identified more than 6000 bridges that were damaged or considered to be dangerous.

During and following these accidents, several news outlets, including The Associated Press (3 August 2007) and *Time* magazine (10 August 2007), featured articles that advocated wireless sensor networks for monitoring bridges and similar structures.

Traditionally, bridges are inspected in different phases and at different levels (Koh and Dyke 2007):

1. visual inspection carried out by road maintenance crews during routine road inspections, normally every day;
2. basic inspections carried out usually at least once a year by local bridge inspectors;
3. detailed inspection, carried out at least every five years on selected bridges by regional bridge inspectors; and
4. special inspections carried out by highly qualified experts and researchers according to technical needs, normally as a consequence of questionable results from basic or detailed inspections.

Fundamentals of Wireless Sensor Networks: Theory and Practice   Waltenegus Dargie and Christian Poellabauer
© 2010 John Wiley & Sons, Ltd

The first phase is a labor-intensive, tedious, inconsistent, and subjective inspection technique (Koh and Dyke 2007), whereas the rest require sophisticated tools, which are usually expensive, bulky, and power consuming. Subsequently, developing auto-mated, efficient, and affordable structural health monitoring techniques is an active research area.

Broadly speaking, tool-based inspection techniques can be classified into local and global inspections (Chintalapudi *et al.* 2006). Local techniques focus on detecting highly localized, imperceptible fractures in a structure. These techniques employ ultrasonic, thermal, X-ray, magnetic or optical imaging techniques, but this type of inspection requires a significant amount of time and the disruption of the normal operation of the structure.

Global inspection techniques, on the other hand, aim to detect a damage or defect that is large enough to affect the entire structure. Often this is carried out by detecting conspicuous changes in the movements of abutments, balustrades and barriers, bridge bearings, decks, towers, expansion joints, railings, etc., to forced or ambient excitations. Global inspection techniques can be considered as an inverse problem, that is, the status of the structure is determined on the basis of its response to an external excitation. The excitation can be ambient (such as an earthquake or a strong wind) or forced (such as a deliberate force produced by a shaker or an impact hammer). In either case, modal parameters, such as natural frequencies, damping ratios, and mode shapes are investigated to identify damage in the form of expansion, de-lamination, corrosion, debonding, cracking, etc.

Modal parameters are determined by several factors including: the magnitude and dura-tion of the excitation; the material from which the structure is made; the size of the structure; the technical restrictions in the construction; the age of the structure; and other surrounding constraints.

More recently, researchers have been developing and testing wireless sensor networks as part of a global inspection mechanism. There are three aspects that make them suitable for the task:

1. The sensor nodes can be placed in areas that are inaccessible to wired and bulky devices.
2. By deploying a large number of nodes, it is possible to establish correlation between different measurements. This facilitates localizing damage.
3. Ideally, the deployment as well as the management (maintenance) of the sensor network does not require disruption of the normal operation of the structure.

2.1.1   Sensing Seismic Events

Seismic responses in large structures are transient by nature and comprise frequencies below a few tens of hertz. The response can be captured by employing acceleration sensors, tilt sensors, and piezoelectric sensors. However, the sensors should be oversampled at high frequency to compensate for noise and imperfect placement.

Some of the challenges pertaining to the analysis of data are: (a) restrictions regarding the characteristics of the excitations; (b) the presence of unreachable degree-of-freedom elements; (c) measurement noise; (d) modeling errors; and (e) environmental constraints. The effectiveness of a technique is measured by its capability to extract a sufficiently large amount of damage-sensitive parameters (stiffness, damping, etc.), given limited and incomplete modal data measured from a real structure.

A damage detection technique can identify a single damage or multiple damages, depending on the model of the structure. Single-damage detection usually employs natural frequencies while the multiple damage detection technique employs mode shapes.

2.1.2 Single Damage Detection Using Natural Frequencies

This technique computes the correlation between the measured and predicted (hypothesis) modal frequencies to determine the damage. The parameter vectors used for evaluating correlation coefficients consist of the ratio of the first n modal frequency changes due to a damage in the structure, that is, $\Delta\omega = (\omega_h - \omega_d)$, where ω_h and ω_d denote the natural frequency vectors of the healthy and damaged elements of the structure, respectively. The hypothesis vector, predicted from an analytic model, is used to infer the location and extent of damage. It is denoted by $\delta\omega$.

Given a pair of parameter vectors, one can estimate the level of correlation in several ways. The simplest way of estimating correlation is to calculate the angle between ω_h and ω_d. A damage localization technique using the pair comparison attempts to find linear correlation of modal frequency variation vectors. One way is to apply Equation (2.1) (Koh and Dyke 2007):

$$C_j = \frac{\Delta\omega^T \delta\omega_j}{|\Delta\omega||\delta\omega_j|} \tag{2.1}$$

where the subscript $j = (1, 2, ..., r)$ indicates the hypothesized location of the damage.

Another possibility is to apply the damage localization assurance criterion, or in short, DLAC, which is expressed as:

$$DLAC_j = \frac{\left|\Delta\omega^T \delta\omega_j\right|^2}{(\Delta\omega^T \Delta\omega_j)(\delta\omega_j^T \delta\omega_j)} \tag{2.2}$$

Equation (2.2), similar to Equation (2.1), compares two frequency change vectors (namely, one based on the measurement obtained from the test structure and the other based on the jth hypothesis of an analytical model of the structure) in order to evaluate the level of correlation between the two parameter vectors.

2.1.3 Multiple Damage Detection Using Natural Frequencies

Damage inference based on a unique pattern in modal frequency changes yields erroneous results when applied to a structure with a multiple or unknown number of defects. Incorporating a sensitivity matrix derived from the analytic model of the structure into Equation (2.1) enables multiple damages to be estimated. The sensitivity matrix consists of the first-order derivatives of the modal frequencies with respect to each foreseeable damage variable, for the ratio of stiffness reduction in each structural element. This is given as (Koh and Dyke 2007):

$$
S = \begin{bmatrix}
\dfrac{\partial \omega_1}{\partial z_1} & \dfrac{\partial \omega_1}{\partial z_2} & \cdots & \dfrac{\partial \omega_1}{\partial z_n} \\[2mm]
\dfrac{\partial \omega_2}{\partial z_1} & \dfrac{\partial \omega_2}{\partial z_2} & \cdots & \dfrac{\partial \omega_2}{\partial z_n} \\[2mm]
\vdots & \vdots & \cdots & \vdots \\[2mm]
\dfrac{\partial \omega_m}{\partial z_1} & \dfrac{\partial \omega_m}{\partial z_2} & \cdots & \dfrac{\partial \omega_m}{\partial z_n}
\end{bmatrix}
\tag{2.3}
$$

where $Z_i, i = 1, 2, ..., n$ is a damage variable.

Consequently, given $\delta\omega = S\delta z$, the multiple damage location assurance criterion (MDLAC) is given as:

$$
\text{MDLAC}_j = \frac{\left| \Delta\omega^T \left[S\delta z_j \right] \right|^2}{\left(\Delta\omega^T . \Delta\omega_j \right) . \left(\left[S\delta z_j \right]^T . \left[S\delta z_j \right] \right)}
\tag{2.4}
$$

2.1.4 Multiple Damage Detection Using Mode Shapes

The problem with Equation (2.4) lies in evaluating all possible combinations of damage variables that maximize the MDLAC. Efficient search algorithms, such as genetic algorithms, can be applied to determine the correct set of damage variables, but this can only be made at the cost of computational complexity.

Multiple damage detection techniques, which use mode shapes instead of natural frequency changes, can avoid the sensitivity matrix thereby bypassing the need to apply search algorithms. There are two types of approach as far as evaluating mode shapes are concerned:

1. The modal assurance criteria (MAC) is determined between paired modes, namely, healthy and damaged. Thus, the number of available MAC values equals the number of measured modes.
2. Instead of individually evaluating the modal assurance criterion value for each mode shape, a single vector of stacked mode shapes can be used for evaluating linear correlation.

In both cases, unlike natural frequencies, mode shapes retain spatial information for a given degree of freedom, that is, correlation between stacked mode shapes can directly indicate a damaged location. On the other hand, mode-based correlation techniques can only weakly capture the extent of the damage.

The relative change in each mode shape between healthy and damaged states can be used to calculate the correlation coefficients. The change of mode shapes $\Delta(m \times r)$ due to damage is transformed into a single vector, vec $[\Delta\Phi](mr \times 1)$ by stacking r columns of damages of the matrix $\Delta\Phi$. Hence, the stack mode shape correlation (SMSC) is:

$$
\text{SMSC}_j = \frac{\text{vec}\,[\Delta\Phi]^T \ \text{vec}\,[\delta\Phi_j]}{\left| \text{vec}\,[\Delta\Phi]^T \ \text{vec}\,[\delta\Phi_j] \right|}
\tag{2.5}
$$

where vec $[\Delta\Phi]$ and vec $[\delta\Phi]$ represent stacked vectors for the variation of the identified and predicted mode shapes, respectively.

2.1.5 Coherence

Another approach to damage detection and localization is by the use of the coherence function. The coherence, $C_{xy}(\omega)$, between two time discrete signals, $x[n]$ and $y[n]$, $0 \le n \le \infty$, is the normalized function of the frequency derived from the cross-spectrum of the two signals:

$$C_{xy}(\omega) = \frac{S_{xy}}{S_{xx}S_{yy}} \qquad (2.6)$$

The coherence function measures the extent to which the two signals are linearly related with each other at each frequency. A unity value indicates that the signals are highly correlated at a given frequency and zero indicates that the signals are uncorrelated at that frequency. Coherence is a complex quantity, but it is often approximated by the magnitude squared coherence (MSC):

$$\left|\gamma_{xy}(w)\right|^2 = \frac{\left|S_{xy}(w)\right|^2}{S_{xx}(w) \cdot S_{yy}(w)} \qquad (2.7)$$

If the two signals are identical, then the coherence gives a unity result for all frequencies. Similarly, if two signals describe completely uncorrelated random processes, the coherence will be zero for all frequencies. For example, a repeated measurement of the seismic response of a healthy structure would exhibit a high coherence (near 1) for most frequencies, whereas the coherence of a seismic response of a damaged structure and a healthy structure would have low coherence (near 0).

Equation (2.7) would always result in a unity magnitude for all frequencies (though the imaginary component may not be 1) if a single window were used to estimate the spectral density. A commonly used multiple windowing method is called weighted overlapped segment averaging (WOSA) and involves splitting two signals, X and Y, into equal lengths of windowed segments. The Fast Fourier Transform (FFT) of these segments is taken and their results are averaged together to estimate the spectral density. The segments can be overlapped to reduce the variance of the spectral estimate (an overlap of 50% is common). However, overlapping is computationally expensive. In general, more windows provide smoother coherences with less variability but require more computation.

Measuring the degree of coherence of two signals at each potential frequency requires that the signals are well represented at each frequency, that is, the power of the signals at each frequency should be appreciably large enough and the two signals should be well synchronized. This requires precise sensors, narrow-band filters to isolate the desired frequency components from higher harmonics, and sampling the signals at high rate. Therefore, instead of evaluating coherence at each frequency, one can integrate the area under the coherence curve for a given frequency range. Domain knowledge of the structural response can be taken into account to limit the range. For large bridges, for example, the range lies below 10 Hz. Subsequently, a normalized integration over the range 0 to 10 Hz to get a 0 to 1 measure of the coherence can be achieved:

$$P_{xy} = \frac{1}{10} \int_0^{10} C_{xy}(\omega) d\omega \qquad (2.8)$$

2.1.6 Piezoelectric Effect

So far the inputs of damage detection techniques are implicitly assumed to be obtained from acceleration sensors or tilt sensors. It is relatively simple to capture seismic response by employing acceleration sensors. It is also possible to employ piezoelectric materials to capture a seismic response. When a mechanical stress is applied to a piezoelectric material, it generates an electric charge; and when an electric field is applied to a piezoelectric material, its dimensions change proportionally to the applied field. This property makes a piezoelectric material suitable for seismic sensing or actuation, or both.

The relation between the mechanical and electrical variables associated with a piezoelectric material can be described as follows (Park $et\ al.$ 2000):

$$S_i = S_{ij}^E T_j + d_{mi} E_m \tag{2.9}$$

$$D_m = d_{mi} T_i + \varepsilon_{mk}^T \tag{2.10}$$

This can be reformulated as:

$$\begin{bmatrix} S \\ D \end{bmatrix} = \begin{bmatrix} s^E & d_t \\ d & \varepsilon^T \end{bmatrix} \begin{bmatrix} T \\ E \end{bmatrix} \tag{2.11}$$

In Equations (2.9)–(2.11), the parameters are defined as follows:

- S is the mechanical strain;
- T is the mechanical stress;
- E is the electric field;
- D is the charge density;
- s is the mechanical compliance;
- d is the piezoelectric strain constant;
- ε is the permittivity; and
- the subscripts i, j, m, and k indicate the direction of stress, strain, or electric field.

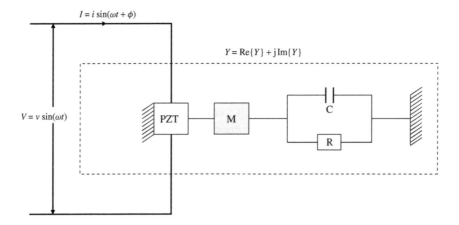

Figure 2.1 A piezoelectric material for capturing mechanical impedance. The PZT is normally bonded directly to the surface of the structure by a high strength adhesive to ensure better mechanical interaction - this is indicated by the gray box, M. The broken line indicates the coupled electromechanical admittance Y (Park $et\ al$ 2000).

Equation (2.9) describes the change in the dimension of a piezoelectric material as a result of an applied electric field, whereas Equation (2.10) describes the electric field that is induced by piezoelectric material in response to an applied mechanical stress.

Equation (2.10) is important for structural health monitoring, because it quantitatively describes the variation in the mechanical impedance due to the presence of structural damage. More precisely, it means that by employing a piezoelectric material, it is possible to couple the mechanical and electrical impedance thereby extracting structural information from electrical impedance measurements. Mathematically, the relationship between the mechanical impedance and the electrical impedance is described as:

$$Y(\omega) = j\omega \frac{w_a l_a}{h_a} \left(\varepsilon_{33}^T (1 - j\delta) - \frac{Z_s(\omega)}{Z_s(\omega) Z_a(\omega)} d_{3x}^2 Y_{xx}^E \right) \qquad (2.12)$$

where

- Y is the electrical admittance;
- Z_a is the mechanical impedance of the piezoelectric material;
- Z_s is the mechanical impedance of the structure;
- Y_{xx} is Young's modulus of piezoelectric material at zero electric field (inverse of compliance);
- d_{3x} is the piezoelectric strain constant at zero stress;
- ε_{33}^T is the permittivity at zero stress;
- d is the dielectric loss tangent to the piezoelectric material; and
- w_a is the width, l_a is the length, and h_a is the thickness of the piezoelectric material, respectively.

The electrical admittance is the inverse of impedance, Z, in electrical engineering. The impedance, Z, can have a resistive, an inductive, or a capacitive component, or a combination of these. If the material is a simple resistor, the impedance is equal to the resistance of the material, R, which can be simplified as:

$$R = \frac{v(t)}{i(t)} \qquad (2.13)$$

where $v(t)$ is a timely varying potential difference measured at the two ends of the resistor and $i(t)$ is a timely varying current flowing through the resistor. R is real, which means that the current is in phase with the voltage. If the material is a pure coil, with no resistive component, the impedance, Z, is expressed as:

$$Z(\omega) = \frac{v(t)}{i(t)} = j\omega L \qquad (2.14)$$

where L is the inductance of the material and $\omega = 2\pi f$ is the angular velocity, signifying the rate of change of the potential difference, $v(t)$. j implies that the voltage and the current are out of phase by 90° as a result of the inductive effect. A positive j indicates that the current is leading the voltage. The inductance, L, of the material depends on the length of the conductor, the number of coils, and the conductivity of the conductor. In reality no coil is purely inductive, it will have some resistive component, R. Hence, for a real coil with a resistive component, the impedance, Z, of Equation (2.14) is modified to:

$$Z(\omega) = \frac{v(t)}{i(t)} = R + j\omega L \qquad (2.15)$$

If the material is capacitive (if two conductors are separated by a dielectric material and there is a potential difference between the two conductors), the impedance of the material is given as:

$$Z(\omega) = \frac{v(t)}{i(t)} = \frac{1}{j\omega C} \qquad (2.16)$$

where C is the capacitance of the material; and j indicates that the voltage and the current are out of phase by $90°$, the voltage leading the current.

The capacitance of a material, C, depends on the cross-sectional area of the two conductors, the distance of separation, and the dielectric material that separates them. In general, the impedance through which an electric charge is flowing as a result of the potential difference v between the two ends of the material is given as:

$$Z(\omega) = \frac{v(t)}{i(t)} = \frac{R + j\omega L}{j\omega C} \qquad (2.17)$$

With this in mind, if the impedance of the piezoelectric material and the magnitude and frequency of the electric field induced inside the piezoelectric material are known, it is possible to determine the mechanical stress in Equation (2.12). Figure 2.1 demonstrates the transformation of mechanical admittance to an electrical admittance through the employment of a piezoelectric material.

Accordingly:

$$Z_s(\omega) = Z_a(\omega) \left(\frac{\varepsilon_{33}^T(1 - j\delta) - \dfrac{Y(\omega)h_a}{j\omega w_a l_a}}{(d_{3x})^2 Y_{xx}^E - \varepsilon_{33}^T(1 - j\delta) + \dfrac{Y(\omega)h_a}{j\omega w_a l_a}} \right) \qquad (2.18)$$

Equation (2.18) illustrates that the mechanical impedance of a structure can be determined from the electrical admittance of the piezoelectric bonding. In other words, structural integrity can be evaluated by measuring the electrical impedance of the piezoelectric sensor.

2.1.7 Prototypes

Wisden is the first prototype to employ wireless sensor networks for monitoring structural health. It is developed at the University of Southern California and deployed (a) on a seismic test structure and (b) in an abandoned, four-story office building in Los Angeles (a victim of the 1994 Northridge earthquake). The seismic test structure used as a platform for conducting seismic experiments is an imitation of a full-scale 28×28 ft^2 hospital ceiling. It supports $10,000$ lb of weight and can be subjected to a uniaxial motion of a peak-to-peak stroke of

10 in. with a 55,000-lb hydraulic actuator with a ±5-in. stroke. The hydraulic pump delivers up to 40 gallons per minute at 3000 lb/in^2. The overall weight of the moving portion of the test structure is approximately 12,000 lb.

The Wisden sensor network (Xu *et al.* 2004; Chintalapudi *et al.* 2006) consists of 25 nodes and a 16-bit vibration card, which is specifically designed for high-quality, low-power vibration sensing. A high-sensitive triaxial accelerometer is attached to the vibration card.

The nodes organize themselves to establish a tree topology WSN. The topology of the network is dynamically adjusted to accommodate incoming nodes and the departure of existing ones (due, for example, to the failure or exhaustion of batteries). The task of the network is to reliably send time-synchronized vibration data to a remote sink over a multi-hop route. To ensure reliable transmission, Wisden implements a negative acknowledgment (NACK), hybrid hop-by-hop/end-to-end transmission scheme. The hop-by-hop scheme enables intermediate nodes to identify and retransmit lost messages by observing gaps between received sequence numbers. For this reason, every node stores transmitted messages in a message cache. Wisden ensures reliable transmission of sensed data in the presence of up to 30% loss rate.

Another prototype was developed at the University of California, Berkeley, and deployed at the Golden Gate Bridge in San Francisco (Kim et al. 2007). The bridge has a center span that sustains a maximum transverse deflection (due to wind or earthquake) of 27.7 ft and maximum upward and downward deflections of 5.8 ft and 10.8 ft, respectively. The towers are 500 ft high above the roadway and 746 ft high above the water. The tops of the towers can have transverse deflections of up to 12.5 in. and toward the shore longitudinal deflections of 22 in. Sixty-four wireless sensor nodes were deployed on this

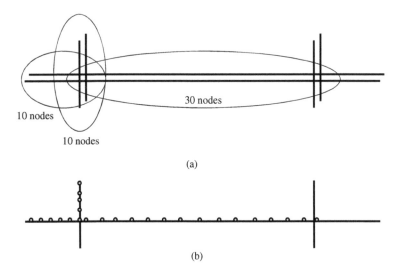

Figure 2.2   The deployment scenario of on the Golden Gate Bridge. (a) The nodes are deployed on both side of the span. (b) A two-dimensional view of the placement of nodes on the bridge.

bridge to establish a structural health monitoring network. The nodes were distributed over the main span and the tower, collecting ambient vibrations synchronously, at a rate of 1 kHz, with less than 10 μs jitter and with an accuracy of 30 μG. Data is collected reliably over a 46-hop network. Figure 2.2 illustrates the deployment at the Golden Gate Bridge.

The goal of the deployment was to determine the response of the structure to both ambient and extreme conditions and to compare actual performance with design predictions. The network measured ambient structural accelerations from wind load at closely spaced locations. It also measured strong shaking from a potential earthquake. The installation as well as the monitoring, was conducted without disrupting the operation of the bridge.

2.2 Traffic Control

Ground transportation is a vital and complex socioeconomic infrastructure. Operationally, it is linked with and provides support for a variety of systems, such as supply-chain, emergency response, and public health. In urban areas, this results in potential congestion. The 2009 Urban Mobility Report, issued by the Texas Transportation Institute, reveals that in 2007, congestion caused urban Americans to travel 4.2 billion hours more and to purchase an extra 2.8 billion gallons of fuel. The total congestion cost is estimated to be $87.2 billion – an increase of more than 50% over the previous decade.

Unfortunately, building new roads is not a feasible solution for many cities of the world owing to the lack of free space and the high cost of demolition of old roads (streets). Many consider better regulation of transportation systems as the only sustainable solution to road congestion.

One approach to dealing with congestions is to put in place distributed sensing systems that reduce congestion. These systems gather information about the density, sizes, and speed of vehicles on roads; infer congestions; and suggest to drivers some alternative routes and emergency exits.

2.2.1 The Sensing Task

A large number of devices are used in traffic control systems. These include video, sonar, radar, inductive loops, magnetometers, microloop probes, pneumatic road tubes, piezoelectric cables, PVDF wire, and pneumatic treadle. Video and sonar-based sensing systems need to be installed on poles, while inductive loops, magnetometers, and pneumatic treadles can be embedded into the transportation infrastructure. Camera-based systems involve human operators to process images, identify incidents, and assign speed rankings. Apparently, this technique is costly and can only be employed in selected streets, such as those that are frequently traveled.

Another way is to fully automate congestion recognition. There are several approaches to do this. For example, automated camera-based systems use machine vision to count and classify vehicles. Alternatively, they target the license numbers of passing vehicles and associate driving history as a means of estimating congestion causes. These approaches are well suited as long as the data from the cameras is reliable. In the presence of fog, smog, dust, snow, or rain, however, roadside cameras are unreliable.

2.2.1.1 Inductive Loops

More recently, in-road sensing devices have been developed as complementary systems. These devices are interesting because they are unaffected by weather and provide direct information with very little ambiguity. One of the most common in-road traffic sensors is the inductive loop (Knaian 2000). This is a coil of wire several meters in diameter and can be buried under the road and connected to a roadside control box that passes an electric current through the coil. By establishing relationship between the current, the magnetic field strength that is induced as a result of the current, and the speed and size of the passing vehicles, it is possible to infer traffic flow. The exact relationship between the current and a vehicle can be defined by using Faraday's induction law.

According to Faraday's law, when an electric current passes through a conductor, it produces a magnetic field around the conductor. The direction of the field is normal to the direction of the current flow. The strength and density of the magnetic field depends on the length and cross-sectional area of the conductor as well as the material from which the conductor is made, that is, the permeability of the conductor, μ. The ratio of the magnetic flux, Φ, to the current is called inductance, L, which is defined as:

$$L = \frac{\Phi}{i} \tag{2.19}$$

If, instead of a straight conductor, the current passes through a solenoid (a long, thin coil, with a length much greater than the diameter of the loop) of N turns and length l, the magnetic flux density, **B**, induced in it is expressed as:

$$\mathbf{B} = \mu_0 \frac{Ni}{l} \tag{2.20}$$

where μ_0 is the permeability of free space; N is the number of turns; i is the current; and l is the length of the coil. The magnetic flux through the coil is obtained by multiplying the flux density **B** by the cross-sectional area, A and the number of turns, N:

$$\Phi = \mu_0 N^2 i \frac{A}{l} \tag{2.21}$$

Reformulating Equation (2.21) will yield:

$$L = \mu_0 N^2 \frac{A}{l} \tag{2.22}$$

The inductance of a solenoid changes when vehicles drive on the road, disturbing the induced magnetic flux. The magnitude of the change depends on the vehicle's speed and size. To determine the speed of the vehicle, two loops separated by a distance, d, of known length are sufficient.

Measuring the change in voltage or current is easier than measuring the change in magnetic field strength or magnetic flux. The induced electromotive force in a closed loop is directly proportional to the rate of change of magnetic flux through the loop. This can be better explained by moving a conductor through a magnetic field, which induces a voltage in that conductor. The induced voltage is proportional to the speed of movement, the length and cross-sectional area of the conductor, and the strength of the magnetic field. If the conductor forms a solenoid, the number of turns of the conductor influences the induced voltage.

The magnetic field, the direction of movement, and the voltage are all orthogonal to each other. To determine their exact direction, Fleming's right-hand rule can be applied. Alternatively, it is also possible to keep the conductor stationary and vary the magnitude and direction of the magnetic field to induce a voltage in a conductor. Mathematically, this is expressed as:

$$\varepsilon = -N\frac{d\Phi_B}{dt} \tag{2.23}$$

where ε is the electromotive force (emf) in volts; N is the number of turns of wire; Φ_B is the magnetic flux in Weber that passes through a single loop. The negative sign in Equation (2.23) indicates that the direction of the electromotive force is opposite to the direction of the magnetic flux. The magnetic flux is a function of the cross-sectional area of the conductor and the magnetic field strength, which is normal to the conductor. Lenz's law can be applied to determine the direction of the induced electromotive force (emf) and current resulting from electromagnetic induction.

$$\frac{d\Phi}{dt} = \frac{A\,dB}{dt} \tag{2.24}$$

More generally, the relation between the rate of change of the magnetic flux through a surface S enclosed by a contour C and the electric field along the contour is expressed as:

$$\oint_C E \cdot dl = -\frac{d}{dt}\int_S \mathbf{B} \cdot d\mathbf{A} \tag{2.25}$$

where E is the electric field; dl is an infinitesimal element of the contour C; and $\mathbf{B}$ is the magnetic field strength. The directions of the contour C and $d\mathbf{A}$ are assumed to be related by the right-hand rule.

Equivalently, the differential form of Faraday's law can be employed:

$$\nabla \times \mathbf{E} = -\frac{\partial \mathbf{B}}{\partial t} \tag{2.26}$$

The only limitation of inductive loops is their physical size. First, deployment requires the complete dismantlement of an entire cross-section of a road. Second, it is difficult to distinguish vehicles in bumper-to-bumper traffic, since two vehicles may cross the loop at the same time.

2.2.1.2 Magnetic Sensors

The presence, direction, and speed of a vehicle can be determined by employing magnetic sensors. The technique requires a magnetic field of known strength and direction. A moving vehicle can disturb the distribution of the magnetic field either by producing its own magnetic field or simply by cutting across it. As the magnitude and direction of the disturbance depends on the speed, size, density, and permeability of the vehicle, it is possible to use magnetic sensors to quantify the disturbance.

Magnetic sensors can be classified into low-field, medium-field, and high-field sensors, according to the range of the strength of the magnetic field they measure (Caruso and Withanawasam 1999). Low-field sensors measure magnetic field strength below $1\,\mu G$

(micro Gauss); medium-field sensors measure between 1 μG and 10 G; high-field sensors can measure above 10 G. The Earth's magnetic field is found in the medium field.

Magnetic fields are set up by the motion of electrical charges. For instance, the magnetic field of a bar magnet is created by the motion of negatively charged electrons within iron atoms. The cause of the Earth's magnetic field is not completely understood, but it is believed to be associated with electrical currents produced by the coupling of convective effects and rotations in the spinning liquid metallic outer core of iron and nickel. It has a uniform distribution over a wide area (several kilometers). It was first measured by Carl Friedrich Gauss in 1835 and has been repeatedly measured since then, showing a relative decay of about 5% over the last 150 years.

Sensors that can measure the Earth's magnetic field comprise an alloy of nickel and iron. Typical examples are anisotropic magnetoresistive (AMR) sensors whose resistive property changes according to the Earth's magnetic field strength. AMR sensors can measure both linear and angular positions and displacement in the Earth's magnetic field.

Almost all road vehicles, including those with polymer body panels, contain a large mass of steel. Since the magnetic permeability of steel is much higher than the surrounding air, it has the capacity to concentrate the flux lines of the Earth's magnetic field. The concentration of magnetic flux (disturbance) at a particular location varies as the vehicle moves and can be detected from a distance of up to 15 m (Weaver 2003). Figure 2.3 demonstrates how an AMR sensor can be used to measure the disturbance in the Earth's magnetic field caused by a moving vehicle.

It is possible to distinguish between different types of vehicles (car, bus, minibus, truck, etc.) by modeling a vehicle as a composition of many dipole magnets (Caruso and Withanawasam 1999). These dipoles have north–south orientations that cause distortions in the Earth's magnetic field. The extent of the distortions of the dipoles depends on, among other things, the permeability of the dipoles. For example, the engine and wheel areas exert stronger distortions than the other parts of a vehicle, and for each vehicle class of interest, it is possible to produce a unique model. When a vehicle passes close to a magnetic sensor, or drives over it, the sensor can detect the different dipole moments of the various parts of the vehicle. The field variation reveals a detailed magnetic signature.

Figure 2.3 Detection of a moving vehicle with an AMR magnetic sensor (Caruso and Withanawasam 1999).

2.2.2 Prototypes

Knaian (2000) proposes the use of wireless sensor networks for traffic monitoring in urban areas. A prototype was deployed in Vassar Street, Cambridge, Massachusetts. The wireless sensor node consists of two AMR magnetic sensors for detecting vehicular activities and a temperature sensor for monitoring road condition (snow, ice, or water). The movement and speed of a vehicle is captured by observing the disturbance it creates in the Earth's magnetic field. This takes the form of an excursion first below and then above a predefined baseline, since the vehicle pulls field lines away from the sensor when it approaches it and then toward the sensor when it drives away from it.

To measure the speed of a vehicle, the node waits until it detects an excursion from the baseline and then starts sampling at a frequency of 2 kHz. An AMR magnetic sensor is placed at the front of the node and another at the back. The waveforms at the outputs of the sensors are identical, except that they are shifted in time and may be affected by noise. When the signal from the rear sensor crosses the baseline, the node begins to count the number of samples until the signal from the forward sensor crosses the baseline. From this count, it computes the speed of the passing vehicle.

In order to detect a vehicle during bumper-to-bumper drive, a minimum of three samples at a frequency of 100 Hz are required for the following vehicle description: average drive speed is 40 mph (or 20 meters per second) and the size of the engine block is 60 cm. In other words, the sampling frequency f_s should be:

$$f_s = 20 \, \text{m/s} \times \frac{3 \times 100}{60 \, \text{cm}} = 100 \, \text{Hz} \tag{2.27}$$

Sampling at this rate enables the node to detect vehicles that travel at a higher speed (as fast as 200 mph), with a minimum separation distance of 3 m. Figure 2.4 displays the block diagram of the sensor node developed at MIT.

Arora et al. (2004) deployed 90 Mica2 sensor nodes at MacDill Air Force Base in Tampa, Florida, to detect the movement of vehicles, soldiers, and people. Seventy-eight of the nodes were magnetic sensor nodes that were deployed in a 60 × 25 square foot area. Additionally, 12 radar sensor nodes were overlaid on the network. The magnetic sensor nodes were distributed uniformly. These nodes form a self-organizing network which connects itself to a remote computer via a base station and a long-haul radio repeater.

The Mica2 nodes were based on a 4 MHz Atmel processor with 4 kbytes random access memory, 128 kbytes of flash program, and 512 kbytes of EEPROM memory (used for data logging). The TinyOS operating system runs on the nodes. Magnetic fields were sensed by using an in-built magnetometer while the TWR-ISM-002 radar motion sensor was used to detect movement of objects.

2.3 Health Care

A wide range of health care applications have been proposed for wireless sensor networks, including monitoring patients with Parkinson's Disease, epilepsy, heart patients, patients rehabilitating from stroke or heart attack, and elderly people. Unlike other types of applications discussed so far, health care applications do not function as stand-alone systems. Rather, they are integral parts of a comprehensive and complex health and rescue system.

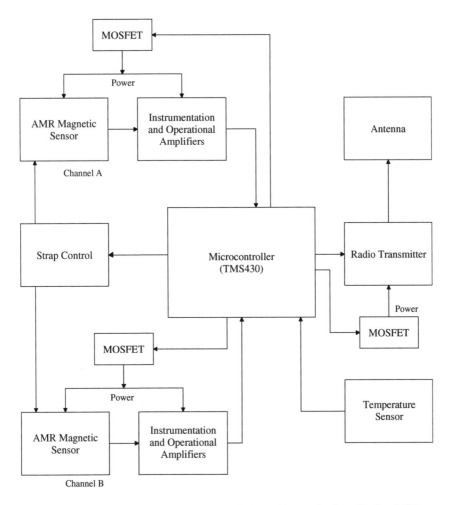

Figure 2.4   Block diagram of the MIT node for traffic monitoring (Knaian 2000).

According to the US Centers for Medicare and Medicaid Services (CMS), the national health spending of the country in 2008 was estimated to be $2.4 trillion. The cost of heart disease and stroke takes around $394 billion. The same report projects a rise in health spending in the US as well as in many western countries. Apparently, this is a concern for policy makers, health care providers, hospitals, insurance companies, and patients.

Interestingly, higher spending does not necessarily correlate with a high quality service or prolonged lifetime (Kulkarni and Öztürk 2007). For example, in 2000, the US spent more on health care than any other country in the world – an average of $4500 per person – but ranked 27th in average life expectancy. Many countries achieved higher life expectancy rates at a lower cost.

While preventive health care has been advocated by many as a means to reduce health spending and mortality rate, studies show that some patients find that certain practices are inconvenient, complicated, and interfere with their daily life (Morris 2007). For example, many miss checkup visits or therapy sessions because of a clash of schedules with established living and working habits, fear of overexertion, or transportation cost.

To deal with these problems, research attempts to provide a comprehensible solution that involves the following tasks:

- building pervasive systems that provide patients with rich information about diseases and their prevention mechanisms;
- seamless integration of health infrastructures with emergency and rescue operations as well as transportation systems;
- developing reliable and unobtrusive health-monitoring systems that can be worn by patients to reduce the responsibilities and presence of medical personnel;
- alerting nurses and doctors when medical intervention is necessary; and
- reducing inconvenient and costly checkup visits by creating reliable links between autonomous health-monitoring systems and health institutions.

2.3.1 Available Sensors

The research community has been very active in developing a plethora of wearable and wireless systems that seamlessly monitor heart rate, oxygen level, blood flow, respiratory rate, muscle activities, movement patterns, body inclination, and oxygen uptake (V_{O_2}). Given below is a concise summary of some of the commercially available wireless sensor nodes for health monitoring:

- pulse oxygen saturation sensors: they measure the percentage of hemoglobin (Hb) saturated with oxygen (SpO_2) and heart rate (HR);
- blood pressure sensors;
- electrocardiogram (ECG);
- electromyogram (EMG) for measuring muscle activities;
- temperature sensors – both for core body temperature and skin temperature;
- respiration sensors;
- blood flow sensors;
- blood oxygen level sensor (oximeter) for measuring cardiovascular exertion (distress).

2.3.2 Prototypes

2.3.2.1 Artificial Retina

Schwiebert *et al.* (2001) developed a microsensor array that can be implanted in the eye as an artificial retina to assist people with visual impairments. The system consists of an integrated circuit and an array of sensors. The integrated circuit is a multiplexer with on-chip switches and pads to support a 10×10 grid of connections; it operates at 40 kHz. Moreover, it has an embedded transceiver for wired and wireless communications. Each connection in the chip interfaces a sensor through an aluminum probe surface. Before the bonding is done, the entire integrated circuit, except the probe areas, is coated with a biologically inert substance.

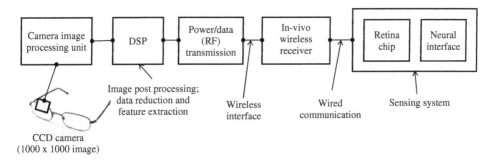

Figure 2.5 The processing components of the artificial retina (Schwiebert *et al.* 2001).

Each sensor is a microbump. It starts with a rectangular shape, but near the end tapers to a point and rests gently on the retina tissue. The sensors are sufficiently small and light to be held in place with relatively little force. The distance between adjacent microbumps is approximately 70 μm. The sensors produce electrical signals proportional to the light reflected from an object being perceived. The ganglia and additional tissues transform the electrical energy into chemical energy, which in turn is transformed into optical signals and communicated to the brain through the optic nerves. The magnitude and wave shape of the transformed energy corresponds to the response of a normal retina to light stimulation.

The system is a full duplex system, allowing communication in a reverse direction. In addition to the transformation of electrical signals into optical signals, neurological signals from the ganglia can be picked up by the microsensors and transmitted out of the sensing system to an external signal processor. In this way, the sensor array is used as a reception and transmission system in a feedback loop.

Two types of wireless communications are foreseen. First, the eventual mapping of an input electrical signal to brain patterns cannot be realized internally with the sensing system alone since signal processing is a computationally intensive process. Second, diagnostic and maintenance operations require the extraction of data from the sensing system. For these reasons, interconnecting the sensing system with an external system is essential. In addition to these requirements, communication is required to transfer data from a charge-coupled device (CCD) camera (embedded in a pair of spectacles) to the sensor array.

Figure 2.5 illustrates the signal processing steps of the artificial retina. A camera embedded in a pair of spectacles directs its output to a real-time digital signal processor (DSP) for data reduction and processing. The camera can be combined with a laser pointer for automatic focusing. The output of the DSP is compressed and transmitted through a wireless link to the implanted sensor array, which decodes the image and produces a corresponding electrical signal.

2.3.2.2 Parkinson's Disease

Lorincz *et al.* (2009) and Weaver (2003) propose the use of WSNs to monitor patients with Parkinson's Disease (PD). The aim is to augment or entirely replace a human observer and to help the physician to fine-tune the medication dosage.

Parkinson's Disease is a degenerative disorder of the central nervous system. It results from degeneration of neurons in a region of the brain that control movements (substantial nigra), creating a deficiency in neurotransmitter dopamine. Deficiency in dopamine causes severe impairment of motor skills and speech, manifesting itself in tremor of the hands, arms, legs, and jaw; unsteady walk and slowness of movements; and lack of balance and coordination.

According to the Parkinson's Disease Foundation (Foundation P.D. 2009), as many as one million Americans live with PD. Approximately 60,000 are diagnosed with the disease each year without considering the thousands of additional cases that go undetected. The foundation estimates that 4 million people worldwide are living with PD. The combined direct and indirect cost of PD, including treatment, social security payments, and lost income from inability to work, is estimated to be nearly $25 billion per year in the United States alone.

Persons under treatment (this is usually an external stimulation given to remnant cells in the substantia nigra to produce more dopamine) can be found in one of the following three phases (Weaver 2003):

1. The exhibition of typical symptoms in the form of tremor and slow movement when the stimulation has worn off. This is known as the *off* state.
2. Normal movements free of tremor when the medication is balanced. This is known as the *on* state.
3. Exaggerated involuntary movements when the medication is at highest concentration. This is known as *dyskinesia*.

The treatment of PD is case-specific, that is, doctors typically monitor patients on an individual basis and provide suitable medication to prolong the duration of the *on* state. This requires a close follow-up of the medication cycle and frequent adjustments. Apparently, the follow-up is costly. According to the Parkinson's Disease Foundation, the medication costs for an individual person are estimated to be $2,500 a year and therapeutic surgery can cost up to $100,000 per patient (Foundation P.D. 2009).

Weaver (2003) developed a wearable system that can reduce personnel cost and help a physician to fine-tune the medication dosage. It consists of a lightweight sensor node with 3D accelerometer sensors, a processor core, and a storage system for logging data for later retrieval. The system could record 17 hours of accelerometer data. The accelerometer sensors can be sampled at a rate of 40 Hz.

The system was deployed on PD patients at the Memorial Hospital's Parkinson Day Center, Cambridge, MA, and a large amount of data was collected while they performed common daily tasks (walking, reading quietly while sitting, sitting in animated conversation, etc.). The patients wore the nodes on their ankles and wrists. The report reveals that the system was able to identify the occurrence of dyskinesia at the rate of 80%.

More recently, Lorincz et al. (2009) at Harvard University employed a more sophisticated wireless sensor node – the Shimmer wireless sensor platform (Sensor platform TSW 2009) – for monitoring patients with PD and epilepsy. The node consists of a TI MSP430 processor, CC2420 IEEE 802.15.4 radio, triaxial accelerometer, and rechargeable Li-polymer battery. It also integrates a MicroSD slot that supports a Flash memory for storing accelerometer data. This way, the node is capable of storing data from the 3D accelerometer sensor continuously for more than 80 days at a sampling frequency of 50 Hz. In addition to the 3D accelerometer sensor, the node platform provides interfaces for gyroscope, ECG, EMG, tilt and vibration sensor, and a passive infrared (PIR) motion sensor.

These nodes were deployed on seven patients for one week. Nine nodes were deployed on the body of each patient, two on each arm and leg, and one at the back. Data from 3D accelerometer sensors and gyroscopes were sampled at a frequency of 100Hz. An initial clinical evaluation was made in various settings, including tuning of deep brain stimulation (DBS) parameters.

2.4 Pipeline Monitoring

Another area of application for wireless sensor networks is the monitoring of gas, water, and oil pipelines. The management of pipelines presents a formidable challenge. Their long length, high value, high risk, and often difficult access conditions require continuous and unobtrusive monitoring. Leakages can occur due to excessive deformations caused by earthquakes, landslides, or collisions with an external force; corrosion, wear, material flaws or even intentional damage to the structure.

To detect leakages, it is vital to understand the characteristics of the substance the pipelines transport. For example, fluid pipelines generate a hot-spot at the location of the leak, whereas gas pipelines generate a cold-spot due to the gas pressure relaxation. Likewise, fluid travels at a higher propagation velocity in metal pipelines than in polyvinyl chloride (PVC). There are a large number of commercially available sensors (fiber optics, temperature sensors, and acoustic sensors) to detect and localize thermal anomalies.

2.4.1 Prototype

The PipeNet prototype was developed as a collaboration project between Imperial College, London, Intel Research, and MIT to monitor water pipelines in urban areas. Its main task is to monitor (1) hydraulic and water quality by measuring pressure and pH, and (2) the water level in combined sewer systems (sewer collectors and combined sewer outflows). Sewerage systems convey domestic sewage, rainwater runoff, and industrial wastewater to sewerage treatment plants. Historically, these systems are designed to discharge their content to nearby streams and rivers in the event of overflow, such as during periods of heavy rainfall. Subsequently, the combined sewer overflows are among the major sources of water quality impairment. Nearly 770 large cities in the US, mainly older communities, have combined sewer systems (Stoianov *et al.* 2007).

PipeNet is deployed in three different settings. In the first setting, pressure and pH sensors are installed on a 12 in. cast-iron pipe which supplies drinking water. Pressure data is collected every 5 min for a period of 5 s at a rate of 100 Hz. The wireless sensor node can locally compute minimum, maximum, average, and standard deviation values and communicate the results to a remote gateway. Likewise, pH data is collected every 5 min for a period of 10 s at a rate of 100 Hz. The sensor nodes use a Bluetooth transceiver for wireless communication.

The pressure sensor is a modified version of the OEM piezoresistive silicon sensor. It has an error compensation mechanism to deal with the effects of nonlinearity and hysteresis. The sensor has a startup time of less than 20 ms and a fast dynamic response. It consumes less than 10 mW. The pH sensor is a glass electrode with an *Ag/AgCl* reference cell.

In the second setting, a pressure sensor is employed to measure the pressure in 8 in. cast iron pipe. The data is collected every 5 min for a period of 5 s at a sampling rate of 300 Hz. For this setting, local processing was not supported; instead, the raw data was transmitted to a remote gateway.

Finally, in the third setting, the water level of a combined sewer outflow collector is monitored. Two pressure transducers were placed at the bottom of the collector and an ultrasonic sensor on the top. The pressure sensors are low-power devices and consume less than 10 mW. The ultrasonic sensor is a high-power device and consumes 550 mW. For efficient power consumption, the pressure sensors are employed for periodic monitoring while the ultrasonic sensor is required only to verify the readings from the pressure sensors when their difference exceeded a set threshold or when the water level exceeded the weir height. In this setting, data collection was carried out at a rate of 100Hz at 5 min intervals for a period of 10 s. Moreover, local data aggregation was performed to reduce the network traffic. The network supported remote configuration to increase the sampling rate up to 600 Hz.

2.5 Precision Agriculture

Another interesting area where wireless sensor networks motivated a large number of researchers is precision agriculture. Traditionally, a large farm is taken as a homogeneous field in terms of resource distribution and its response to climate change, weeds, and pests. Accordingly, farmers administer fertilizers, pesticides, herbicides, and water resources. In reality, a large field exhibits wide spatial diversity in soil types, nutrient content, and other important factors. Therefore, treating it as a uniform field can cause inefficient use of resources and loss of productivity.

Precision agriculture is a method of farm management that enables farmers to produce more efficiently through a frugal use of resources. This encompasses different aspects, such as micro-monitoring soil, crop, and climate change in a field, and providing a decision support system (DSS). Precision agriculture uses Geographic Information System management tools; GPS, radar, aerial images, etc., to accurately diagnose a field and apply vital farming resources.

A large number of technologies have been developed over the last several years to facilitate and automate precision agriculture. Some of these are:

- *Yield monitors:* These are devices that use, among other things, mass flow sensors, moisture sensors, and a GPS receiver to monitor instantaneous yield based on time and distance. The sensors enable measurement of the mass or the volume of grain flow (grain flow sensors), separator speed, ground speed, grain moisture, and header height.
- *Yield mapping:* Couples GPS receivers with yield monitors to provide spatial coordinates for the yield monitor data.
- *Variable rate fertilizer:* Manages the application of liquid and gaseous fertilizer materials.
- *Weed mapping:* Enables a farmer to map weeds while combining, seeding, spraying, or field scouting.
- *Variable spraying:* By knowing weed locations from weed mapping, spot control can be implemented. This enables booms to be turned on and off electronically and alter the amount (and blend) of herbicide applied.

- *Topography and boundaries:* Enable the production of very accurate topographic maps that can be used to interpret yield maps and weed maps as well as planning for grassed waterways and field divisions. Field boundaries, roads, yards, tree stands, and wetlands can all be accurately mapped to aid farm planning.
- *Salinity mapping:* This is used to map fields that are affected by salinity. Salinity mapping is valuable in interpreting yield maps and weed maps as well as tracking the change in salinity over time.
- *Guidance systems:* These are devices that can accurately position a moving vehicle within a 12 in. radius (or less). They are useful for spraying and seeding as well as field scouting.

The main challenge in applying precision agriculture technologies is the need to collect amount of data over several days that is large enough to characterize the entire field. In this regard, wireless sensor networks can be excellent tools as large-scale sensing technologies.

2.5.1 Prototypes

Several prototype deployments have already been carried out in Spain (López Riquelme *et al.* 2009), the US (Pierce and Elliott 2008), Canada (Beckwith *et al.* 2004), The Netherlands (Baggio 2005), India (Panchard *et al.* 2007), and Italy (Matese *et al.* 2009), to mention just a few. In the following subsections, a brief summary of some of the prototypes is given.

2.5.1.1 Wine Vineyard

Beckwith *et al.* (2004) deployed a wireless sensor network in Okanagan Valley, British Columbia, to monitor and characterize significant variation in temperature – heat summation and periods of freezing temperatures – over one management block of a wine vineyard. In the vineyard, temperature is the predominant parameter that affects the quality as well as the quantity of harvest. For example, wine grapes see no real growth until the temperature goes above $10\,°C$. Moreover, different wine grapes have different requirements for heat units (in other words, different sites will be able to support different grapes). Subsequently, the deployment aims to measure the temperature over a $10\,°C$ baseline that a site accumulates over the growing season.

The network consisted of 65 nodes distributed in a grid-like pattern 10 to 20 meters apart, covering about 2 acres. The experience shows that due to the self-configuring nature of the network and the inherent structured layout of vineyard fields, the planning and execution of the deployment process was easy – it took the researchers approximately 24 man-hours to deploy the network.

The network topology was determined by two essential constraints: placement of nodes in an area of viticulture interest and the support for multi-hop communication. The network was reliably established and functional for the period from the onset of grape maturity through the second major Arctic outflow (i.e., cold front) in the region. The data was used to investigate several aspects:

- the existence of covariance between the temperature data collected by the network with known agriculturally significant data;

- growing degree day differences; and
- potential frost damage.

The mean data enabled the observation of the relative differences in the accumulation of heat units during that period. According to the authors' report, the mean ranged from 7.95 to 11.94 °C. Moreover, the cooler areas accumulated two-thirds of the heat units of the warmest areas. By overlaying the temperature data onto a topological map, it was possible to observe that while the temperature covaried to some degree with elevation and aspect, these did not predict temperature value very well. An interesting conclusion was the extent of variation in this vineyard – there was a measured difference of over 35% of heat summation units (HSUs) in as little as 100 meters.

2.5.1.2 Lofar Agro

Baggio (2005) reports the deployment of a wireless sensor network at Lofar Agro, The Netherlands. The network was tasked to monitor phytophthora, a fungal disease, in a potato field. Whereas many factors contribute to phytophthora, climatological conditions are supposed to be the main causes. Thus, by observing the humidity and temperature conditions in the field as well as the wetness of the potato leaves, the researchers attempted to determine the potential risk of the disease and the requirements for fungicide.

With this goal in mind, a potato field was instrumented with a wireless sensor network. The network consisted of 150 wireless sensor nodes, each of which integrates temperature and humidity sensors. Because the radio range of these nodes dramatically reduced when the potato crop was flowering, an additional 30 nodes were deployed as relaying nodes to ensure the network's connectivity. These relaying nodes were installed at a height of 75 cm to enhance communication, while the sensing nodes were installed at a height of 20, 40, and 60 cm.

Furthermore, the field was equipped with a weather station to measure luminosity, air pressure, precipitation, and wind strength and direction. Since the humidity of the soil was considered to be the major factor in the development of the micro climate, a large number of soil humidity sensors were used.

The nodes sampled surrounding temperature and humidity at a rate of 1 sample per minute and stored the results temporarily. Afterwards, the data was communicated to a remote base station every 10 min. To efficiently utilize energy, delta encoding (in which 10 samples were encoded in a single packet) and periodic sleeping (with a 7% duty cycle) techniques were used.

The sampled data were logged to a server through a gateway and a backbone network. The server logged the data, filtered out erroneous readings, and handed the accumulated data to the phytophthora decision support system (DSS) server. Finally, the DSS combined the field data with a detailed weather forecast to determine the treatment policy.

2.6 Active Volcano

Monitoring active volcanoes is another application domain for wireless sensor networks (WSNs).

Volcanoes occur when broken slabs of the Earth's outermost shell, known as lithosphere, float on the hotter and softer layer in the Earth's mantle. This phenomenon causes occasional collision between the lithosphere plates and is attributed to be the cause of most volcanoes.

In most cases, the Earth's volcanoes are hidden from view, occurring on the ocean floor along spreading ridges. Scientists attempt to capture and study the nature of active volcanoes by employing seismic and acoustic sensors and by collecting seismic and infrasonic signals. At present, typical active volcanoes are monitored by expensive devices that are difficult to move or require an external supply voltage. The deployment and maintenance of these devices require vehicle or helicopter assistance. Data storage is also a concern since, in typical scenarios, stations should log data to a Compact Flash card or a hard drive, which must be retrieved on a periodic basis.

WSNs can be very useful for active volcano monitoring. First, a large number of small, cheap, and self-organizing nodes can be deployed to cover a vast field. In contrast to the expensive and bulky equipments presently used, the deployment of WSNs is fast and economical. Second, through a high density and wide coverage, it is possible to achieve high spatial diversity. Third, the networks can operate without requiring stringent maintenance routines.

2.6.1 Prototypes

Werner-Allen *et al.* (2006) at Harvard University proposed (the use of WSNs) and deployed two prototype networks on two sites – namely, on Volcán Tungurahua in central Ecuador and Volcán Reventador in northern Ecuador. The first deployment (2004) in central Ecuador consisted of three wireless sensor nodes that integrated microphones. The second deployment (2005) consisted of a larger network, with 16 sensor nodes that integrated seismoacoustic sensors. The network had a linear topology and extended over a length of 3 km.

An important task in active volcano monitoring is to capture discrete events, such as eruptions, earthquakes, or tremor activities. Typically, these events are transient, lasting less than 60 s and occurring several times a day. Consequently, the nodes were tasked to gather data pertaining to these events and to collaborate with each other to support a multi-hop communication link.

The researchers employed the raw data to investigate volcanic activities. As a result, they were able to capture 230 volcanic events just over a period of three weeks. Interestingly, the prototype deployment was also used to investigate the performance of large-scale sensor networks for collecting high-resolution volcanic data. The researchers observed that the study of active volcanoes necessitates high data rates and data fidelity; and sparse arrays with high spatial separation between nodes.

It was reported that a single missed or corrupted sample could invalidate an entire record and small differences in sampling rates between two nodes could frustrate analysis. This implies that samples must be accurately time stamped to allow comparisons between correlated measurements.

The sensor architecture of the nodes deployed consisted of an 8-dBi 2.4-GHz external omnidirectional antenna, a seismometer, a microphone, and a custom hardware interface board. Fourteen of the 16 nodes were fitted with Geospace Industrial GS-11 geophones and single-axis seismometers with a corner frequency of 4.5 Hz, oriented vertically. The remaining two nodes integrated 3D Geospace Industries GS-1 seismometers with corner frequencies of 1 Hz, each axis producing a separate set of data.

The node's hardware consisted of a Texas Instruments MSP430 microcontroller, 48 kbytes of program memory, 10 kbytes of static RAM, 1 Mbyte of external flash memory, and a 2.4-GHz Chipcon CC2420 IEEE 802.15.4 radio. TinyOS was used as an operating environment. The flash memory was used for buffering raw data. The seismoacoustic sensors were interfaced with the node through an external board that provided up to four Texas Instruments AD7710 analog-to-digital converters (ADCs) to achieve a high resolution (24 bits per channel) data. Even though the microcontroller provided 16 bits on-board ADCs, they were considered to be inadequate for two main reasons. First, a resolution of at least 20 bits was required, whereas the on-board ADCs provided only 16 bits of resolution. Second, the seismoacoustic signals required a filter whose center frequency should be 50 Hz. This type of filter could not be realized with analog elements, but could be approximated digitally. The process, however, required oversampling. Thus, the AD7710 sampled at more than 30 kHz and provided a programmable output word rate of 100 Hz. Consequently, the high sample rate and computation that digital filtering required were delegated to a specialized device.

The deployment enables the researchers to clearly distinguish the tasks that can be carried out by WSNs from those that can be carried out by conventional techniques. For instance, due to good spatial distribution, WSNs enable large inter-node separations to obtain widely separated views of seismic and infrasonic signals as they propagate. Moreover, they are suitable for capturing triggered events. On the other hand, WSNs are inadequate for capturing complete waveforms for long duration.

2.7 Underground Mining

Finally, another application domain for which wireless sensor networks have been proposed is underground mining.

Underground mining is one of the most dangerous working environments in the world. Perhaps the incident of 3 August 2007 at the Crandall Canyon mine, Utah, USA, is a good example of the danger associated with underground mining. It also highlights some of the contributions of wireless sensor networks to facilitate safe working conditions and rescue operations.

In this fateful incident, six miners were trapped inside the coal mine. Though their precise location was not known, experts estimated that the men were trapped 457 m below ground, 5.5 km away from the mine entrance. There were different opinions about the exact cause of the accident. The owners of the mine claimed that a natural earthquake was the cause. Seismologists at the University of Utah observed that seismic waves of 3.9 magnitude were recorded on the same day in the area of the mine, leading scientists to suspect that mine operations were the cause of the seismic spikes.[1]

Following the accident, a costly and irksome rescue attempt was undertaken. This included the drilling of 6.4 cm and 26 cm holes into the mine cavity through which an omnidirectional microphone and a video camera were lowered, and an air sample was taken. The air sample indicated the presence of sufficient oxygen (20%), a small

---

[1] It has been claimed that workers were using a risky technique known as "retreat mining", with which the last standing pillars of coal were deliberately pulled down and the roof was allowed to fall in after an area was exhaustively mined.

concentration of carbon dioxide, and no trace of methane. The microphone detected no sound and the video camera revealed some equipment, but not the six missing miners.

This evidence caused a mixed anticipation. If the miners were alive, the amount of oxygen was sufficient to sustain life for some additional days. Moreover, the absence of methane gave hope that there would be no immediate danger of explosion as a result of the drilling of holes. However, the absence of carbon dioxide and the evidence from the camera and the microphone undermined the expectation of finding the missing persons alive. More than six labor-intensive days were required to collect the above evidence.

Unfortunately, despite the efforts, the rescue mission had to be suspended because an additional seismic shift in the mountain collapsed another part of the mine, resulting in a horizontal coal explosion onto the rescuers. The downward motion of the seismic waves indicated a further settling and collapse within the coal mine. This fact strengthened the seismologists' proposition that man-made causes produced the first incident. Three rescuers were killed and several were injured.

2.7.1 Sources of Accidents

Seismic shifts are not the only danger in underground mining. Explosions sparked by methane gas and coal-dust can cause significant damage. The following are some of the sources responsible for methane production:

1. A large portion of methane is created during a coalification process. [This is a process by which plant biomass is converted by biological and geological forces into coal. The methane is then stored in coal seams and the surrounding strata, most likely to be released during coal mining.]
2. Inadequate ventilation.
3. Methane from a fallen coal.
4. Methane from the mining faces.
5. Methane from the walls and ceilings of coal and rock roadways.
6. Methane from the gob of a coal mine.

Besides being the cause of mining explosions, methane emission from underground mining is a serious ecological threat. For example, the EPA[2] expects methane emissions from US coal mines to reach 28.0 MMTCE (4.9 Tg) by 2010, excluding possible Climate Change Action Plan (CCAP) reductions. This is because underground coal production – mined at increasingly greater depths – is projected to grow faster than surface production. At present, methane emission from underground (coal) mines accounts for around 10% of the total US anthropogenic methane emissions.

Coal dust is produced at every step of the mining process and accumulates through the movement of air and the transportation of coal: on the floors, walls, and ceilings of the mine, all the way from the mine entrance to the deepest shafts. Methane gas explosions create carbon monoxide when the density of the gas is high, but if there is not much gas, it is dispersed in the air. When a coal-dust explosion occurs, the coal dust does not burn completely, since the dust is a solid substance. The explosion forms a high-density

---

[2] *Source:* http://www.epa.gov/methane/sources.html

coal-dust cloud which prevents adequate air circulation. This contributes to the production of carbon monoxide. Even if a coal-dust explosion does not spread throughout the length and breadth of the mine, the resulting carbon monoxide gas does. In this way, all the workers can be exposed to poisonous gas.

2.7.2 The Sensing Task

Several sensing tasks can be defined for wireless sensor networks. First, they can be deployed to locate individuals under normal or abnormal (such as during an entrapment) situations. Second, they can be used to locate collapse holes. Third, they can be used to measure and forecast seismic shifts due to internal (mining operations) as well as external (earthquake) causes. Fourth, they can measure the concentration of gases, including methane, oxygen, and carbon dioxide.

As of November 2008, the US Mine Safety and Health Administration (MSHA) estimates the methane emissions from ventilation systems on a quarterly basis. Based on these measurements, MSHA estimates the average daily methane emissions for each underground mine. With MSHA's admission, there is an apparent measurement and reporting error associated with this method, as the average of the four quarterly measurements cannot be representative of the true average at a given mine. The average emissions at a particular mine may be over- or underestimated.[3]

Having said this, there are formidable challenges to network deployment in underground mining, namely, the extreme hostile environment for radio communication. First, because of the turns and twists of underground tunnels, it is impossible to maintain a line-of-sight communication link, and signals arrive at their destination after being highly reflected, refracted, and scattered. Second, because of the high percentage of relative humidity, signal absorption and attenuation are extremely high.

Consequently, even though there are some reports of prototype deployments (see, for example, Li and Liu 2009 and Chehri *et al.* 2009), the formidable communication challenges undermine the results.

Exercises

2.1 Most applications in wireless sensor networks extract time and frequency domain features to detect interesting events. Define the following features:

(a) Autocorrelation function
(b) Correlation coefficients
(c) Cross-correlation function
(d) Autoregression function
(e) Coherence

2.2 Explain the difference between time domain and frequency domain features.

2.3 A 2D accelerometer sensor measures the movement of a structure to an ambient excitation. The normalized raw data that is collected for 1 second from the x- and y-axes

---

[3] *Source:* http://www.msha.gov/. Last visited November 20, 2009

are given below. In each case, the measurement is one-dimensional and should be read from left to right and top to bottom.

$$
x = \begin{bmatrix}
0.13 & 0.13 & 0.13 & 0.11 & 0.09 & 0.08 & 0.06 & 0.05 & 0.04 & 0.02 \\
-0.01 & -0.02 & -0.01 & -0.02 & -0.04 & -0.06 & -0.11 & -0.12 & -0.13 & -0,10 \\
0.12 & 0.00 & -0.06 & -0.03 & 0.00 & 0.02 & 0.02 & 0.03 & 0.03 & 0.03 \\
0.03 & 0.03 & 0.03 & 0.02 & 0.03 & 0.03 & 0.02 & 0.03 & 0.02 & 0.02 \\
0.03 & 0.02 & 0.02 & 0.03 & 0.03 & 0.02 & 0.01 & 0.05 & 0.05 & 0.03 \\
0.08 & -0.04 & 0.02 & -0.03 & -0.07 & 0.06 & 0.18 & 0.14 & 0.08 & 0.04 \\
0.03 & 0.03 & 0.02 & 0.00 & -0.03 & -0.07 & -0.13 & -0.21 & -0.31 & -0.31 \\
-0.42 & -0.37 & -0.28 & 0.31 & -0.01 & -0.28 & 0.12 & -0.12 & 0.04 & -0.01 \\
0.03 & 0.03 & 0.02 & 0.03 & 0.03 & 0.03 & 0.03 & 0.02 & 0.02 & 0.02 \\
0.03 & 0.02 & 0.03 & 0.03 & 0.03 & 0.03 & 0.02 & 0.02 & 0.03 & 0.12
\end{bmatrix}
$$

$$
y = \begin{bmatrix}
-0.01 & -0.02 & -0.02 & -0.02 & -0.04 & -0.04 & -0.03 & -0.02 & -0.02 & -0.02 \\
-0.03 & -0.03 & 0.01 & 0.02 & 0.02 & 0.03 & 0.02 & 0.03 & 0.05 & 0.13 \\
-0.01 & 0.04 & -0.02 & -0.06 & 0.02 & -0.01 & 0.01 & 0.00 & 0.01 & 0.01 \\
0.01 & 0.01 & 0.01 & 0.01 & 0.01 & 0.01 & 0.01 & 0.01 & 0.01 & 0.01 \\
0.01 & 0.02 & 0.02 & 0.01 & 0.01 & 0.01 & 0.01 & 0.01 & -0.02 & -0.07 \\
0.03 & -0.09 & -0.05 & -0.06 & -0.14 & -0.18 & -0.03 & 0.05 & 0.01 & -0.05 \\
-0.04 & -0.02 & -0.02 & -0.03 & -0.04 & -0.05 & -0.07 & -0.04 & 0.00 & 0.01 \\
0.02 & 0.11 & 0.00 & -0.07 & 0.40 & -0.06 & -0.09 & 0.17 & -0.03 & 0.04 \\
0.01 & 0.01 & 0.01 & 0.01 & 0.01 & 0.00 & 0.01 & 0.02 & 0.01 & 0.01 \\
0.01 & 0.02 & 0.02 & 0.02 & 0.01 & 0.01 & 0.01 & 0.02 & 0.00 & -0.02
\end{bmatrix}
$$

(a) Calculate the autocorrelation for both sequences.

(b) Calculate the correlation coefficients of the sequences.

(c) Calculate the Fast Fourier Transform (FFT) of both sequences.

2.4 To improve the expressiveness of frequency domain features, it is preferred to compute the Short Time Fourier Transform (STFT) of a time series sequence instead of the FFT of the entire frame.

(a) Divide the 1 s frame into 10 subframes such that there is an overlap of 50% between each subframe, except the first and last.

(b) Calculate the STFT for each window.

(c) Now reduce the overlap to 25%, compute the STFT, and compare the results with the results obtained from the 50% overlap subframes.

2.5 How can oversampling of sensor data overcome the effect of noise?

2.6 One of the time domain features used to recognize interesting events is the zero-crossing rate, which can be expressed as:

$$
\text{ZCR}(s) = \frac{1}{T} \sum_{i=0}^{T-1} F(s(i) \cdot s(i-1) < 0)
$$

where s is a discrete, time-series sequence; and $s(i)$ and $s(i-1)$ are two consecutive samples. $F = 1$ if the evaluation is true, $F = 0$ otherwise.

(a) Compute the zero-crossing rates for the two time series measurements given above.

(b) What conclusion can be drawn from the zero-crossing rate?

2.7 Another interesting feature is the spectral centroid, a frequency domain feature which represents the balancing point of the spectral power distribution:

$$C_t = \frac{\sum_{n=1}^{N} M_t[n] \cdot n}{\sum_{n=1}^{N} M_t[n]}$$

where $M_t[n]$ is the magnitude value of the spectrum at position (frequency) n. Calculate the spectral centroid for the two time series sequences given above.

2.8 In structural health monitoring, inspection techniques are classified into global and local. Explain the difference between these techniques. For which of these techniques are wireless sensor networks suitable?

2.9 Explain how the property of a pipeline changes at a location where gas and oil leakages occur.

2.10 Explain how an acoustic sensor can be used to monitor the content of a pipeline.

2.11 Explain the principle of a piezoelectric sensor to measure movement?

2.12 How can a magnetic sensor be employed to measure the movement of vehicles?

2.13 What is an electromyograph and for what application can it be used?

2.14 Describe the three phases of Parkinson's Disease.

2.15 What is a heat unit?

References

Arora, A., Dutta, P., Bapat, S., Kulathumani, V., Zhang, H., Naik, V., Mittal, V., Cao, H., Demirbas, M., Gouda, M., Choi, Y., Herman, T., Kulkarni, S., Arumugam, U., Nesterenko, M., Vora, A., and Miyashita, M. (2004) A line in the sand: A wireless sensor network for target detection, classification, and tracking. *Comput. Netw.* **46** (5), 605–634.

Baggio, A. (2005) Wireless sensor networks in precision agriculture. *ACM Workshop Real-World Wireless Sensor Networks*.

Beckwith, R., Teibel, D., and Bowen, P. (2004) Report from the field: Results from an agricultural wireless sensor network. *LCN'04: Proceedings of the 29th Annual IEEE International Conference on Local Computer Networks* (pp. 471–478). IEEE Computer Society, Washington, DC, USA.

Caruso, M., and Withanawasam, L. (1999) Vehicle detection and compass applications using AMR magnetic sensors. *Sensors Expo Proceedings* (pp. 477–489).

Chehri, A., Fortier, P., and Tardif, P.M. (2009) Uwb-based sensor networks for localization in mining environments. *Ad Hoc Netw.* **7** (5), 987–1000.

Chintalapudi, K., Fu, T., Paek, J., Kothari, N., Rangwala, S., Caffrey, J., Govindan, R., Johnson, E., and Masri, S. (2006) Monitoring civil structures with a wireless sensor network. *IEEE Internet Computing* **10** (2), 26–34.

Foundation, P.D. (2009) http://www.pdf.org/. Last updated, 15 November 2009.

Kim, S., Pakzad, S., Culler, D., Demmel, J., Fenves, G., Glaser, S., and Turon, M. (2007) Health monitoring of civil infrastructures using wireless sensor networks. *IPSN'07: Proceedings of the 6th International Conference on Information Processing in Sensor Networks* (pp. 254–263). ACM, New York, NY, USA.

Knaian, A. (2000) *A wireless sensor network for smart roadbeds and intelligent transportation systems*. Master's Thesis, Massachusetts Institute of Technology.

Koh, B.H., and Dyke, S.J. (2007) Structural health monitoring for flexible bridge structures using correlation and sensitivity of modal data. *Comput. Struct.* **85** (34), 117–130.

Kulkarni, P., and Öztürk, Y. (2007) Requirements and design spaces of mobile medical care. *SIGMOBILE Mob. Comput. Commun. Rev.* **11** (3), 12–30.

Li, M., and Liu, Y. (2009) Underground coal mine monitoring with wireless sensor networks. *ACM Trans. Sen. Netw.* **5** (2), 1–29.

López Riquelme, J.A., Soto, F., Suardíaz, J., Sánchez, P., Iborra, A., and Vera, J.A. (2009) Wireless sensor networks for precision horticulture in southern Spain. *Comput. Electron. Agric.* **68** (1), 25–35.

Lorincz, K., Chen, Br., Challen, G.W., Chowdhury, A.R., Patel, S., Bonato, P., and Welsh, M. (2009) Mercury: A wearable sensor network platform for high-fidelity motion analysis. *SenSys'09: Proceedings of the 7th ACM Conference on Embedded Networked Sensor Systems* (pp. 183–196). ACM, New York, NY, USA.

Matese, A., Di Gennaro, S.F., Zaldei, A., Genesio, L., and Vaccari, F.P. (2009) A wireless sensor network for precision viticulture: The NAV system. *Comput. Electron. Agric.* **69** (1), 51–58.

Morris, M. (2007) Technologies for heart and mind: new directions in embedded assessment. *Intel Technology Journal*. **11** (1), 67–75.

Panchard, J., Rao, S.T.V.P., Hubaux, J.P., and Jamadagni, H.S. (2007) Commonsense net: A wireless sensor network for resource-poor agriculture in the semi-arid areas of developing countries. *Inf. Technol. Int. Dev.* **4** (1), 51–67.

Park, G., Cudney, H., and Inman, D. (2000) Impedance-based health monitoring of civil structural components. *Journal of Infrastructure Systems* **6** (4), 153–160.

Pierce, F.J., and Elliott, T.V. (2008) Regional and on-farm wireless sensor networks for agricultural systems in eastern Washington. *Comput. Electron. Agric.* **61** (1), 32–43.

Schwiebert, L., Gupta, S.K., and Weinmann, J. (2001) Research challenges in wireless networks of biomedical sensors. *MobiCom'01: Proceedings of the 7th Annual International Conference on Mobile Computing and Networking* (pp. 151–165). ACM, New York, NY, USA.

Sensor platform TSW (2009) http://shimmer-research.com/wordpress/?page_id = 20. Last visited 16 November 2009.

Stoianov, I., Nachman, L., Madden, S., and Tokmouline, T. (2007) Pipenet: A wireless sensor network for pipeline monitoring. *IPSN'07: Proceedings of the 6th International Conference on Information Processing in Sensor Networks* (pp. 264–273). ACM, New York, NY, USA.

Weaver, J. (2003) *A wearable health monitor to aid Parkinson disease treatment*. Master's Thesis, MIT.

Werner-Allen, G., Lorincz, K., Welsh, M., Marcillo, O., Johnson, J., Ruiz, M., and Lees, J. (2006) Deploying a wireless sensor network on an active volcano. *IEEE Internet Computing* **10** (2), 18–25.

Xu, N., Rangwala, S., Chintalapudi, K.K., Ganesan, D., Broad, A., Govindan, R., and Estrin, D. (2004) A wireless sensor network for structural monitoring. *SenSys'04: Proceedings of the 2nd International Conference on Embedded Networked Sensor Systems* (pp. 13–24). ACM, New York, NY, USA.

3

Node Architecture

The wireless sensor nodes are the central element in a wireless sensor network (WSN). It is through a node that sensing, processing, and communication take place. It stores and executes the communication protocols and the data-processing algorithms. The quality, size, and frequency of the sensed data that can be extracted from the network are influenced by the physical resources available to the node. Therefore, the design and implementation of a wireless sensor node is a critical step.

The node consists of sensing, processing, communication, and power subsystems. The designer has a plethora of options in deciding how to build and put together these subsystems into a unified, programmable node. The processor subsystem is the central element of the node and the choice of a processor determines the tradeoff between flexibility and efficiency – in terms of both energy and performance. There are several processors as options: microcontrollers, digital signal processors, application-specific integrated circuits, and field programmable gate arrays.

There are a number of ways to connect the sensing subsystem with the processer. Connecting two or more analog sensors with a multichannel ADC system that integrates multiple high-speed ADCs into a single IC design is one way. However, these types of ADCs are known to produce crosstalk and to increase uncorrelated noise, reducing the signal-to-noise ratio (SNR) on individual channels. Moreover, the coupled signals can create spurs that are similar to harmonic terms, reducing spurious free dynamic range (SFDR) and total harmonic distortion (THD). For low-frequency signals, however, the effect is not significant. Some sensors have their own built-in ADC which can be directly connected with the processor through a standard chip-to-chip protocol. Most microcontrollers have one or more internal ADCs to interface analog devices.

Likewise, the communication subsystem can be interfaced with the processor subsystem in different ways. One way is to use the SPI serial bus. Some transceivers have their own processor board to perform low-level signal processing pertaining to the physical and the data link layer, thereby relieving the main processor from these concerns. The communication subsystem is the most energy intensive subsystem and its power consumption should be regulated. Almost all commercially available transceivers provide a controlling functionality to switch the transceiver between various active operation levels; idle and sleep state.

The power subsystem provides DC power to all the other subsystems to bias their active components such as crystal oscillators, amplifiers, registers, and counters. Moreover, it provides DC–DC converters so that each subsystem can obtain the right amount of bias voltage.

Fundamentals of Wireless Sensor Networks: Theory and Practice   Waltenegus Dargie and Christian Poellabauer
© 2010 John Wiley & Sons, Ltd

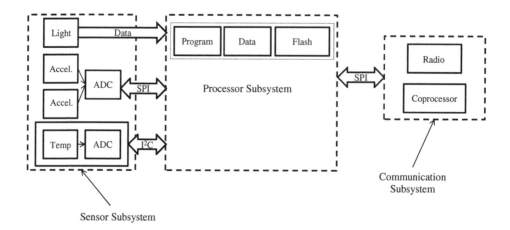

Figure 3.1 Architecture of a wireless sensor node.

Figure 3.1 shows the different subsystems of a wireless sensor node and the different integration techniques. The power subsystem and its relationship with the other subsystems are not shown here.

3.1 The Sensing Subsystem

The sensing subsystem integrates one or more physical sensors and provides one or more analog-to-digital converters as well as the multiplexing mechanism to share them. The sensors interface the virtual world with the physical world. Sensing physical phenomena is not something new. The Chinese astronomer, Zhang Heng, invented the *Houfeng Didong Yi* – a seismoscope – in the year 132 AD to measure the magnitude of seasonal winds and the movement of the Earth. Likewise, magnetometers have been in use for more than 2000 years.

But the advent of *microelectromechanical systems* (MEMS) has made sensing a ubiquitous process. Nowadays, there are a plethora of sensors that measure and quantify physical attributes at a cheap price. A physical sensor contains a transducer, a device that converts one form of energy into another form of energy, typically into an electrical energy (voltage). The output of this transducer is an analog signal having a continuous magnitude as a function of time. Therefore, an analog-to-digital converter is required to interface a sensing subsystem with a digital processor.

Table 3.1 provides a detailed summary of the types of sensors that have been employed in WSNs. It also provides a concise summary of the events they capture and the aspects of these events. The list is by no means exhaustive, but it highlights the scope and usefulness of WSNs, and the wealth of sensors that can be employed.

3.1.1 Analog-to-Digital Converter

The analog-to-digital converter (ADC) converts the output of a sensor – which is a continuous, analog signal – into a digital signal. This process requires two steps:

1. The analog signal has to be quantized (i.e., converted from a continuous valued signal into a discrete valued signal; discrete both in time and magnitude). The most important

Table 3.1 Summary of the sensors employed in wireless sensor networks

Sensor	Application area	Sensed event	Remark
Accelerometer	AVM (Werner-Allen et al. 2006)	2D and 3D acceleration of movements of people and objects	Volcanic activities by capturing primary and secondary seismic waves
	SHM (Xu et al. 2004)		Stiffness of a structure due to modal changes in the structure
	Health care (Benbasat and Paradiso 2007)		Stiffness of bones, limbs, joints; inclination, exertion
			Motor fluctuation in Parkinson's Disease (detection of bradykinesia, hyperkinesia)
	Transportation (Department 2000)		Irregularities in rail, axle-box or wheels of a train system
	SCM (Malinowski et al. 2007)		Defect of fragile objects during transportation (during loading and unloading of goods; irregular drive)
Acoustic emission sensor	SHM (Staszewski et al. 2004)	Elastic waves generated by the energy released during crack propagation	Measures microstructural changes or displacements
Acoustic sensor	Transportation (Chellappa et al. 2004)	Acoustic pressure vibration	Vehicle detection
	Pipelines (Sinha 2005)		Measure structural irregularities
			Gas contamination at ppm level
Capacitance sensor	PA	Solute concentration	Measure the water content of a soil
ECG	Health care (Lorincz et al. 2009)	Heart rate	
EEG		Brain electrical activity	
EMG		Muscle activity	
Electrical/electro-magnetic sensors	PA	Electrical resistivity/conductivity capacitance or inductance affected by the composition of tested soil	Measure nutrient contents and distribution
Gyroscope	Health care (Jovanov et al. 2005)	Angular velocity	Detection of gait phases (step detection)

(continued overleaf)

Table 3.1 Summary of the sensors employed in wireless sensor networks *(continued)*

Sensor	Application area	Sensed event	Remark
Humidity sensor	PA (Szewczyk et al. 2004) HM	Relative as well as absolute humidity	
Infrasonic sensor	AVM (Werner-Allen et al. 2006)	Concussive acoustic waves produced as a result of earth quake or volcanic eruption	
Magnetic sensor	Transportation (Haoui et al. 2008)	Presence, intensity, direction, rotation, and variation of a magnetic field	Presence, speed, and density of a vehicle on a street; congestion
Oximeter	Health care (Morris and Guilak 2009)	Blood oxygenation of patient's hemoglobin	Cardiovascular exertion (stress) and trending of exertion relative to activity
pH sensor	Pipeline (water) (Stoianov et al. 2007)	Concentration of hydrogen ions	Indicates the acid and alkaline content of water – measure of cleanliness
Photo acoustic spectroscopy	Pipeline (Sinha 2005)	Gas sensing	Detects gas leak in a pipeline
Piezoelectric cylinder	Pipeline (Sinha 2005)	Gas velocity	A leak produces a high-frequency noise that produces vibration
Soil moisture sensor	PA (López Riquelme et al. 2009)	Soil moisture	Fertilizer and water management
Temperature sensor	PA HM (López Riquelme et al. 2009)	Temperature	
Barometer sensor	PA HM (Szewczyk et al. 2004)	Pressure exerted on a fluid	
Passive infrared sensor	Health care (Shnayder et al. 2005) HM (Szewczyk et al. 2004)	Infrared radiation from objects	Motion detection
Seismic sensor	AVM (Werner-Allen et al. 2006)	Measure primary and secondary seismic waves (body wave, ambient vibration)	Detection of earthquake
Oxygen sensor	Health care (Murphy and Heinzelman 2002)	Amount and proportion of oxygen in the blood	
Blood flow sensor	Health care (Murphy and Heinzelman 2002)	The Doppler shift of a reflected ultrasonic wave in the blood	

decision at this stage is to determine the number of allowable discrete values. This decision in turn is influenced by two factors: (a) the frequency and magnitude of the signal; and (b) the available processing and storage resources.

2. The sampling frequency. In communication engineering and digital signal processing, this frequency is decided by the Nyquist rate.[1] In wireless sensor networks, however, the Nyquist rate does not suffice. Oversampling is required because of noise.

The prevailing consequence of the first step is the quantization error while the second is aliasing.

An ADC is specified, among other things, in terms of its resolution, which is an expression of the number of bits that can be used to encode the digital output. For example, an ADC with a resolution of 24 bits can represent 16,777,216 distinct discrete values. The resolution of an ADC can also be expressed in volts – since the output of most MEMS sensors is analog voltage. The voltage resolution of an ADC is equal to its overall voltage measurement range divided by the number of discrete intervals. In other words:

$$Q = \frac{E_{pp}}{2^M} \tag{3.1}$$

where Q is the resolution in volts per step (volts per output code); E_{pp} is the peak-to-peak analog voltage; and M is the ADC's resolution in bits.

Here Q suggests that the interval between the discrete steps (values) is uniform. But in reality this is not so. In most ADCs, the least significant bit changes as a function of 0.5 times Q and the most significant bit changes as a function 1.5 times Q. Those bits in the middle have a resolution of Q voltage.

In selecting an ADC, knowledge of the process or activity being monitored is important. Consider an industrial process whose thermal property ranges from -20 to $+80°$ C. The choice of the physical sensor as well as the ADC depends on the type of thermal change that is of interest. If, for example, a change of 0.5 °C is required, an ADC with a resolution of 8 bits is sufficient. If, on the other hand, a change of 0.0625 °C is required, then the ADC should have a resolution of 11 bits.

3.2 The Processor Subsystem

The processor subsystem brings together all the other subsystems and some additional peripherals. Its main purpose is to process (execute) instructions pertaining to sensing, communication, and self-organization. It consists of a processor chip, a nonvolatile memory (usually an internal flash memory) for storing program instructions, an active memory for temporarily storing the sensed data, and an internal clock, among other things.

Whereas a wide range of off-the-shelf processors are available for building a wireless sensor node, one has to make a careful choice, as it affects the cost, flexibility, performance, and energy consumption of the node. If the sensing task is well defined from the outset and does not change over time, a designer may choose either a field programmable gate array or a digital signal processor. These processors are very efficient in terms of their energy consumption; and for most simple sensing tasks, they are quite adequate. However, as these

---

[1] For a band-limited signal, the Nyquist rate sets a lower bound on the sampling frequency. Hence, the minimum sampling rate should be twice the bandwidth of the signal.

are not general-purpose processors, the design and implementation process can be complex and costly.

In many practical cases, however, the sensing goal changes or a modification may be required. Moreover, the software that runs on the wireless sensor node may require occasional updates or remote debugging. Such tasks require a considerable amount of computation and processing space at runtime. In which case, special-purpose, energy-efficient processors are not suitable.

Most existing sensor nodes at present use microcontrollers. There are some justifications besides those just mentioned. WSNs are emerging technologies; and the research community is still active with research for developing energy-efficient communication protocols and signal-processing algorithms. As this requires dynamic code installation and update, the microcontroller is the best option.

3.2.1 Architectural Overview

A major concern in resource-constrained processors is the efficient execution of algorithms, since this requires the transferring of information from and to memory. This includes program instructions and the data to be processed or manipulated. For example, in WSNs, the data stem from the physical sensors and the program instructions relate to communication, self-organization, data compression, and aggregation algorithms.

The processor subsystem can be designed by employing one of the three basic computer architectures: Von Neumann, Harvard, and Super-Harvard (SHARC). The Von Neumann architecture provides a single memory space that is used by program instructions and data. It provides a single bus to transfer data between the processor and the memory. This architecture has a relatively slow processing speed because each data transfer requires a separate clock. Figure 3.2 illustrates a simplified view of the Von Neumann architecture.

The Harvard architecture modifies the Von Neumann architecture by providing separate memory spaces for storing program instructions and data. Each memory space is interfaced with the processor with a separate data bus. In this way, program instructions and data can be accessed at the same time. Additional to this feature, the architecture supports a special single instruction, multiple data (SIMD) operation, a special arithmetic operation and a bit reverse addressing. It can easily support multitasking operating systems, but it has no virtual memory or memory protection. Figure 3.3 displays the Harvard architecture.

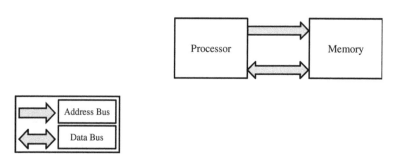

Figure 3.2   The Von Neumann architecture.

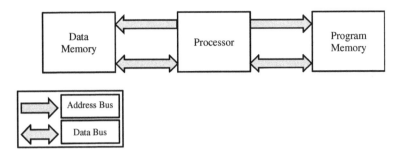

Figure 3.3   A simplified view of the Harvard architecture.

The next generation of processor architecture is the Super-Harvard architecture, best known as SHARC. It is an extension of the Harvard system; it adds two essential components to its predecessor and provides alternatives for accessing I/O devices from within the processor subsystem. One of the components is an internal instruction cache that enhances the performance of the processor unit. It can be used to temporarily store frequently used instructions, thereby reducing the need for repeatedly fetching them from the program memory. Moreover, the architecture enables an underutilized program memory to be used as a temporary relocation place for data.

In SHARC, external I/O devices can directly be connected with the memory unit through an I/O controller. The configuration enables a direct data streaming from an external hardware into the data memory, without the need to involve the microcontroller. This is known as Direct Memory Access (DMA).

DMA is desirable for two reasons: (1) the costly CPU cycles can be invested in a different task; and (2) it makes program memory bus and data memory bus accessible from outside the chip, providing an additional interface to off-chip memory and peripherals. Figure 3.4 illustrates an overview of the SHARC architecture.

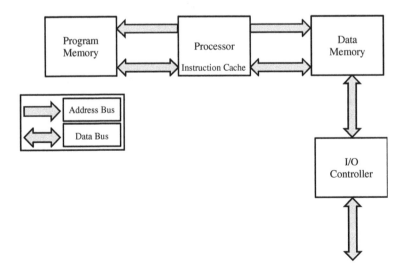

Figure 3.4   An overview of the Super-Harvard architecture.

3.2.2 Microcontroller

A microcontroller is a computer on a single integrated circuit, consisting of a comparatively simple central processing unit and additional components such as high-speed buses, a memory unit, a watchdog timer, and an external clock. Microcontrollers are integrated in many products and embedded devices. Today such simple systems as elevators, ventilators, office machines, household appliances, power tools, and toys ubiquitously employ microcontrollers.

3.2.2.1 Structure of a Microcontroller

Typically, a microcontroller integrates the following components:

- a CPU core that ranges from small and simple 4-bit to complex 32- or 64-bit processors;
- a volatile memory (RAM) for data storage;
- a ROM, EPROM, EEPROM, or flash memory for storing relatively simple instruction program code;
- parallel I/O interfaces;
- discrete input and output bits, allowing control or detection of the logic state of an individual package pin;
- a clock generator – which is often an oscillator with a quartz timing crystal;
- one or more internal analog-to-digital converters; and
- serial communications interfaces such as Serial Peripheral Interface and Controller Area Network for interconnecting system peripherals such as event counters, a timer, and a watchdog.

3.2.2.2 Advantages and Disadvantages

A microcontroller can be chosen over other types of small-scale processors because of the programming flexibility it offers. Its compact construction, small size, low power consumption, and low cost make it suitable for building computationally less intensive, standalone applications. Most of the commercially available microcontrollers can be programmed with assembly language and the C programming language.

The use of higher-level programming languages increases the programming speed and eases debugging. There are development environments that offer an abstraction of all the functionalities of a microcontroller. This enables application developers to program microcontrollers without the need to have a low-level knowledge of the hardware.

However, microcontrollers are not as powerful and as efficient as some custom-made processors such as digital signal processors (DSPs) and field programmable gate arrays (FPGAs). Moreover, for applications which demand simple sensing tasks but large-scale deployments (such as in precision agriculture and active volcano monitoring), one may prefer to use architecturally simple but energy- and cost-efficient processors such as application-specific integrated circuits.

3.2.3 Digital Signal Processor

A comprehensive understanding of DSPs requires a knowledge of digital signal processing. Broadly speaking, digital signal processing deals with processing discrete signals with digital filters. These filters minimize the effect of noise on a signal or selectively enhance or modify the spectral characteristics of a signal.

While analog signal processing requires complex hardware components, digital signal processing, on the contrary, requires mainly simple adders, multipliers, and delay components. A DSP is a specialized microprocessor designed to carry out complex mathematical operations at an extremely high efficiency, processing hundreds of millions of samples every second and providing real-time performance. Most commercially available DSPs are designed with the Harvard architecture.

3.2.3.1 Advantages and Disadvantages

Powerful and complex digital filters can be realized with commonplace DSPs. These filters perform remarkably well in signal detection and estimation, both of which require significant numerical computations. This is particularly interesting for multimedia WSN applications in which in-network audio and video signal processing may be required to compress or aggregate large size data. DSPs are also useful for applications that require the deployment of nodes in harsh physical settings where signal transmission may suffer corruption due to noise and interference.

Having said this, a WSN carries out other tasks (tasks pertaining to network management, self-organization, multi-hop communication, topology control, etc.) in addition to numerical computations. These tasks require protocols which are not necessarily characterized as numerical operations. Furthermore, the protocols may require periodical upgrade or modifications, which means flexibility in network reprogramming is vital.

3.2.4 Application-Specific Integrated Circuit

An application-specific integrated circuit (ASIC) is an integrated circuit (IC) that can be customized for a specific application. There are two types of design approaches: fully customized and half-customized. To understand the difference between these two, it is useful to understand the basic building blocks of an ASIC.

The ASIC architecture consists of cells and metal interconnects. A cell is an abstraction of a logical functionality that is physically implemented by active components (transistors). When several of these cells are interconnected by the metal interconnects, they make up an application-specific integrated circuit. The manufacturing of cells has reached such a maturity that there is a standard library of cells consisting of a collection of low-level logic functions, including basic gates (AND, OR, and INVERT), multiplexers, adders, and flip-flops. As the standard cells have identical size, they can be arranged in rows to ease the process of automated digital layout. Using predefined cells from a cell library makes the ASIC design process much easier.

In a fully customized IC, some (possibly all) logic cells, circuits, or layouts are custom made. The aim is to optimize cell performance (for example, execution speed) and to include features that are not defined or supported by the standard cell library. Fully customized ASICs are expensive and their design time is lengthy. On the other hand, a half-customized ASIC is built with logic cells that are available in the standard library.

In both cases, the final logic structure is configured by the end user. This reduces time to market and financial risks by eliminating the need to cycle through an integrated circuit production. An ASIC is a cost-efficient solution, since the interconnections as well as the logic structure can be specified according to the user's needs. This also offers great flexibility and reusability.

3.2.4.1 Advantages and Disadvantages of ASICs

Unlike a microcontroller, an ASIC can easily be designed and optimized to meet a specific customer demand. Even with a half-customized design, it is possible to design multiple microprocessor cores and embedded software in a single cell. Moreover, even though a fully customized ASIC is costly, with a hybrid approach (full custom and standard cell design), developers can achieve control over size and execution speed. Thus, it is possible to design with optimum performance and cost. Typical disadvantages include difficulties in designing, the lack of reconfigurability, and the usually high development costs.

3.2.4.2 Application of ASIC

Perhaps the most suitable role of ASICs in WSNs is not to replace microcontrollers or DSPs but to complement them. As briefly mentioned in the introduction section of this chapter, some of the subsystems may integrate customized processors to handle rudimentary and low-level tasks and to decouple these tasks from the main processing subsystem. For example, some communication subsystems are shipped with an embedded processor core to enhance the quality of received signals, cancel noise, and perform cyclic redundancy checks. These types of special-purpose processors can be efficiently realized by making use of ASICs.

3.2.5 Field Programmable Gate Array

The distinction between ASICs and FPGAs is not always clear. In fact, it is not unusual for companies that manufacture programmable ASICs to call their products FPGAs. While the basic anatomy of both architectures is essentially the same, FPGAs are more complex in design and more flexible to program. As the emphasis is on the (re)programming and reconfigurability aspect, typical features of a FPGA are summarized as follows:

- in a FPGA, none of the mask layers is customized;
- a FPGA includes some programmable logic components, or *logic blocks* – these are: a 4-input lookup table (LUT), a flip-flop, and an output block;
- there is a well-defined and formal method for programming the basic logic cells and the interconnect;
- there is a matrix of programmable interconnects surrounding the basic logic cells producing a configuration instance; and
- there are programmable I/O cells that surround the core.

FPGAs are programmed electrically, by modifying a packaged part. This process may take from a few milliseconds to a few minutes, depending on the programming technology and the size of the part. The programming is done with the support of circuit diagrams and hardware description languages, such as VHDL and Verilog.

3.2.5.1   Advantages and Disadvantages

FPGAs have a higher bandwidth compared to DSPs; they are more flexible in their application and can support parallel processing. While DSPs and microcontrollers can incorporate an internal ADC, a FPGA does not. Similar to a DSP, a FPGA has the capability to work with floating point representation. Additionally, a FPGA exposes its processing speed to application developers, thereby giving them a greater flexibility of control. On the other hand, FPGAs are complex; and the design and realization process is costly.

3.2.6   Comparison

Working with a microcontroller is preferred if the design goal is to achieve flexibility. Working with all the others is preferred if power consumption and computational efficiency is desired. Whereas microcontrollers have limited memory, steady progress is being made to increase memory size. Recently, more and more microcontrollers are becoming available on the market with attractive features; for example, the TI MSP430F2618 and MSP430F5437 offer an active memory (RAM) of 8 kilobytes (KB) and 16 KB, as well as 116 KB and 256 KB flash memory respectively. Both consume less power and perform better than earlier models. The Atmel ATMega1281 and its next version ATMega2561 also have good architecture and better memory and performance. Both of them have 8 KB active memory and 128 KB and 256 KB flash memory, respectively. The Jennic architecture – JN5121 and JN5139 – integrates a microprocessor and a radio subsystem into a single package to enhance processing speed. It has 96 KB and 192 KB RAM, respectively; and 128 KB flash memory.

In comparison, DSPs are expensive, large in size, and less flexible. For example the PIC 16F873 and Sx28AC microcontrollers have 5 million instructions per second (MIPs) and 75 MIPS processing power; 24 and 20 general purpose I/O; 4 KB and 2 KB RAM, respectively; and 28-pin each. The devices cost around $5.81 and $4.05, respectively. Compared to that, the DSP56364 has 100 MIPS processing power, 16 general-purpose I/O, 1 KB RAM, 100 or 112-pin and costs $11.00. Moreover, DSPs are best for signal processing, with specific algorithms.

FPGAs are faster than both microcontrollers and digital signal processors and support parallel computing. In wireless sensor networks, since sensing, processing, and communication should take place at the same time, FPGSs can be useful. However, their production cost and the difficulty with programming make them less desirable.

ASICs have higher bandwidths; they are the smallest in size, perform much better, and consume less power than any of the other processing types. Their main disadvantage is the high cost of production owing to the complex design process, usually with lower production quantity and a reduced reusability. Performance can be improved with the application of multicore systems where several applications could run in parallel. This enables the integration of ASICs into the other subsystems, so that when the main processor subsystem is idle and therefore should be turned off, elementary and rudimentary tasks can be carried out by the more efficient ASICs.

3.3   Communication Interfaces

As the selection of the right type of processor is vital to the performance as well as the energy consumption of a wireless sensor node, the way the subcomponents are interconnected with the processor subsystem is also vital.

Fast and energy-efficient data transfer between the subsystems of a wireless sensor node is critical to the overall efficiency of the network it sets up. However, the practical size of the node puts a restriction on system buses. Whereas communication via a parallel bus is faster than a serial bus, a parallel bus needs more space. Moreover, it requires a dedicated line for every bit that should be transmitted simultaneously while the serial bus requires a single data line only. Owing to the size of the node, parallel buses are never supported in node design.

The choice, therefore, is often between serial interfaces such as the serial peripheral interface (SPI), the general purpose input/output (GPIO), the secure data input/output (SDIO), the inter-integrated circuit (I^{2}C), and the Universal Serial Bus (USB). Among these, the most commonly used buses are the SPI and the I^{2}C.

3.3.1   Serial Peripheral Interface

The Serial Peripheral Interface (SPI – pronounced as "spy") is a high-speed, full-duplex synchronous serial bus. It was developed at Motorola in the mid-1980s. It does not have an official standard as such, but manufacturers building devices that use the SPI should conform to the implementation specification of other manufacturers in order to support correct communication (for example, devices should agree on whether to transmit the most significant bit (MSB) or the least significant bit (LSB) first).

The SPI bus defines four pins: (Master-Out/Slave-In) MOSI, (Master-In/Slave-Out) MISO, (Serial Clock) SCLK, and (Chip Select) CS. Some manufacturers refer to MOSI as SIMO and to MISO as SOMI, but the semantics is the same. Likewise, CS is sometimes referred to as (Slave Select) $\overline{\text{SS}}$. As the name suggests, MOSI is used to transmit data from the master to the slave when a device is configured as a master. In case it is configured as a slave, this port is used to receive data from the corresponding master. The semantics are reversed for the MISO port. SCLK is used by the master to send the clock signal that is needed to synchronize transmission; and by the slave to read this signal. Every communication is initiated by the master. A master device signals a slave with which it wants to communicate via the CS port. Since SPI is a single master bus, the microcontroller is by default the master in a wireless sensor node. Thus, components cannot communicate directly with each other but only via the microcontroller – for example, with this configuration, an ADC cannot send a sampled data directly to a RAM. Figure 3.5 illustrates two types of configurations. In (a), a single master communicates with a single slave device, while in (b), a master is connected with multiple slave devices.

Both master and slave devices hold shift registers. In most cases these are 8-bit registers, but heterogeneous sizes are also allowed. Both registers are connected in a ring-forming 16-bit shift register. This is the common mode of connection. Assuming that the MSB is transferred first, during a transmission cycle, the MSB that is sent by the master is inserted to the slave's LSB register while, in the same cycle, the slave's MSB is shifted to the master's

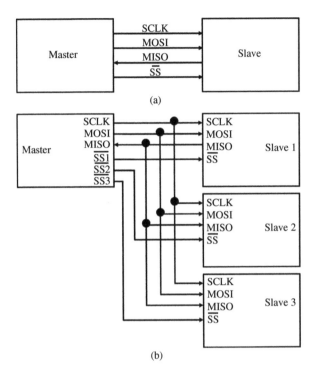

Figure 3.5 Connecting devices with the SPI bus. (a) A single master connecting with a single slave. (b) A single master connecting with multiple slaves.

LSB. After all bytes have been sent, the slave's register contains the master's word, while the master holds the slave's word.

Since master and slave form a commonly used shift register, every device in every transmission must read and send data. For devices that do not provide feedback (for example, LC displays do not offer status or bug messages) or do not require input data (some devices may not accept any commands at all), this means adding pseudo bytes into the shift register.

SPI supports a synchronous communication protocol. Consequently, the master and the slave must agree on the timing. To do so, the master sets the clock according to the slave's maximum clock speed – the baud generator of the master reads the slave's clock and calculates the master's clock by dividing the read speed with an internally defined value. Furthermore, master and slave should agree on two additional parameters, namely, clock polarity (CPOL) and clock phase (CPHA). CPOL defines whether a clock is used in high- or low-active mode. CPHA determines the times when the data in the registers are allowed to change and when the written data can be read. There are four different combinations (shown in Table 3.2) which are all incompatible with each other.

3.3.2 Inter-Integrated Circuit

The inter-integrated circuit (I^2C) is a multi-master half-duplex synchronous serial bus (see Figure 3.6). It was developed by Philips Semiconductors, which is also the owner of the official standard. I^2C uses only two bidirectional lines (unlike SPI, which uses four). The

Table 3.2 Common SPI modes

SPI mode	CPOL	CPHA	Description
0	0	0	SCLK is low-active. Sampling is allowed on odd clock edges. Data changes on even clock edges.
1	0	1	SCLK is low-active. Sampling is allowed on even clock edges. Data changes on odd clock edges.
2	1	0	SCLK is high-active. Sampling is allowed on odd clock edges. Data changes on even clock edges.
3	1	1	SCLK is high-active. Sampling is allowed on even clock edges. Data changes on odd clock edges.

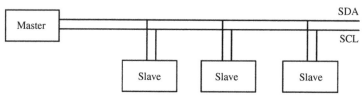

Figure 3.6 Connecting devices with the I²C serial bus.

aim of I²C is to minimize costs for connecting devices within a system by accommodating lower transmission speeds. I²C defines two speed modes: a Fast-mode, with a bit rate of up to 400 kbps and a High-speed-mode (referred to as Hs-mode) that supports a transmission rate of up to 3.4 Mbps. The 100 kbps rate (Standard-mode) was defined in earlier versions. Nevertheless, Fast-mode and Hs-mode components are downwards compatible to ensure communication with older components.

As the standard does not specify a CS or $\overline{SS}$ port, every device type that uses I²C must have a unique address that will be used to communicate with a device. In earlier versions, a 7-bit address was used, allowing 112 devices to be uniquely addressed (4 bits are reserved). This address space turned out to be insufficient due to an increasing number of devices. Currently I²C uses 10-bit addressing.

In the old protocol, a master device flags the start condition(s) and transmits the slave's 7-bit address. Then the master expresses *read or write* interest. At this time, the slave sends an acknowledgment (ACK). Afterwards, the data transmitter sends a 1-byte (8-bits) data, which is then acknowledged by the receiver. If there is still data to be sent, the transmitter keeps on sending and the receiver keeps on acknowledging in return. Finally, the master raises the stop flag (stop condition) to signify the end of a communication.

In the new protocol, after the start condition (S), a leading 11110 introduces the 10-bit addressing scheme. The last two address bits of the first byte concatenated with the eight bits of the second byte form the whole 10-bit address. Devices that only use 7-bit addressing simply ignore messages with the leading 11110.

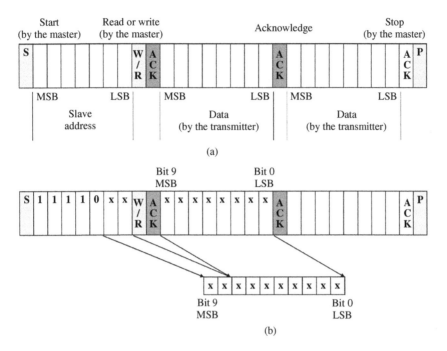

Figure 3.7   Communication protocols in the I²C serial bus.

Figure 3.7 shows the old and the new I²C protocol. In (a) the old protocol is displayed. In (b) the new protocol is displayed with the first two bytes of a transfer and the 10-bit address.

As already mentioned, I²C provides two lines; these are Serial Clock (SCL) and Serial Data Analyzer (SDA). Hs-mode devices have additional ports called SDAH and SCLH. Since each master generates its own clock signal, communicating devices must synchronize their clock speeds. In case they do not, a slower slave device could wrongly detect its address on the SDA line while a faster master device is sending data to a third device. Besides clock synchronization, I²C requires arbitration between master devices wanting to send or receive data at the same time. I²C does not explicitly define any fair arbitration algorithm. Rather the master that holds the SDA line low for the longest time wins the medium. Additionally, I²C enables a device to read data at a byte level for fast communication. This, however, may raise the need for more time to store the received bytes, in which case the device can hold the SCL low until it completes reading or sending the next byte. This type of clock synchronization is called handshaking.

Table 3.3 gives a comparison between SPI and I²C.

3.3.3   Summary

Buses are essential highways to transfer data between the processor subsystem and the other subsystems. Due to the concern for size, only serial buses can be used by a wireless sensor node. These buses demand high clock speeds to gain the same throughput that can be

Table 3.3 Comparison of SPI and I^{2}C

SPI	I^{2}C
• Four lines enable full-duplex transmission.	• Two lines reduce space and simplify circuit layout. Lowers costs.
• No addressing is required due to CS. This reduces overhead and increases throughput. However, one needs additional hardware configurations to connect more than one slave.	• Addressing enables multimaster mode, which in turn enables more than one device to initiate communication.
• Allowing only one master avoids conflicts.	• Multimaster mode is prone to conflicts when two or more master devices communicate simultaneously. Arbitration is required.
• Hardware requirement support increases with an increasing number of connected devices and therefore raises costs.	• Hardware requirement is independent of the number of devices using the bus.
• The master's clock is configured according to the slave's speed. This frees the slaves from requiring clocking a device. However, speed adaptation slows down the master.	• Slower devices may stretch the clock thereby increasing latency and keeping other devices waiting for accessing the bus.
• Speed depends on the maximum speed of the slowest device.	• Speed is limited to 3.4 MHz and all devices need to support the highest speed that is used in the system, otherwise a slower device may wrongly detect its device address.
• Heterogeneous register size allows flexibility in the devices that are supported.	• Homogeneous register size reduces overhead, since no additional control bits are required to be transmitted.
• Combined registers imply every transmission should be read.	• Devices that do not read or provide data are not forced to provide potentially useless bytes.
• The absence of an official standard leads to application specific implementations.	• Official standard eases integration of devices since developers can rely on a certain implementation.

achieved with parallel buses. However, they can also be bottlenecks; this is particularly the case with the Von Neumann architecture, since the same bus is used for data as well as instructions. They also do not scale well with processor speed. For example I^{2}C in its latest version is limited to 3.4 MHz while the clock speed of one of the most commonly used microcontroller family, the TI MSP430x1xx series, has a clock frequency of 8 MHz.

Delays due to contention for bus access become critical if some of the devices act unfairly and keep the bus occupied. For example, I^{2}C allows slave devices to stretch clock signals if it is deemed appropriate to "packet" communication and give priority to components that need to exchange time-critical data.

3.4 Prototypes

In this section, some example prototype node architectures are presented. The architectures are not chosen because of their commercial success or energy efficiency, but because they demonstrate the different node realization possibilities discussed in the preceding section.

3.4.1   The IMote Node Architecture

The IMote sensor node architecture (Figure 3.8) is a multipurpose architecture that consists of a power management subsystem, a processor subsystem, a sensing subsystem, a communication subsystem, and an interfacing subsystem.

The sensing subsystem (Figure 3.9) provides an extensible platform to connect multiple sensor boards. One realization of the sensor board contains a 12-bit, 4-channel ADC; a high-resolution temperature/humidity sensor; a low-resolution digital temperature sensor; and a light sensor. These devices are interfaced to the processing subsystem through the SPI and I^2C buses. As can be seen in the figure, the I^2C bus is chosen to connect low data rate sources whereas the SPI bus is used to interface high data rate sources.

The processing subsystem provides a main processor (microprocessor) and a digital signal processor (DSP). The main processor has the ability to operate at a low-voltage (0.85 V) and a low-frequency (13 MHz) mode, thus enabling low power operation. Likewise, the frequency can be scaled to 104 MHz at the lowest voltage level and can be increased up to 416 MHz with Dynamic Voltage Scaling (DVS). Moreover, it has many low power modes, including sleep and deep sleep modes. The coprocessor is intended to accelerate multimedia operations, which are computation intensive.

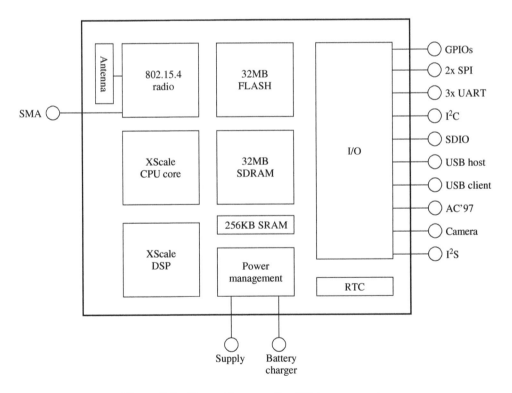

Figure 3.8   The architecture of an IMote sensor node.

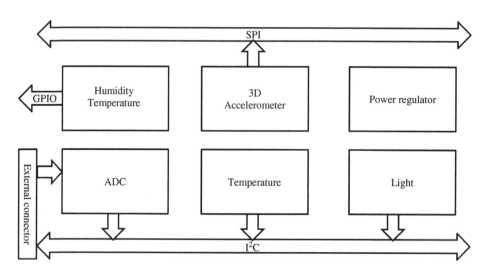

Figure 3.9   The sensing subsystem of the IMote architecture.

Similar to the sensor subsystem, the communication subsystem provides an extensible interface to accommodate different types of radios. One realization is based on the Chipcon (CC2420) transceiver, which implements the IEEE 802.15.4 radio specification. The transceiver provides a transmission rate of 250 kbps over 16 channels in the 2.4-GHz band.

3.4.2   The XYZ Node Architecture

The XYZ architecture consists of four subsystems. Figure 3.10 shows the schematic diagram of the node architecture. The processor subsystem is based on the ARM7TDMI core microcontroller, which is capable of operating at a maximum frequency of 58 MHz. The microcontroller can operate at two different modes, depending on the application requirement: at 32 bits and 16 bits. The processing subsystem provides an on-chip memory of 4 KB boot ROM and a 32 KB RAM, which can be extended by up to 512 KB of flash memory.

The peripheral components that interface the processing subsystem with the remaining subsystem include an embedded DMA controller, four 10-bit ADC inputs, serial ports (RS232, SPI, I²C, SIO), and 42 multiplexed general-purpose I/O pins. Most of the multiplexed GPIO pins are available on two 30-pin headers together with the DC voltage provided by the power subsystem or directly by an on-board voltage regulator.

The communication subsystem is based on the Chipcon CC2420 radio, which is connected to the processing subsystem through a SPI interface. The CC2420 is a 2.4 GHz IEEE 802.15.4 compliant single-chip RF transceiver. The processor subsystem controls the communication subsystem by either turning it off or putting it in sleep mode. The communication through the SPI interface enables the radio to wake up a sleeping processor when an RF message has been successfully received.

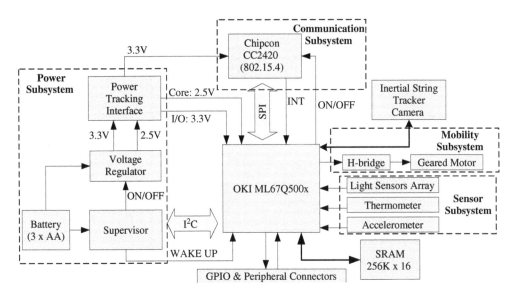

Figure 3.10   The XYZ node architecture (Lymberopoulos and Savvides 2005).

While in the architecture the mobility subsystem appears to be a separate subsystem, it should be considered as a part of the sensor subsystem.

3.4.3   The Hogthrob Node Architecture

The Hogthrob node architecture (Bonnet *et al.* 2006) is designed for a specific application, namely, to monitor the activities of sows in a large-scale pig production. The basic assumption behind the main sensing task is that there is a direct correlation between the movement of a sow and the onset of estrus. Therefore, a network that is established by nodes worn by sows monitors movements to capture this vital state, so that appropriate care can be given for pregnant sows. For example, in Denmark a law is already in place that requires pregnant sows to move freely in a large pen. Apart from this, other vital conditions, such as illness (by detecting cough or limping), are also monitored by the sensor network.

The node architecture consists of the usual subsystems. Unlike many existing architectures, the Hogthrob node's processing subsystems consists of two processors, a microcontroller, and a field programmable gate array (FPGA). The microcontroller performs less complex, less energy-intensive tasks, such as controlling the communication subsystem and other peripherals. It also initializes the FPGA and functions as an external timer and an ADC converter to it. The FPGA executes the sow monitoring application and coordinates the functions of the sensor node.

Figure 3.11 displays a partial view of the node architecture and the various interfacing buses. There are a number of interfaces supported by the processing subsystem, including the I^{2}C interface for the sensing subsystem, the SPI for the communication subsystem, the JTAG

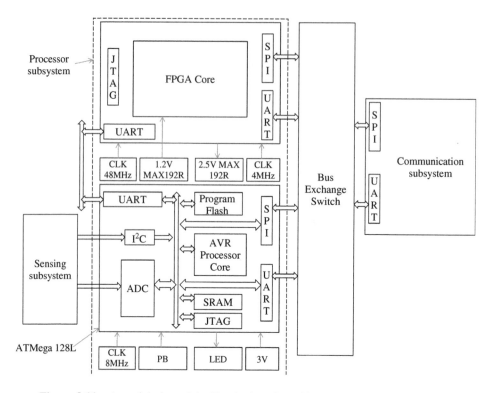

Figure 3.11 A partial view of the Hogthrob node architecture (Bonnet *et al*. 2006).

interface for in-system programmability and debugging, and the serial (RS-232) interface for interaction with a PC.

Exercises

3.1 A vibration sensor outputs an analog signal with a peak-to-peak voltage of 5 V at a frequency of 100 Hz.

(a) What should be the minimum sampling frequency, so that no information is lost during the digitization process?

(b) Suppose a resolution of 0.025 V is required to detect an interesting event. What should be the resolution of the ADC in terms of bits to convert the analog signal to a digital signal?

3.2 What is the drawback of using a multichannel ADC?

3.3 What is aliasing?

3.4 Define each of the following terms as applied to discrete time signal processing systems:

(a) Linearity

(b) Time-invariance

(c) Causality

(d) Stability

3.5 While they are not the most energy-efficient solutions, microcontrollers are the predominant processors in wireless sensor networks. Explain some of the reasons.

3.6 Explain the reason why using the Von Neumann architecture is not efficient for a wireless sensor node.

3.7 Why are parallel buses not desirable in a wireless sensor node?

3.8 What is the side effect of using a serial bus that supports full-duplex communication?

3.9 Explain the following terms in the context of the serial bus, SPI:

(a) Serial Data Out

(b) Serial Data In

(c) Serial Clock

3.10 State how a master component can communicate with multiple slaves in:

(a) I^2C

(b) SPI

3.11 Explain, with the help of diagrams, how the data transfer protocol of the I^2C bus functions.

3.12 Explain the basic similarities and differences between a FPGA and an ASIC.

3.13 Explain some of the distinct features of the Super-Harvard architecture.

3.14 A large number of commercially available wireless sensor nodes integrate three types of memory architectures: EEPROM (flash memory), RAM, and ROM. Explain the purpose of each of them.

3.15 The communication subsystem of a wireless sensor node is usually interfaced with the processor subsystem through a SPI bus instead of an I^2C bus. Why is this?

3.16 While memory management is very useful, it cannot be supported in wireless sensor networks. Why?

3.17 What is a virtual memory?

3.18 In most communication systems, the last stage in the reception process requires digital-to-analog converters (DAC). But in this book, the DAC is not discussed. What do you think is the reason?

3.19 Explain two different ways of interfacing an analog temperature sensor with a processor subsystem.

3.20 How can two hardware components with different speeds communicate with each other through a serial bus?

References

Benbasat, A.Y., and Paradiso, J.A. (2007) A framework for the automated generation of power-efficient classifiers for embedded sensor nodes. *SenSys '07: Proceedings of the 5th International Conference on Embedded Networked Sensor Systems* (pp. 219–232). ACM, New York, NY, USA.

Bonnet, P., Leopold, M., and Madsen, K. (2006) Hogthrob: Towards a sensor network infrastructure for sow monitoring (wireless sensor network special day). *DATE '06: Proceedings of the Conference on Design, Automation and Test in Europe* (p. 1109). European Design and Automation Association, 3001 Leuven, Belgium.

Chellappa, R., Qian, G., and Zheng, Q. (2004) Vehicle detection and tracking using acoustic and video sensors. *Proc. of IEEE Conf. on Acoustics, Speech, and Signal Processing.*

Department UT (2000) *Developed wheel and axle assembly monitoring system to improve passenger train safety.*

Haoui, A., Kavaler, R., and Varaiya, P. (2008) Wireless magnetic sensors for traffic surveillance. *Transportation Research Part C: Emerging Technologies.* **16** (3), 294–306.

Jovanov, E., Milenkovic, A., Otto, C., and de Groen, P.C. (2005) A wireless body area network of intelligent motion sensors for computer assisted physical rehabilitation. *Journal of Neuro-Engineering and Rehabilitation.* **2**: 6.

López Riquelme, J.A., Soto, F., Suardíaz, J., Sánchez, P., Iborra, A., and Vera, J.A. (2009) Wireless sensor networks for precision horticulture in southern Spain. *Comput. Electron. Agric.* **68** (1), 25–35.

Lorincz, K., Chen, Br., Challen, G.W., Chowdhury, A.R., Patel, S., Bonato, P., and Welsh, M. (2009) Mercury: A wearable sensor network platform for high-fidelity motion analysis. *SenSys '09: Proceedings of the 7th ACM Conference on Embedded Networked Sensor Systems* (pp. 183–196). ACM, New York, NY, USA.

Lymberopoulos, D., and Savvides, A. (2005) XYZ: A motion-enabled, power aware sensor node platform for distributed sensor network applications. *IPSN '05: Proceedings of the 4th International Symposium on Information Processing in Sensor Networks* (p. 63). IEEE Press, Piscataway, NJ, USA.

Malinowski, M., Moskwa, M., Feldmeier, M., Laibowitz, M., and Paradiso, J.A. (2007) Cargonet: A low-cost micropower sensor node exploiting quasi-passive wakeup for adaptive asynchronous monitoring of exceptional events. *SenSys '07: Proceedings of the 5th International Conference on Embedded Networked Sensor Systems* (pp. 145–159). ACM, New York, NY, USA.

Morris, M., and Guilak, F. (2009) Mobile heart health: Project highlight. *IEEE Pervasive Computing* **8** (2), 57–61.

Murphy, A.L., and Heinzelman, W.B. (2002) *MiLAN: Middleware linking applications and networks.* Technical Report, Rochester, NY, USA.

Shnayder, V., Chen, Br., Lorincz, K., Jones, T.R.F.F., and Welsh, M. (2005) Sensor networks for medical care. *SenSys '05: Proceedings of the 3rd International Conference on Embedded Networked Sensor Systems* (p. 314). ACM, New York, NY, USA.

Sinha, D.N. (2005) *Acoustic sensor for pipeline monitoring.* Technical Report, Los Alamos National Laboratory.

Stoianov, I., Nachman, L., Madden, S., and Tokmouline, T. (2007) Pipenet: A wireless sensor network for pipeline monitoring. *IPSN '07: Proceedings of the 6th International Conference on Information Processing in Sensor Networks* (pp. 264–273). ACM, New York, NY, USA.

Szewczyk, R., Mainwaring, A., Polastre, J., Anderson, J., and Culler, D. (2004) An analysis of a large scale habitat monitoring application. *SenSys '04: Proceedings of the 2nd International Conference on Embedded Networked Sensor Systems* (pp. 214–226). ACM, New York, NY, USA.

Werner-Allen, G., Lorincz, K., Welsh, M., Marcillo, O., Johnson, J., Ruiz, M., and Lees, J. (2006) Deploying a wireless sensor network on an active volcano. *IEEE Internet Computing* **10** (2), 18–25.

Staszewski, W.J., Boller, G., and Tomlinson, G. (eds) (2004) *Health Monitoring of Aerospace Structures: Smart Sensor Technologies and Signal Processing.* John Wiley & Sons Ltd.

Xu, N., Rangwala, S., Chintalapudi, K.K., Ganesan, D., Broad, A., Govindan, R., and Estrin, D. (2004) A wireless sensor network for structural monitoring. *SenSys '04: Proceedings of the 2nd International Conference on Embedded Networked Sensor Systems* (pp. 13–24). ACM, New York, NY, USA.

4

Operating Systems

An operating system (OS) in a WSN is a thin software layer that logically resides between the node's hardware and the application and provides basic programming abstractions to application developers. Its main task is to enable applications to interact with hardware resources, to schedule and prioritize tasks, and to arbitrate between contending applications and services that try to seize resources. Additional features include:

- memory management;
- power management;
- file management;
- networking;
- a set of programming environments and tools – commands, command interpreters, command editors, compiler, debuggers, etc. – to enable users to develop, debug, and execute their own programs; and
- legal entry points into the operating system for accessing sensitive resources such as writing to input components.

Traditionally, operating systems are classified as single-task/multitasking and single-user/multi-user operating systems. A single-task operating system processes one task at a time while a multitasking operating system can execute multiple tasks simultaneously. Multitasking operating systems require a large amount of memory to manage the states of multiple tasks but they enable tasks with different complexity to execute in parallel. For example, in a wireless sensor node, the processor subsystem may interact with the communication subsystem while aggregating data that arrive from the sensing subsystem. A multitasking OS is the best candidate for this type of environment. However, due to the limited resources, the overhead of concurrent processing may not be affordable. In a single-task OS one task is executed at a time, therefore, as a rule tasks should have a short duration. In a single-user OS, one user (the owner of the resources) is active at a time, whereas a multi-user operating system allows multiple users to share the resources of a single system at the same time.

The choice of a particular operating system depends on several factors. In the following sections, typical functional and nonfunctional aspects will be discussed.

Fundamentals of Wireless Sensor Networks: Theory and Practice   Waltenegus Dargie and Christian Poellabauer
© 2010 John Wiley & Sons, Ltd

4.1 Functional Aspects

4.1.1 Data Types

In wireless sensor networks (WSNs), communication between the different subsystems is vital. These subsystems communicate with each other for various reasons such as to exchange data, delegate functionalities, and signaling. Interactions take place through well-formulated protocols and data types that are supported by the OS. Complex data structures have strong expression power but consume resources, while simple data types are resource efficient but have limited expression capability. Almost all of the existing operating systems or runtime environments in WSNs support the native data types of the C programming language and some of the complex data types such as *struct* and *enum*.

4.1.2 Scheduling

Task scheduling is one of the basic functions of an OS. How efficiently tasks can be organized, prioritized, and executed determines the efficiency of the OS.

Broadly speaking, there are two scheduling mechanisms: queuing-based and round-robin scheduling. In a queuing-based scheduling, tasks originating from the various subsystems are temporarily stored in a queue and executed serially according to a predefined rule. Some operating systems enable tasks to specify priority levels so that they can be given precedence.

Queuing-based scheduling can be further classified into first-in-first-out (FIFO) and sorted queue. In a FIFO scheme, tasks are processed according to their arrival time: a task that arrives first will be executed first as soon as the processor is free. A nonpreemptive OS will execute the task to the end before another task is admitted for execution. In a preemptive OS, however, a task of higher priority may interrupt a task of low priority. In a sorted queue scheme, tasks in a queue are sorted according to some criteria. One way is to sort tasks according to their estimated execution duration. This approach prevents long-duration tasks from blocking short-duration tasks. The approach is also known as the shortest job first (SJF) rule.

The FIFO scheme is the simplest and the most economical as it incurs minimum system overheads. However, the FIFO scheme may not treat tasks fairly, since tasks of long duration may block short-duration tasks for a long time. In the SJF scheme, sorting incurs system overhead, since each task in the queue must be evaluated to estimate execution duration and to sort tasks accordingly.

Round-robin scheduling is a time-sharing scheduling technique in which several tasks can be processed concurrently. The scheduler defines a time frame by dividing time into slots and tasks will be given slots in a multiplexed manner. This way, all the tasks advance toward their completion.

Regardless of how tasks are executed, a scheduler can be either a nonpreemptive or preemptive scheduler. In strictly nonpreemptive scheduling, a task is executed to the end and will not be interrupted by another task. On the contrary, in strictly preemptive scheduling, the scheduler decides how time is shared between tasks and allows a task of higher priority to interrupt a task of lower priority. There is also the so-called "politely-preemptive" scheduling in which, even if tasks are interruptible, the scheduler will not interrupt a process if it is in a critical section.

4.1.3 Stacks

A stack is a data structure that is used to temporarily store data objects in memory by piling one upon another. The objects are accessed on a last-in first-out (LIFO) basis. The processor core uses stacks to store system state information when it begins executing subroutines. This way it "remembers" where to return after the subroutine is completed. Subroutines can also call other subroutines by storing the state of the current subroutine on top of the previous state information in the stack. When the subroutine is completed, the processor pulls the first address it finds on the top of the stack and jumps to that location.

In a multithreaded OS, each thread requires its own stack to manage state information. This is one of the reasons why multithreaded operating systems are expensive in WSNs.

4.1.4 System Calls

The OS provides a number of basic functions which enable the separation of concern, namely, the need to decouple the concern of accessing hardware resources and additional low-level services from the implementation details of the access mechanisms. Users invoke these operations whenever they wish to access a hardware resource such as a sensor, watchdog timer, or the radio without the need to concern themselves how the hardware is accessed. In a UNIX environment, these basic functions are commonly known as system calls.

4.1.5 Handling Interrupts

An interrupt is an asynchronous signal generated by a hardware device (a sensor, a watchdog timer, a radio) and causes the processor to interrupt executing the present instruction and to call for an appropriate interrupt handler. The processor stores the state of the interrupted process in a stack and gives control to the interrupt handler. For example, an interrupt signal may be raised by the communication subsystem when it receives a packet that should be processed immediately. The processor subsystem must then suspend the execution of the present instruction and calls for the appropriate module in the OS that should handle radio packets. Additional to the hardware devices, the OS can define several system events that can flag interrupt signals. In some cases, the OS itself can generate periodic interrupts to allow the processor to monitor the state of hardware resources and inform corresponding event handlers in case there is an interest in a specific hardware state.

In the same way that tasks can have different priority levels, interrupt signals can also have different priority levels. A high-priority interrupt can interrupt a low-priority interrupt. In such systems, programs can choose whether or not they wish to be interrupted by setting an interrupt mask. The mask enables them to "evade" low-level interrupts that they have nothing to do with. Masking interrupts can be dangerous and can corrupt data. Some operating systems have nonmaskable interrupts for the most crucial operations.

4.1.6 Multithreading

A thread is the path taken by a processor or a program during its execution. In a single-task, nonpreemptive operating systems, tasks are monolithic and there is only a single thread of execution. In a multithreaded environment, a task can be divided into several logical pieces which can be scheduled independently from each other and executed concurrently. Likewise, multiple tasks originating from different sources can be executed concurrently in multiple threads. Threads of the same task share a common data and address space and can communicate with each other if necessary. There are two main advantages of a multithreaded OS:

1. tasks do not block other tasks; this is particularly important to deal with tasks pertaining to I/O systems;
2. short-duration tasks can be executed along with long-duration tasks.

While threads are resource conservative by nature, they cannot be created endlessly. The creation of threads slows down the processor and there may not be sufficient resources to divide among a large number of threads. Therefore, some operating systems support only a limited number of threads and keep them in a "pool". Each thread in a pool awaits a task assignment. Once a request is received, it is assigned to an available thread in the pool. Upon completion of the task, the thread returns to the pool and awaits the next assignment. If all the threads in the pool are used, the system holds in a queue an upcoming request until the next thread returns to the pool. In this way the OS can keep the number of threads to a manageable size.

4.1.7 Thread-Based vs Event-Based Programming

In wireless sensor networks, it is vital to support concurrent tasks, particularly tasks related to I/O systems, and the choice is between thread-based and event-based execution paradigms.

The decision whether threads or events should be supported in an OS must take several factors into account, including the need for separate stacks and the need to estimate their maximum size for saving context information. Thread-based programs use multiple threads of control within a single program and a single address space. This way, a thread blocked by an I/O device can be suspended while other tasks are executed in different threads. However, the programmer must carefully protect shared data structures with locks and use condition variables to coordinate the execution of threads. To deal with all these problems, the OS needs to synchronize program execution. In general, program code written for multithreading environments is complex, bug-prone, and may lead to deadlocks and race conditions.

In event-based programming, there are events and event handlers. Event-handlers register at the OS scheduler to be notified when a named event occurs. The kernel typically implements a loop function that polls for events and calls the appropriate event-handlers when events occur. An event is processed to completion unless its handler reaches a blocking operation, in which case it registers a new callback and returns control to the scheduler.

4.1.8 Memory Allocation

The memory unit is a precious resource. That is where the OS resides. Additionally, data and the application's program code are temporarily stored there. How, and for how long, memory is allocated for a piece of program determines the speed of task execution.

Memory can be allocated to a program either statically or dynamically. A static memory allocation is a frugal way of using the memory, but it can only be used if the program's memory requirement is known in advance. The memory allocation takes place when the program starts – as part of the execution operation – and is never freed. Since the program's memory requirement is precisely known at the time of compilation, memory is used efficiently. On the other hand, static memory allocation does not allow runtime adaptation.

Dynamic memory allocation is used when the size and duration of the required memory are not known at the time of the program's compilation. This includes programs that use dynamic data structures whose memory requirement cannot be determined when the program starts. Such programs often use memory on a transient basis. They allocate some memory, use it for a while, but then reach a point where they no longer need that particular piece. Because memory is not inexhaustible, memory that is no longer used can be released or assigned to a different owner. Dynamic memory allocation enables flexibility in programming but produces a considerable management overhead.

As a strategy to increase the memory capacity of a node, most architectures use EEP-ROM or flash memory to store program code. Consequently, it is possible to deploy relatively complex applications and communication protocols. However, reading and writing to flash memory is costly with respect to energy consumption.

4.2 Nonfunctional Aspects

4.2.1 Separation of Concern

Operating systems designed to support resource constrained devices and the networks which are established by these devices are different from general-purpose operating systems, mainly due to the extremely tight resource budget. In general-purpose operating systems, there is a clear separation between the operating system and the applications that run on top of it. These two interact with each other through well-defined interfaces and system calls. The operating system itself has several distinct services that can be upgraded, debugged, or altogether removed independently. Such a distinction is difficult to support in wireless sensor networks.

In most cases, the operating system consists of a number of lightweight modules which can be "wired" together in order to create a monolithic program code that is responsible for the sensing, processing, and communication tasks. The wiring takes place at compilation time, producing a single system image that can be installed on the individual nodes. Some operating systems provide an indivisible system kernel along with a set of library components for building an application. There are also other operating systems that provide a kernel and a set of reconfigurable (reprogrammable) low-level services that abstract the hardware components of a node. The services can be "wired" together to make up an application and since the kernel functions independently of these services – even though its role is limited – there is a separation of concern to some extent. Separation of concern enables flexible and efficient reprogramming and reconfiguration. An update or an upgrade can be made as a whole or in part as required. The latter enables the efficient use of communication bandwidth as well as memory space during a software update.

4.2.2 System Overhead

An operating system executes program code and, therefore, requires its own share of resources. How much resource it consumes depends on its size and the type of services it provides to the higher-level services and applications. The resources consumed by the operating system are the system's overhead.

Presently available wireless sensor nodes have resources that are measured in terms of kilobytes and a few megahertz. These resources have to be shared by programs carrying out sensing, data aggregation, self-organization, network management, and communication. The operating system's overhead should be understood in view of these tasks.

4.2.3 Portability

In Chapter 3, it has been shown that different hardware architectures can be applied to develop a wireless sensor node. Ideally, nodes of heterogeneous architectures and operating systems should be able to coexist and collaborate with each other. At present, however, the existing operating systems do not provide this type of support.

A related issue is the portability of an operating system to deal with the rapid evolution of the hardware architecture. Wireless sensor networks (WSNs) are still emerging technologies. In the past decade, architectural design has undergone a remarkable evolution and as more and more application domains are studied, this evolution is expected to continue in order to accommodate unforeseen requirements. Hence, operating systems should be portable and extensible.

4.2.4 Dynamic Reprogramming

Once a WSN is deployed, it may be necessary to reprogram some part of the application or the operating system for the following reasons:

- complete knowledge of the deployment setting may not be known at the time of deployment and, as a result, the network may not perform optimally;
- both the application requirements and the properties of the physical environment in which the networks operate can change over time; and
- it may be necessary to detect and fix bugs while the network is still operating.

Manual replacement of a piece of software may not be feasible because of the large number of nodes in question. The alternative is to develop an operating system that provides dynamic reprogramming support. Apparently, if there is no clear separation between the application and the operating system, dynamic reprogramming cannot be supported.

If, on the other hand, there is a separation between the two, dynamic programming can be supported in principle, but its practical implementation depends on several factors.

First, the operating system should be able to receive the software update piece by piece and assemble and store it temporarily in the active memory. Second, the OS should make sure that this is indeed an updated version. Third, it should be able to remove the piece of software that should be updated and install and configure the new version. All these consume resources and may cause their own bugs.

Software reprogramming (update) requires robust code dissemination protocols which are responsible for splitting and compressing the code; ensuring code consistency and version controlling; and providing a robust dissemination strategy to deliver the code over a wireless link.

Current developments in programming tools and environments will be discussed in Chapter 12.

4.3 Prototypes

4.3.1 TinyOS

TinyOS (Gay *et al.* 2007; 2.x Working Group 2005) is the most widely used, richly documented, and tool-assisted runtime environment in WSNs. Moreover, it has undergone a long design and evolution process, which makes its operating principle understandable.

TinyOS's compact architecture makes it suitable for supporting many applications. Conceptually, the architecture consists of a scheduler and a set of components which can be connected with each other through well-defined interfaces. Components are classified into configuration components and modules. A configuration component specifies how two or more modules are connected with each other (this is called "wiring"), whereas modules are the basic building blocks of a TinyOS program. The composition of multiple configurations into a single executable code produces a TinyOS application. TinyOS does not provide a clear separation of concern between the operating system and the application.

A component is made up of a frame, command handlers, event handlers, and a set of nonpreemptive tasks. A component is similar to an object in object-based programming languages in that it encapsulates a state and interacts through well-defined interfaces. An interface can define commands, event handlers, and tasks. Each of them executes in the context of the frame and operates on its state. Therefore, a component has to formally declare the commands it uses and the events it signals. In this way it is possible to determine at the time of compilation the resources required by an application.

Components are structured hierarchically and communicate with each other through commands and events: higher-level components issue commands to lower-level components and lower-level components signal events to higher-level components. Consequently, higher-level components implement event handlers and lower-level components command processors (or function subroutines). The physical hardware is found at the base of the component hierarchy. Figure 4.1 illustrates the logical boundary between an application and the operating system.

As can be seen in Figure 4.1, there are two components at the highest level, namely, the routing component and the sensor application. The routing component is responsible for establishing and maintaining the network while the sensor application is responsible for sensing and processing. The two components communicate with each other and

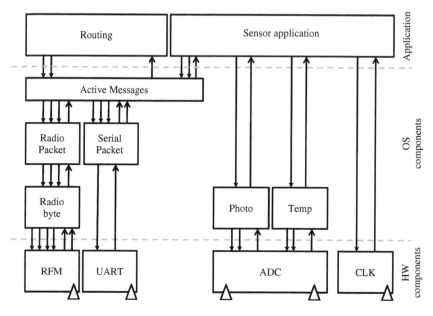

Figure 4.1   Logical distinction between the low-level and high-level components (Hill *et al.* 2000).

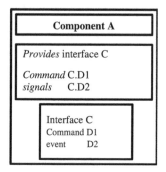

Figure 4.2   A TinyOS component providing an interface.

with the lower-level components asynchronously, through active messages. Additionally, a higher-level component can communicate with a low-level component by issuing non-blocking commands and by expressing an interest in named events.

Figures 4.2 to 4.4 demonstrate the logical structure of components and component configurations. In Figure 4.2, component A declares its service by providing interface C, which in turn provides command D1 and signals event D2. In Figure 4.3, component B expresses interest in interface C by declaring a call to command D1 and by providing an event handler to process event D2.

In Figure 4.4, a binding between Component A and Component B is established through configuration E.

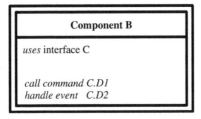

Figure 4.3   A TinyOS component that uses an interface.

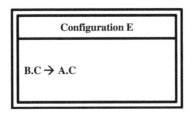

Figure 4.4   A TinyOS configuration that wires an interface provider and an interface user.

4.3.1.1   Tasks, Commands, and Events

TinyOS defines tasks, commands, and events as fundamental building blocks of a TinyOS runtime environment. These are responsible for enabling effective communication between the components of a single frame. Tasks are monolithic processes that should be executed to completion. In other words, they cannot be preempted by other tasks, though they can be interrupted by events. This is how TinyOS supports concurrency and ensures that tasks do not interfere with each other or corrupt each other's data.

Because tasks should execute to completion, it is possible to allocate a single stack to store context information. Tasks can call lower-level commands; signal higher-level events; and post (schedule) other tasks, including themselves. For example, a task responsible for reading packets from a communication subsystem can schedule itself repeatedly until it has completed reading all the packets. In TinyOS, scheduled tasks are based on the FIFO principle, and TinyOS architecture is effective for tasks that are of short duration.

Commands are nonblocking requests made by higher-level components to lower-level components. To deal with potential long-running operations, TinyOS introduces the concept of *split-phase* operation. In a split-phase system, a function call returns immediately, and the called function notifies the caller when the task is completed. It is called split-phase because it splits invocation and completion into two separate phases of execution. A typical example is a packet transmission task. A packet transmission can be a blocking task, since a receiver should wait for a timeout, $t_{timeout}$, before retransmitting a packet. However, if an ACK packet is received before $t_{timeout}$ expires, then the receiver gives up control. In TinyOS, this task is decomposed into two events: a timeout event and a packet received event.

Every component interested in a named event should provide an event handler to process it. Event handlers will be called when hardware events occur. The lowest level components have handlers that are directly connected to hardware interrupts, such as external interrupts, timer events, or counter events. An event handler may react to the occurrence of an event in different ways. For example, it can deposit information into its frame, post tasks, signal higher-level events, or call lower-level commands.

Resource allocation in TinyOS is optimized by adopting a static memory allocation. Since the memory requirement of an application is known at the time of its composition, it avoids the extra overhead associated with dynamic allocation. The lack of true separation of concern limits the adaptability of TinyOS. Moreover, without additional support outside of TinyOS, there is no mechanism to dynamically load and remove components.

As an event-based system, TinyOS does not directly support execution contexts. Hence, a complex program typically requires a state machine. State machines can be less expressive and may be perceived by many programmers as difficult to manage. A typical example mentioned in the literature is dealing with cryptographic operations. These operations require several seconds to complete, monopolizing the processor's precious cycles and making the system unable to respond to external events. A thread-based operating system handles this type of situation by preempting the task on behalf of time critical or short-duration tasks.

4.3.2 SOS

SOS (Han *et al.* 2005) attempts to establish a balance between flexibility and resource efficiency. Unlike TinyOS, it supports runtime reconfiguration and reprogramming of program code. The OS consists of a kernel and a set of modules that can be loaded and unloaded. In functionality, a module is similar to a TinyOS component – it implements a specific task or function. Moreover, in the same way that TinyOS components can be "wired" to build an application, a SOS application is composed of one or more interacting modules. Unlike a TinyOS component, which has a static place in memory, a module in SOS is a position-independent binary. This typical feature enables SOS to dynamically link modules with each other.

The SOS kernel provides interfaces to the underlying hardware. Additionally, it provides a priority-based scheduling mechanism and supports dynamic memory allocation.

4.3.2.1 Interaction

Interaction with a module takes place through messages (asynchronous communication) and direct calls to registered functions (synchronous communication). A message that originates from module A to module B should first go through the scheduler which places it in a priority queue. Then the kernel calls the appropriate message handler in module B and passes the message to it.

Modules implement message handlers that are specific to their purpose of existence. A module can also interact with another module by directly calling one of its registered functions. Interaction through a function call is faster than the message-based communication. This approach requires modules to explicitly register their public functions at the kernel. All modules that are interested in these functions have to subscribe to them. A function registration takes place by calling a system function called `ker_register_fn` at the time of module initialization. The call enables the module to inform the kernel where

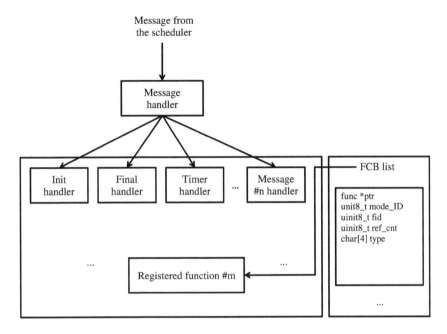

Figure 4.5   Interaction between modules in SOS system (Han *et al*. 2005).

in its binary image the function is implemented. The kernel completes the registration by creating a *function control block* (FCB), which stores key information about the function. This information is used to handle function subscription and to support dynamic memory management and runtime module update (replacement). Figure 4.5 illustrates the two basic types of interactions between modules.

Modules subscribe to a named function by calling the `ker_get_handle` system function. In doing so, they provide the kernel with the module's and function's IDs, which will be used to locate the FCB of interest. If the lookup is successful, the kernel returns a pointer to the function pointer of the subscribed function. The subscriber accesses the subscribed function by dereferencing the pointer. This enables the kernel to replace the function with a newer version by changing the function pointer in the FCB. The process is transparent to the subscribers.

4.3.2.2   Dynamic Reprogramming

Five basic features enable SOS to support dynamic reprogramming. First, modules are position-independent binaries – they use relative addresses rather than absolute addresses, hence, they are relocatable. Second, every SOS module implements two types of handlers – the *init* and *final* message handlers. The *init* message handler will be called by the kernel when first a module is loaded. Its purpose is to set the module's initial state, including initial periodic timers, function registration, and function subscription. The kernel calls the *final* message handler before a module is unloaded. The aim is to release all resources the module owns, including timers, memory and registered functions and to enable the module to gracefully exit the system. After the *final* message, the kernel performs garbage collection.

Third, during compilation, SOS uses a linker script to place the *init* handler of a module at a known offset in the binary. The script will enable easy linking during module insertion. Fourth, SOS keeps the state of a module outside of it. This enables the newly inserted module to inherit the state information of the module it replaces. Fifth, whenever a module is inserted, SOS generates and keeps metadata that contains information pertaining to the ID of the module, the absolute address of the *init* handler, and a pointer to the dynamic memory holding the module state.

In SOS, dynamic module replacement (update) takes place in three steps.

1. When a new module is available, a code distribution protocol advertises it in the network. The advertisement contains the module's ID, version number, and the required size of memory. When the local distribution protocol receives the advertisement, it evaluates the packet to decide whether the module is an updated version of a module which already exists locally or a new module in which the node is interested. In both cases, it also makes sure that there is sufficient space in the program memory to download the module.
2. Once a decision to download the module is made, the protocol proceeds with downloading the module and examines the metadata that is contained in the first packet it receives. The metadata contains, among other things, the size of the memory required to *store the local state of the module*. Module insertion is immediately *aborted* if the SOS kernel determines that it does not have sufficient RAM for the local state of the module.
3. If, on the other hand, everything is all right, module insertion takes place. During module insertion, the kernel creates metadata to store the absolute address of the handler, a pointer to the dynamic memory holding the module state and the identity of the module. Then the SOS kernel invokes the handler of the module by scheduling an *init* message for the module.

4.3.3   Contiki

Contiki (Dunkels *et al.* 2004) is a hybrid operating system. By default, its kernel functions as an event-driven kernel but multithreading support is implemented as an application library. There is a dynamic linking strategy to couple the multithreading library with applications that explicitly require it.

Like SOS, Contiki realizes the separation of the basic system support (by the kernel) from the rest of dynamically loadable and reprogrammable services, which are called processes. The services communicate with each other through the kernel by posting events. The kernel itself does not provide any hardware abstraction; instead it allows device drivers and applications to communicate directly with the hardware. This limited scope of the kernel makes it easy to reprogram and replace services.

Each Contiki service manages its own state in a private memory and the kernel keeps a pointer to the process state. However, a service shares with other services the same address space. It also implements an event handler and an optional poll handler. Figure 4.6 illustrates Contiki's memory assignment in ROM and RAM.

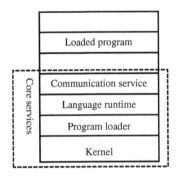

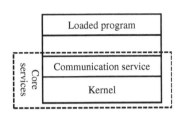

Figure 4.6   The Contiki operating system: the system programs are partitioned into core services and loaded programs (Dunkels *et al.* 2004).

As can be seen in Figure 4.6, Contiki has two main parts: the services inside the broken lines consist of the core services; those outside of the broken lines are the dynamically loadable services. The partition is made at compilation time. The core is made up of the kernel, the program loader, a communication stack with device drivers for the communication hardware, and additional frequently used services. These services are compiled into a monolithic binary image and deployed on a wireless sensor node. This part of the operating system cannot be modified dynamically unless a special boot loader is used to overwrite or patch it.

The program loader is responsible for loading runtime programs into the active memory. It can obtain the binaries either from a remote source through the communication service or directly from a local storage (EEPROM). Normally, program binaries are stored in EEPROM.

The kernel is the central element of the OS. Its basic assignment is to dispatch events and to periodically call polling handlers. Subsequently, a program execution in Contiki is triggered by either events that are dispatched by the kernel or through the polling mechanism. Event handlers process an event to completion, unless they are preempted by interrupts or other mechanisms – such as a thread preempting another thread when Contiki operates in a multithreaded environment.

The kernel supports synchronous and asynchronous events. Synchronous events are dispatched to the target process as soon as possible and control is returned to the posting process once the event is processed to the end. Asynchronous events, on the other hand, are dispatched at a convenient time. In addition to these events, the kernel provides a polling mechanism, in which the status of hardware components is sampled periodically. During this time, polling handlers that express interest in a named hardware device are notified in accordance with their priority.

4.3.3.1   Service Structure

One of the interesting features of the Contiki OS is its support of dynamic loading and reconfiguration of services. This is achieved by defining services, service interfaces, service stubs, and a service layer. Services are to Contiki what modules are to TinyOS.

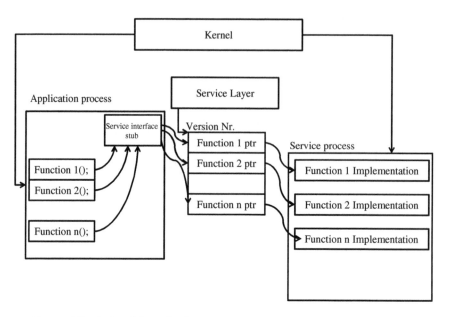

Figure 4.7   A Contiki service interaction architecture (Dunkels *et al*. 2004).

A Contiki service consists of a service interface and its implementation, which is also called a process. The service interface consists of a version number and the list of functions with pointers to the functions that implement the interface. A service stub enables an application program to dynamically communicate with a service through its service interface. A service layer is similar to a lookup service or a registry service. Active services register by providing the description of their service interface and ID and version number. This way, the service layer keeps track of all active services. Figure 4.7 illustrates how application programs interact with Contiki services.

Because programs call services through their service interface stubs, there is no need for them to know about the implementation details or the location in memory of the services. When a service is called, the service interface stub queries the service layer and obtains a pointer to the service interface. Upon obtaining a service whose interface description as well as version number matches with the service stub, the interface stub calls the implementation of the requested function. The loose coupling of a service with the program that uses the service enables the OS to update the service without the need to modify the application's program.

4.3.3.2 Protothreads

Contiki introduces the concept of protothreading by combining some of the features of events and threads. Protothreads can be viewed as lightweight (stackless) threads, but they can also be viewed as interruptible tasks in event-based programming (Dunkels *et al*. 2006). A protothread provides a conditional blocking **wait** statement, PT_WAIT_UNTIL(), which takes a conditional statement and blocks the protothread until the statement is evaluated as true. If the conditional statement is true by the time the protothread reaches the `PT_WAIT_UNTIL()` statement, then it continues executing without an interruption. The `PT_WAIT_UNTIL()` condition can be any conditional statement, including complex Boolean expressions.

Because a protothread is stackless, only an explicit `PT_WAIT_UNTIL()` statement can block. From the scheduler's point of view, all protothreads in the system run on the same stack and context switching is achieved by stack rewinding. The beginning and end of a protothread are explicitly declared by using the statements `PT_BEGIN` and `PT_END`, respectively. A protothread can exit prematurely with the statement `PT_EXIT`.

The protothread concept does not specify when or how a protothread should be invoked or scheduled. In the Contiki implementation, processes are implemented as protothreads that run on top of the event-driven kernel and, therefore, a protothread is invoked whenever the process receives an event – for example, when the process receives a message from another process or a timer event. Likewise, the protothread concept does not predefine how memory is appropriated to manage the states of protothreads. As with scheduling, this is also implementation-specific. If, for example, the operating system is based on a fixed set of protothreads, memory for state management can be statically allocated in advance. But memory can also be allocated in a dynamic fashion if the number of protothreads is not known in advance. In the Contiki implementation, static memory allocation is the typical setting and the state of a protothread is held in the process control block.

Protothreads simplify the design of state machines in event-driven programming since they reduce the number of explicit state machines and state transitions. Their cost is in terms of a memory overhead and a few processor cycles. To illustrate the usefulness of protothreads, consider a MAC protocol that turns off the radio subsystem on a periodic basis, but enables the radio subsystem to complete communication before it enters into a sleep state. This behavior is summarized as follows:

1. Initially (at $t = t_0$) the radio is turned on.
2. The radio remains on for a period of t_{awake} seconds.
3. Once t_{awake} is over, the radio has to be switched off, but it has to complete an ongoing communication.
4. If the communication is not completed, the MAC protocol has to wait for a period, t_{wait_max} before switching off the radio.
5. If the communication is completed or the maximum wait period is over, then the radio should be off and remain in the off state for a period t_{sleep}.
6. The same process is repeated.

Figures 4.8 and 4.9 display the implementations of the sleeping schedule with event-based programming and protothreads, respectively. The state machine implementation requires an explicit state variable that can take on the value ON, WAITING, or OFF. A conditional IF statement is used to perform different actions depending on the value of the state variable. The code can be placed in an event handler function that can be called whenever an event occurs. Possible events in this case are an expiration of a timer and the completion of communication. As can be seen in Figure 4.8, the code that controls the state machine constitutes more than one-third of the total lines of code. Also, the six-step structure of the mechanism is not immediately evident from the code.

```
state: {ON, WAITING, OFF}

radio_wake_eventhandler:
    if (state = ON)
        if (expired(timer))
            timer ← t_sleep
            if (not communication_complete())
                state ← WAITING
                wait_timer ← t_wait_max
            else
                radio_off()
                state ← OFF
    elseif (state = WAITING)
        if (communication_complete() or
                expired(wait_timer))
            state ← OFF
            radio_off()
    elseif (state = OFF)
        if (expired(timer))
            radio_on()
            state ← ON
            timer ← t_awake
```

Figure 4.8 A sleep schedule for the communication subsystem implemented with events (pseudocode) (Dunkels *et al.* 2006).

```
radio_wake_protothread:
    PT_BEGIN
    while (true)
        radio_on()
        timer ← t_awake
        PT_WAIT_UNTIL(expired(timer))
        timer ← t_sleep
        if (not communication_complete())
            wait_timer ← t_wait_max
            PT_WAIT_UNTIL(communication_complete() or
                            expired(wait_timer))
        radio_off()
        PT_WAIT_UNTIL(expired(timer))
    PT_END
```

Figure 4.9 A sleep schedule for the communication subsystem implemented with protothreads (pseudocode) (Dunkels *et al.* 2006).

The sleeping schedule implemented with the protothreads is evidently shorter and more intuitive.

4.3.4 LiteOS

LiteOS (Cao *et al.* 2008; Cao and Abdelzaher 2006) is a thread-based operating system and supports multiple applications. It is based on the principle of a clean separation between the OS and the applications that run on top of it. Unlike all the other operating systems, LiteOS does not provide components or modules that should be "wired" together in order to build an application. As far as LiteOS is concerned, developing building blocks and determining the way they interact with each other is entirely the task of application developers.

Instead, LiteOS provides several system calls: a shell that isolates the system calls from a user; a hierarchical file management system; and a dynamic reprogramming technique.

In LiteOS, the entire network is modeled as a distributed file system. A user at the side of the base station can identify, interact with, and reprogram a named node using a shell that is installed on a resource-rich computer. Each node in the network runs a multithreaded kernel which has three main components: a scheduler, a set of system calls, and a binary installer. The kernel's system calls enable a remote user to access and manage local files and directories. The local files are classified into sensor data, device drivers, and application binaries. Within the system hierarchy, a node is a stateless component. The user's interaction history is maintained by the shell at the remote computer. Figure 4.10 illustrates the LiteOS operating system architecture.

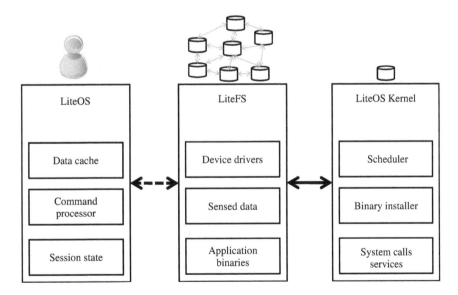

Figure 4.10   The LiteOS operating system architecture (Cao *et al*. 2008).

4.3.4.1 Shell and System Calls

The shell inherits several features from the Linux operating environment. It provides a mounting mechanism to a wireless node which is one-hop away from it; so that the entire network can be considered as a distributed and hierarchical file system. A user can access the resources of a named node as if they were locally available. The shell supports a large number of Linux commands that can be executed on the distributed file system. This way, LiteOS provides a familiar interface to Linux users.

The commands are classified into five categories: file commands, process commands, debugging commands, environment commands, and device commands. File commands are useful to navigate through the hierarchical file system and to move, copy, and delete files and directories. Below is given an example operation using file commands (Cao *et al.* 2008):

```
$ pwd
Current directory is /sn01/node101/apps
$ cp /c/Blink.lhex Blink.lhex
Copy complete
$ exec Blink.lhex
File Blink.lhex successfully started
$ ps
Name State
Blink Sleep
```

In this example, the `pwd` command prints the working directory in a node identified as sn01. This is node101/apps. Next, using the `cp` command, the file `Blink.lhex` is copied from the root directory of the resource-rich computer to the specified directory in node sn01. Then, using the `exec` command, the file is executed. The command ps reports the process status, which, in this case, indicates a sleeping thread.

The process commands are useful to manage threads, that is, creating, suspending, and killing threads. Up to eight threads can run simultaneously in LiteOS. The debugging commands enable a debugging environment to be set up to debug a program code. The environment commands provide support for managing the operating system's environment (at the side of the user) – displaying interaction history and providing command reference (manual). Finally, the device commands provide direct access to hardware devices such as sensors and the radio subsystem.

4.3.4.2 LiteFS

The LiteFS is a distributed file system and an essential feature of LiteOS. Through it, a user has access to the entire sensor network and can program and manage individual nodes. Similar to files in a Linux environment, files in LiteOS represent data, application binaries, and device drivers. Locally, the file system is organized as follows: the RAM contains the list of active (opened) files as well as information pertaining to the memory allocation in EEPROM and the flash memory. The file system structure is stored in the EEPROM memory, and the actual files are stored in a flash memory. This is displayed in Figure 4.11.

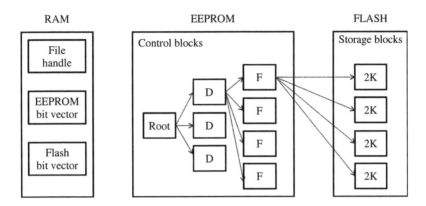

Figure 4.11   The file system structure of LiteFS (Cao *et al*. 2008).

In RAM, up to eight files can be simultaneously opened. LiteFS uses two-bit vectors to keep track of EEPROM and flash memory allocation. Eight bytes are used for the former and 32 bytes are used for the latter. This amounts to a total of 104 bytes of RAM. In EEPROM, each file is represented by a 32-byte control block. The space available for the control blocks is partitioned into 65 blocks. The first block is the root block, which is initialized every time the file system is formatted. The remaining blocks are either directory blocks (specified as D in the figure) or file blocks (specified as F in the figure). A file control block addresses at most 10 logical flash pages, each page holding 2 kbytes of data (or 8 physical flash pages). When a file occupies more than 20 kbytes, LiteFS allocates another control block for this file and stores the address of the new block in the previous block.

4.3.4.3   Dynamic Reprogramming

LiteOS supports the dynamic replacement and reprogramming of user applications. This can be done with or without the availability of the original source code. If the original source code is available to the OS, it will be recompiled with a new memory setting and all references and pointers to the old version will be redirected accordingly. If the original source code is not available to the OS, LiteOS uses a *differential patching* mechanism to replace an older version binary. The approach is to directly encode relocation information into application binaries by inserting differential patches and distributing them along with the binary images.

A mathematical model that has three parameters has been proposed for the differential patching. The model's parameters are the start address of the binary executable in the flash memory (S), the start address of allocated memory in RAM (M), and the stack top (T). The difference between T and M is the actual memory space allocated for the program code. Once these parameters are known, it is possible to insert the updated piece into the old binary image. The model parameters are obtained empirically and require knowledge of the node architecture. This limits the usefulness of the patching scheme.

4.4   Evaluation

Ranking the strength of an operating system, like all ranking assignments, is a difficult assignment. One has to rank in context to provide an appropriate perspective. In wireless sensor networks, there are several contexts pertaining to development, deployment, run-time performance, and code evolution. If one is concerned with design issues, perhaps the prevailing aspects are the following. How rich are the interfaces provided by the OS to access hardware devices? How flexible and expressive is the programming environment that can be supported by the OS? Is there a rich choice of modules, components, and library files that can be supported by the OS and that are useful to build an application? How portable is the OS? How manageable can the application's program code be?

If one is concerned with deployment issues, the most prevailing aspects are dynamic code installation and dynamic code propagation. The code has to be installed and tested on a large number of nodes and doing so manually is an onerous task. Likewise, if one is concerned with code evolution, dynamic code propagation and reprogramming are the most important factors. If one is concerned with runtime behavior, the most prevailing concern is the efficiency of the OS, particularly in terms of its compactness and power consumption.

In view of these aspects, TinyOS is compact in size and efficient in its use of resources, since the cost of managing separate entities (operation system and application) is lumped

Table 4.1   Comparison of functional aspects of existing operating systems

OS	Programming paradigm	Building blocks	Scheduling	Memory allocation	System calls
TinyOS	Event-based (split-phase operation, active messages)	Components, interfaces, and tasks	FIFO	Static	Not available
SOS	Event-based (active messages)	Modules and messages	FIFO	Dynamic	Not available
Contiki	Predominantly event-based, but it provides optional multithreading support	Services, service interface stubs, and service layer	FIFO, poll handlers with priority scheduling	Dynamic	Runtime libraries
LiteOS	Thread-based (based on thread pool)	Applications are independent entities	Priority-based scheduling with optional round-robin support	Dynamic	A host of system calls available to the user (file, process, environment, debugging, and device commands)

into a single assignment of managing a monolithic binary. But replacement or reprogramming cost is high. SOS, Contiki, and LiteOS provide flexible support for dynamic reprogramming and hence are well suited to applications which potentially undergo intensive updating and upgrading processes. However, the cost of image propagation should not be underestimated. LiteOS's approach of viewing the network as a distributed file system is interesting, since it provides the user an intuitive way of navigating the network. However, since nodes are stateless and all update history should be stored at the user's side, it results in extra traffic in the network to disseminate commands and state information.

Generally, the field of wireless sensor networks is relatively young. The operation environments as well as the application requirements are likely to evolve and to be made more compact and refined. Subsequently, the tradeoff is between dynamic reprogramming and code replacement on the one hand, and code execution efficiency on the other.

Tables 4.1 and 4.2 provide summaries of the functional and nonfunctional aspects of the four operating systems presented in this chapter.

Table 4.2 Comparison of nonfunctional aspects of existing operating systems

OS	Minimum system overhead	Separation of concern	Dynamic reprogramming	Portability
TinyOS	332 bytes	There is no clean distinction between the OS and the application. At compilation time a particular configuration produces a monolithic, executable code.	Requires external software support	High
SOS	ca. 1163 bytes	Replaceable modules are compiled to produce an executable code. There is no clean distinction between the OS and the application.	Supported	Medium to low
Contiki	ca. 810 bytes	Modules are compiled to produce a reprogrammable and executable code, but there is no separation of concern between the application and the OS.	Supported	Medium
LiteOS	Not available	Applications are separate entities; they are developed independent of the OS.	Supported	Low

Exercises

4.1 What is a process in the context of operating systems?

4.2 What is intra-process communication and how does it differ from inter-process communication?

4.3 Explain the difference between a system program and an application program?

4.4 What are system calls?

4.5 Explain the following terms and what are some of the mechanisms to avoid them?

    (a) race condition

    (b) deadlock

    (c) starvation

4.6 Compare the following scheduling mechanisms:

    (a) FIFO scheduling

    (b) sorted queue

    (c) round-robin

4.7 What are interrupts and interrupt handlers?

4.8 Why do most operating systems in WSNs define a kernel?

4.9 What is a preemptive process? Provide an example.

4.10 How is concurrency supported in TinyOS?

4.11 What is split-phase programming and how is it useful in WSNs?

4.12 Explain the difference between configuration components and modules in TinyOS.

4.13 Why do threads require their own separate stacks and what is the problem with this approach in WSNs?

4.14 Give three reasons for supporting dynamic reprogramming in WSNs.

4.15 Explain the difference between event-based and thread-based operating systems. Discuss some of the advantages and disadvantages of the two approaches in the context of WSNs.

4.16 Explain the difference between static and dynamic memory allocation.

4.17 State how the separation of concern is supported in the following operating systems:

    (a) Contiki

    (b) SOS

    (c) LiteOS

4.18 Explain the following concepts in TinyOS:

    (a) commands

    (b) tasks

    (c) events

4.19 What is the difference between a TinyOS command and a SOS message?

4.20 Why is the state of a module stored in a separate memory space (outside of the module) in SOS?

4.21 Explain how SOS supports dynamic reprogramming.

4.22 How is multithreading supported in a Contiki environment?

4.23 What is the function of a program loader in Contiki and why is it important?

4.24 How is module replacement supported in Contiki?

4.25 What is the advantage of considering a WSN as distributed file system in LiteOS?

4.26 What is differential patching in LiteOS?

4.27 Explain the functions of the following message handlers in SOS:

(a) *init-handler*

(b) *final-handler*

4.28 State the type of scheduling strategy that the following operating systems employ:

(a) TinyOS

(b) SOS

(c) Contiki

(d) LiteOS

4.29 How does TinyOS deal with dynamic reprogramming?

4.30 Why is separation of concern in TinyOS not a priority?

References

2.x Working Group TT (2005) Tinyos 2.0. *SenSys '05: Proceedings of the 3rd International Conference on Embedded Networked Sensor Systems* (p. 320). ACM, New York, NY, USA.

Cao, Q., Abdelzaher, T., Stankovic, J., and He, T. (2008) The LiteOS operating system: Towards Unix-like abstractions for wireless sensor networks. *IPSN '08: Proceedings of the 7th International Conference on Information Processing in Sensor Networks* (pp. 233–244). IEEE Computer Society, Washington, DC, USA.

Cao, Q., and Abdelzaher, T. (2006) LiteOS: A lightweight operating system for C++ software development in sensor networks. *SenSys '06: Proceedings of the 4th International Conference on Embedded Networked Sensor Systems* (pp. 361–362). ACM, New York, NY, USA.

Dunkels, A., Gronvall, B., and Voigt, T. (2004) Contiki: A lightweight and flexible operating system for tiny networked sensors. *LCN '04: Proceedings of the 29th Annual IEEE International Conference on Local Computer Networks* (pp. 455–462). IEEE Computer Society, Washington, DC, USA.

Dunkels, A., Schmidt, O., Voigt, T., and Ali, M. (2006) Protothreads: Simplifying event-driven programming of memory-constrained embedded systems. *Proceedings of the 4th International Conference on Embedded Networked Sensor Systems (SenSys)*, Boulder, CO. SenSys '06. ACM, New York, NY, USA (pp. 29–42).

Gay, D., Levis, P., and Culler, D. (2007) Software design patterns for TinyOS. *ACM Trans. Embed. Comput. Syst.* **6** (4), 22.

Han, C.C., Kumar, R., Shea, R., Kohler, E., and Srivastava, M. (2005) A dynamic operating system for sensor nodes. *MobiSys '05: Proceedings of the 3rd International Conference on Mobile Systems, Applications, and Services* (pp. 163–176). ACM, New York, NY, USA.

Hill, J., Szewczyk, R., Woo, A., Hollar, S., Culler, D., and Pister, K. (2000) System architecture directions for networked sensors. *ASPLOSIX: Proceedings of the 9th International Conference on Architectural Support for Programming Languages and Operating Systems* (pp. 93–104). ACM, New York, NY, USA.

Part Two

Basic Architectural Framework

5

Physical Layer

One of the desirable aspects of wireless sensor nodes is their ability to communicate over a wireless link. Because of it, mobile applications can be supported; flexible deployment of nodes is possible; and the nodes can be placed in areas that are otherwise inaccessible to wired nodes. Once the deployment is carried out, it is possible to rearrange node placement in order to attain optimal coverage and connectivity; and the rearrangement can be made without disrupting the normal operation of the structure or process the nodes monitor.

However, wireless communication poses some formidable challenges. Some of these challenges are limited bandwidth, limited transmission range, and poor packet delivery performance because of interference, attenuation, and multipath scattering. In order to tackle these challenges, it is vital to understand their properties and some of the mitigation strategies that are already in place.

This chapter provides a fundamental introduction to point-to-point wireless digital communication.

5.1 Basic Components

The basic components of a digital communication system are the transmitter, the channel, and the receiver. Since wireless sensor nodes are placed close to each other in a wireless sensor network, here short range communication is of interest. For a more comprehensive treatment of wireless and digital communications, the reader is referred to Proakis (2000) and Wilson (1995).

Figure 5.1 provides a block diagram of a digital communication system. The communication source in the context of this book represents one or more sensors and produces a message signal, an analog signal. The signal is a baseband signal having dominant frequency components near zero. The message signal has to be converted to a discrete signal (discrete both in time and amplitude) in order to be processed by the processor subsystem. The conversion requires sampling the signal at least at Nyquist rate,[1] so that no information will be lost. After sampling, the discrete signal is converted to a binary stream. This process is called *source encoding*. It is essential to implement an efficient source-coding technique so that the channel's bandwidth and signal power requirements are satisfied. One way to

---

[1] For a band-limited signal, the Nyquist rate sets a lower bound on the sampling frequency. Hence, the minimum sampling rate should be twice the bandwidth of the signal.

Fundamentals of Wireless Sensor Networks: Theory and Practice   Waltenegus Dargie and Christian Poellabauer
© 2010 John Wiley & Sons, Ltd

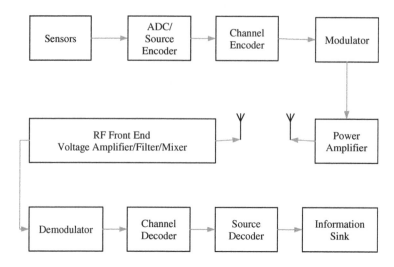

Figure 5.1   Components of a digital communication system.

achieve this is by defining a probability model of the information source, so that the length
of each information symbol depends on its probability of occurrence.

The next step is channel encoding and its aim is to make the transmitted signal robust
to noise and interference. Moreover, in case of signal corruption, it enables an error to be
recognized and the original data to be recovered. There are two essential approaches: to
transmit symbols from a predetermined codebook, and to transmit redundant symbols.

After channel encoding, modulation takes place. This is a process by which the baseband
signal is transformed into a bandpass signal. Modulation is useful for various reasons, but
the main reason is to transmit and receive signals with short antennas. In general, the shorter
the wavelength of the transmitted signal, the shorter is the length of the antenna. Finally, the
modulated signal has to be amplified and the electrical energy is converted into electromagnetic energy (electromagnetic radiation) by the transmitter's antenna, and the signal is
propagated over a wireless link to the desired destination.

The components of the receiver block carry out the reverse process to retrieve the message signal from the electromagnetic waves. The receiver antenna induces a voltage that
is, ideally, similar in shape, frequency, and phase with the modulated signal. Due to various types of losses and interferences, the magnitude and shape of the signal is changed and
has to pass through a series of amplification and filtering processes. It is then transformed
back to a baseband signal through the process of demodulation and detection. Finally, the
baseband signal undergoes a pulse-shaping process and two additional stages of decoding
(channel and source) in order to extract the sequence of symbols that represent the original
analog signal, which is the message.

5.2   Source Encoding

A source encoder transforms an analog signal into a digital sequence. The process consists
of sampling, quantizing, and encoding.

To explain these stages, suppose a sensor produces an analog signal that can be expressed as $s(t)$. During the sampling process, $s(t)$ will be sampled and quantized by the analog-to-digital converter (ADC) that has a resolution of Q distinct values. As a result, a sequence of samples, $S = (s[1], s[2], ..., s[n])$ are produced. The difference between the sampled $s[j]$ and its corresponding analog value at time t_j is the quantization error. As the signal varies over time, the quantization error also varies and can be modeled as a random variable with a probability density function, $P_s(t)$.

The aim of the source encoder is to map each quantized element, $s[j]$ into a corresponding binary symbol of length r from a codebook, C. If all the binary symbols in the codebook are of equal length, the codebook is called a Block Code. Often, however, the symbol length as well as the sampling rate is not uniform. It is customary, therefore, to assign short-sized symbols and high sampling rates to the most probable sample values and long-sized symbols and low sampling rates to less probable sample values. Figure 5.2 illustrates the input–output relationship of a source encoder.

A codebook, C, can be uniquely decoded, if each sequence of symbols, $(C(1), C(2), ...)$ can be mapped back to a corresponding value in $S = (s[1], s[2], ..., s[n])$. A binary codebook has to satisfy Equation (5.1) to be uniquely decoded.

$$\sum_{i=1}^{u} \left(\frac{1}{r}\right)^{l_i} \leq 1 \tag{5.1}$$

where u is the size of the codebook and l_i is the size of the codeword $C(i)$.

A codebook can be instantaneously decoded if each symbol sequence can be extracted (decoded) from a stream of symbols without taking into consideration previously decoded symbols. This will be possible if and only if there does not exist a symbol in the codebook, such that the symbol $\mathbf{a} = (a_1, a_2, ..., a_m)$ is not a prefix of the symbol $\mathbf{b} = (b_1, b_2, ..., b_n)$, where $m < n$ and $a_i = b_i$, $\forall i = 1, 2, ..., m$ within the same codebook. Table 5.1 lists different types of codebooks.

$$a \in A : A = q - ray \longrightarrow \boxed{\text{Source Encoder}} \xrightarrow{\ c \in C : C = M - ray\ }$$

Figure 5.2   Input – output relationship of a source encoder.

Table 5.1   Source-encoding techniques

	C^1	C^2		C^3	C^4		C^5	C^6
s_1	0	00		0	0		0	0
s_2	10	01		100	10		01	10
s_3	00	10		110	110		011	110
s_4	01	11		11	1110		111	111
Block code	No	Yes		No	No		No	No
Uniquely decoded	No	Yes		No	Yes		Yes	Yes
$\sum_{i=1}^{n} \left(\frac{1}{2}\right)^{l_i}$	$1\frac{1}{4}$	1		1	$\frac{15}{16} < 1$		1	1
Instantly decoded	No	Yes (block code)		No	Yes (comma code)		No	Yes

5.2.1 The Efficiency of a Source Encoder

The efficiency of a source encoder is a quantity that expresses the average length, $L(C) = E[l_i(C)]$ of symbols used to represent the sampled analog signal.

Suppose the probability of a q-ary source – that is, it has q distinct symbols – producing the symbol s_i is P_i and the symbol C_i in a codebook is used to encode s_i. The expected length of the codebook is given by:

$$L(C) = \sum_{i=1}^{q} P_i \cdot l_i(C) \tag{5.2}$$

Sometimes, it is necessary to express efficiency in terms of the information entropy or Shannon's entropy. In information theory, Shannon's entropy is defined as the minimum message length necessary to communicate information. It is related to the uncertainty associated with the information. If the symbol s_i can be expressed by a binary symbol of n bits, the information content of s_i is:

$$I(s_i) = -\log_2 P_i = \log_2 \frac{1}{P_i} \tag{5.3}$$

The entropy (in bits) of a q-ary memoryless source encoder is expressed as:

$$H_r(A) = E[I_r(s_i)] = \sum_{i=1}^{q} P(s_i) \cdot I_r(s_i) = \sum_{i=1}^{q} P(s_i) \cdot \log_2 \frac{1}{P(s_i)} \tag{5.4}$$

The efficiency of a source encoder in terms of entropy reveals the unnecessary redundancy in the encoding process. This can be expressed by:

$$\eta(C) = \frac{H(S)}{L(C)} \tag{5.5}$$

The redundancy of the encoder is:

$$\frac{L - H(S)}{L} = 1 - \eta \tag{5.6}$$

5.2.1.1 Example

Suppose the analog signal in Figure 5.3 is quantized into four distinct values, 0, 1, 2, 3. As can be seen in the figure, some values (2) occur more frequently than others (0 and 3). If the probability of occurrence of these values can be expressed as $P(0) = 0.05$, $P(1) = 0.2$, $P(2) = 0.7$, $P(3) = 0.05$, then, it is possible to compute the efficiency of two of the codebooks given in Table 5.1, namely C^2 and C^3.

For $P_1 = 0.05$, $\log_2 \left(\frac{1}{0.05} \right) = 4.3$. Because l_i has to be a whole number and there should be no loss of information, l_1 must be 5. Likewise, $l_2 = 3$; $l_3 = 1$; and $l_4 = 5$. Hence:

$$E[L(C^2)] = \sum_{j} l_j \cdot P_j = (5 \times 0.05) + (3 \times 0.2) + (1 \times 0.7) + (5 \times 0.05) = 1.8 \tag{5.7}$$

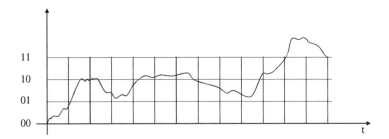

Figure 5.3   An analog signal with four possible values.

Using Equation (5.4), the entropy of C^2 is calculated as:

$$H(C^2) = 0.05 \log_2 \left(\frac{1}{0.05} \right) + 0.2 \log_2 \left(\frac{1}{0.2} \right)$$

$$+ 0.7 \log_2 \left(\frac{1}{0.7} \right) + 0.05 \log_2 \left(\frac{1}{0.05} \right) = 1.3 \tag{5.8}$$

Therefore, the encoding efficiency of the codebook, C^2 (see Table 5.2) is:

$$\eta(C^2) = \frac{1.3}{1.8} = 0.7 \tag{5.9}$$

The redundancy in C^2 is:

$$\text{rdd}_{C^2} = 1 - \eta = 1 - 0.67 = 0.3 \tag{5.10}$$

In terms of energy efficiency, this implies that 30% of the transmitted bits are unnecessarily redundant, because C^2 is not compact enough.

In the same way l_j is computed for C^2, the expected symbol length (in bits) for C^3 (see Table 5.3) is given as:

$$E[L(C^3)] = \sum_j l_j \cdot P_j$$

$$= (3 \times 0.05) + (2 \times 0.2) + (1 \times 0.7) + (3 \times 0.05)$$

$$= 1.4 \tag{5.11}$$

Table 5.2   Description of the compactness of C^2

j	a_j	P_j	l_j
1	00	0.05	5
2	01	0.2	3
3	10	0.7	1
4	11	0.05	5

Table 5.3 Description of the compactness of C^3

j	a_j	P_j	l_j
1	100	0.05	3
2	11	0.2	2
3	0	0.7	1
4	110	0.05	3

Because the probabilities of the symbols are unchanged, entropy also remains unchanged. The encoding efficiency of C^3 is therefore:

$$\eta(C^3) = \frac{1.3}{1.4} = 0.9 \tag{5.12}$$

The redundancy, rdd, in C^3 is:

$$\text{rdd}_{C^3} = 1 - \eta = 1 - 0.9 = 0.1 \tag{5.13}$$

5.2.2 Pulse Code Modulation and Delta Modulation

Pulse code modulation (PCM) and delta modulation (DM) are the two predominantly employed source encoding techniques. In digital pulse code modulation, the signal is first quantized and then each sample is represented by a binary word from a finite set of words. The size of the individual words as well as the number of words in the set determines the resolution of a PCM technique and the source encoder bit rate.

In PCM information is conveyed in the presence or absence of pulses and not in the amplitude or the location of the edges of the pulses. Because of this property, PCM greatly enhances (almost noise free) the transmission and regeneration of binary words. The associated cost with this form of source encoding is the quantization error and the energy and bandwidth required to transmit multiple bits for each sampled output. Figure 5.4 illustrates a PCM technique that uses two bits to encode a single sample. Four distinct levels are permissible during sampling.

Delta modulation is a digital pulse modulation technique which has found widespread acceptance in low bit rate digital systems. It is a differential encoder and transmits bits of information which describes the difference between successive signal values, as opposed to the actual values of a time-series sequence. The difference signal, $V_d(t)$, is produced by first estimating the signal's magnitude based on previous samples ($V_i(t_0)$) and comparing this value with the actual input signal, $V_{in}(t_0)$. The polarity of the difference value indicates the polarity of the pulse transmitted. The difference signal is a measure of the slope of the signal, which can be achieved by first sampling the analog signal and then by varying the amplitude, width, or the position of the digital signal in accordance with the amplitude of the sampled signal. Figure 5.5 illustrates delta modulation.

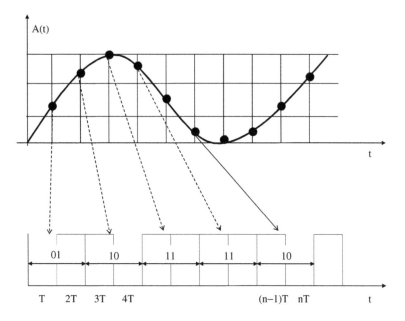

Figure 5.4   A PCM based source encoding.

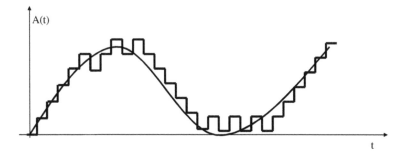

Figure 5.5   Delta encoding.

5.3   Channel Encoding

The main purpose of a channel encoder is to produce a sequence of data that is robust to noise and to provide error detection and forward error correction mechanisms. In simple and cheap transceivers, forward error correction is costly and, therefore, the task of channel encoding is limited to the detection of errors in packet transmission.

The physical channel sets limits to the magnitude and the rate of signal transmission. Figure 5.6 illustrates these restrictions. According to the Shannon–Hartley theorem, the

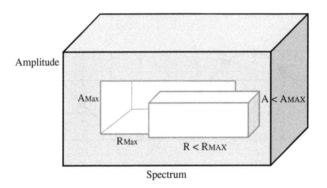

Figure 5.6 Stochastic model of a channel.

capacity of a channel to transmit a message without an error is given as:

$$C = B \cdot \log_2 \left(1 + \frac{S}{N} \right) \tag{5.14}$$

where C is the channel capacity in bits per second; B is the bandwidth of the channel in hertz; S is the average signal power over the entire bandwidth, measured in watts; and N is the average noise power over the entire bandwidth, measured in watts.

Equation (5.14) states that for data to be transmitted free of errors, its transmission rate should be below the channel's capacity. It also indicates how the signal-to-noise (SNR) ratio, can improve the channel's capacity. The equation reveals two independent reasons why errors can be introduced during transmission:

1. Information will be lost if the message is transmitted at a rate higher than the channel's capacity. This type of error is called *equivocation* in information theory. It is characterized as a subtractive error.
2. Information will be lost because of noise, which adds irrelevant information into the signal.

A stochastic model of the channel helps to quantify the impact of these two sources of errors.

Suppose an input sequence of data x_l that can have j distinct values, $x_l \in X = (x_1, x_2, ..., x_j)$, is transmitted through a physical channel. Let $P(x_l)$ denote $P(X = x_l)$. The channel's output can be decoded with a k-valued alphabet to produce $y_m \in Y = (y_1, y_2, ..., y_k)$. Let $P(y_m)$ denote $P(Y = y_m)$. At time t_i, the channel generates an output symbol y_i for an input symbol x_i.

Assuming that the channel distorts the transmitted data, it is possible to model distortion (or transmission probability) as a stochastic process:

$$P(y_m|x_l) = P(Y = y_m|X = x_m) \tag{5.15}$$

where, $l = 1, 2, ..., j$ and $m = 1, 2, ..., k$.

In the subsequent analysis of the stochastic characteristic of the channel, the following assumptions hold:

- The channel is discrete, namely, X and Y have finite sets of symbols.
- The channel is stationary, namely, $P(y_m|x_l)$ are independent of the time instance, i.
- The channel is memoryless, namely, $P(y_m|x_l)$ are independent of previous inputs and outputs.

One way of describing transmission distortion is by using a channel matrix, P_C.

$$P_C = \begin{bmatrix} P(y_1|x_1) & \cdots & P(y_k|x_1) \\ \vdots & & \vdots \\ P(y_1|x_j) & \cdots & P(y_k|x_j) \end{bmatrix} \tag{5.16}$$

where

$$\sum_{m=1}^{k} p(y_m|x_j) = 1 \, \forall j \tag{5.17}$$

Moreover:

$$P(y_m) = \sum 1 = 1 P(y_m|x_l).P(x_l) \tag{5.18}$$

or, more generally:

$$(\overrightarrow{P_y}) = (\overrightarrow{P_x}) \cdot [P_C] \tag{5.19}$$

where both $(\overrightarrow{P_y})$ and $\overrightarrow{P_x}$ are row matrices.

5.3.1 Types of Channels

5.3.1.1 Binary Symmetric Channel

A binary symmetric channel (BSC) is a channel model through which bits of information (0 and 1) can be transmitted. The channel transmits a bit of information correctly (regardless of whether 0 or 1 is transmitted) with a probability p and incorrectly (by flipping 1 to 0 and 0 to 1) with a probability $1 - p$. Such a model is displayed in Figure 5.7.

The conditional probabilities for correct and incorrect transmissions are given as:

$$P(y_0|x_0) = P(y_1|x_1) = 1 - p \tag{5.20}$$

$$P(y_1|x_0) = P(y_0|x_1) = p \tag{5.21}$$

The channel matrix of a binary symmetric channel is, therefore given as:

$$P_{BSC} = \begin{bmatrix} (1-p) & p \\ p & (1-p) \end{bmatrix} \tag{5.22}$$

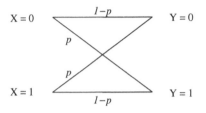

Figure 5.7 A binary symmetric channel model.

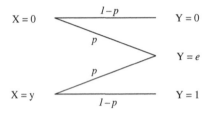

Figure 5.8 A stochastic model of a binary erasure channel.

5.3.1.2 Binary Erasure Channel

In a binary erasure channel (BEC), there is no guarantee that the transmitted bit of information can be received at all (correctly or otherwise). Therefore the channel is characterized as a binary input and a ternary output channel. The probability of erasure is p and the probability that the information is correctly received is $1 - p$. In an erasure channel the probability of error is zero. Figure 5.8 illustrates a binary erasure channel.

The channel matrix for a binary erasure channel is given as:

$$P_{\text{BEC}} = \begin{bmatrix} (1-p) & p & 0 \\ 0 & p & (1-p) \end{bmatrix} \tag{5.23}$$

Equation (5.23) states that a bit of information is either transmitted successfully with $P(1|1) = P(0|0) = 1 - p$ or is erased altogether by the channel with a probability of p. The probability that 0 is received by transmitting 1 or vice versa is 0.

5.3.2 Information Transmission over a Channel

Given the input message, $X : (X, \overset{Px}{\to}, H(X))$, the channel matrix, $[P_C]$ and the output message, $Y : (Y, \overset{Py}{\to}, H(Y))$, it is possible to describe the impact of irrelevance and equivocation as well as the percentage of information that can be transmitted over the channel without an error, which is also called transinformation or mutual information.

5.3.2.1 Irrelevance

The content of information that can be introduced into the channel due to noise is described as the conditional information content, $I(y|x)$. It is the information content of y that can be observed provided that x is known. The conditional entropy is given as:

$$H(y|x) = E_y[I(y|x)] = \sum_{y \in Y} P(y|x) \cdot \log_2 \left(\frac{1}{P(y|x)} \right) \tag{5.24}$$

$P(y|x)$ can be known from the channel matrix $[P_C]$. The average conditional entropy over all input message symbols, $x \in X$, is given by:

$$H(Y|X) = E_x[H(Y|x)] = \sum_{x \in X} P(x) \cdot \sum_{y \in Y} P(y|x) \cdot \log_2 \left(\frac{1}{P(y|x)} \right) \tag{5.25}$$

which is also equal to:

$$H(Y|X) = E_x[H(Y|x)] = \sum_{x \in X} \sum_{y \in Y} P(y|x) \cdot P(x) \cdot \log_2 \left(\frac{1}{P(y|x)} \right) \tag{5.26}$$

From Baye's law, it is clear that:

$$p(x, y) = P(y|x) \cdot P(x) \tag{5.27}$$

According to Equation (5.26), a good channel encoder is one that reduces the irrelevance entropy.

5.3.2.2 Equivocation

The content of information that can be lost because of the channel's inherent constraints can be quantified by observing the input x given that the output y is known:

$$H(X|Y) = \sum_{x \in X} \sum_{y \in Y} P(x|y) \cdot P(y) \cdot \log_2 \left(\frac{1}{P(x|y)} \right) \tag{5.28}$$

Once again, applying Baye's conditional probability:

$$P(x|y) = \frac{P(y|x) \cdot P(x)}{P(y)} = \frac{P(y|x) \cdot p(x)}{\sum_{x \in X} P(y|x) \cdot P(x)} \tag{5.29}$$

The conditional probability of Equation (5.29) is also known as the probability of inference or posterior probability. Therefore, equivocation is sometimes called inference entropy. A good channel encoding scheme is one that has a high inference probability. This can be achieved by introducing redundancy during channel encoding.

5.3.2.3 Transinformation

The information content $I(X; Y)$ that overcomes the channel's constraints to reach the destination (the receiver) is called transinformation. Given the input entropy, $H(X)$, and equivocation, $H(X|Y)$, the transinformation is computed as:

$$I(X; Y) = H(X) - H(X|Y) \tag{5.30}$$

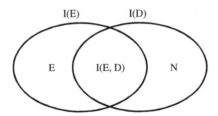

Figure 5.9   Irrelevance, equivocation, and transinformation.

Expanding Equation (5.30) yields:

$$\sum_{x \in X} P(x) \cdot \log_2 \left(\frac{1}{P(x)} \right) - \sum_{x \in X} \sum_{y \in Y} P(x, y) \cdot \log_2 \left(\frac{1}{P(x|y)} \right) \tag{5.31}$$

Rearranging the terms in Equation (5.31) also yields:

$$H(Y) - H(Y|X) = I(Y; X) \tag{5.32}$$

Irrelevance, equivocation, and transinformation, are summarized in Figure 5.9.

5.3.3 Error Recognition and Correction

Apart from improving the transinformation of a channel, it is also essential to recognize and correct errors during transmission. Error recognition can be achieved by permitting the transmitter to transmit only specific types of words. If a channel decoder recognizes unknown words, it attempts to correct the error or requests for retransmission (known as automatic repeat request, ARQ). In principle, a decoder can correct only m number of errors, where m depends on the size of the word. Error correction, or more precisely, forward error correction, can be achieved by sending n bits of information together with r control bits. The problem with forward error correction is that it slows down transmission.

5.4 Modulation

Modulation is a process by which the characteristics (amplitude, frequency, and phase) of a carrier signal are modified according to the message (a baseband) signal. Modulation has several advantages:

- the message signal will become resilient to noise;
- the channel's spectrum can be used efficiently; and
- signal detection will be simple.

5.4.1 Modulation Types

The message signal is a baseband signal. That means, its dominant frequency components are in the vicinity of zero. If this signal were to be transmitted over a wireless link without any form of modulation, one would need a receiver antenna whose size should be approximately equal to one-fourth of the size of the signal's wavelength. This is very long and it is impractical to deploy such an antenna on wireless devices.

The alternative is to superimpose the message signal on a bandpass carrier signal whose wavelength is very much smaller than the baseband signal. Due to several practical reasons, sinusoidal carrier signals are used for modulation. The properties of a sinusoidal carrier signal can be described by Equation (5.33).

$$s_c(t) = S_C \sin(2\pi ft + \phi(t)) \tag{5.33}$$

where S_C is the peak amplitude of the signal; f is the frequency; and $\phi(t)$ is the phase (relative position of the signal with respect to a reference signal). A radio frequency signal can also be described in terms of its wavelength, which is a function of the propagation speed and the frequency. Figure 5.10 shows two sinusoidal signals that have the same frequency and amplitude, but are also out of phase by ϕ degrees.

It is also customary to use polar presentation to describe the relationship between two sinusoidal signals that have the same frequency. Figure 5.11 illustrates the relationship between the two sinusoidal signals shown in Figure 5.10.

A message signal, $s_m(t)$, can change either the amplitude, the phase, or frequency of $s_c(t)$. If $s_m(t)$ changes the amplitude of $s_c(t)$, the modulation is known as amplitude modulation (AM). If $s_m(t)$ changes the frequency of $s_c(t)$, the modulation is known as frequency modulation (FM). If $s_m(t)$ changes the phase of $s_c(t)$, the modulation is known as phase modulation. Likewise, $s_m(t)$ can be a digital (binary) signal, in which case, the corresponding modulation types are called, amplitude shift keying (ASK), frequency shift keying (FSK), and phase shift keying (PSK).

A modulation process can further be classified into coherent or noncoherent; binary or q-ary; and power-efficient or spectrum-efficient. In a coherent modulation technique a carrier signal of the same frequency (and ideally, of the same phase) is required to demodulate (detect) the received signal. In a noncoherent modulation technique, no additional carrier signal is required to demodulate the received signal. In a binary modulation, the modulating (message) signal is binary, whereas in a q-ary modulation, the modulating signal can have m discrete values. In a power-efficient modulation technique, the aim is to optimize the power of the modulated signal, whereas in a spectrum-efficient modulation technique, the aim is to optimize the bandwidth of the modulated signal.

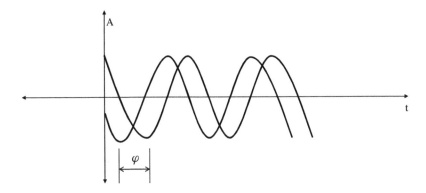

Figure 5.10   Two signals having a phase difference of ϕ.

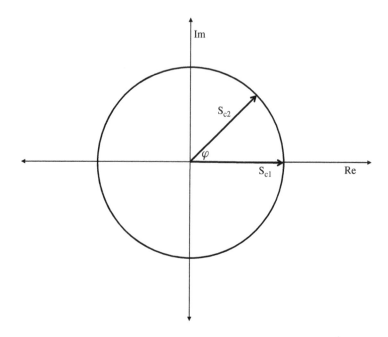

Figure 5.11   Representation of a relationship between signals with a polar diagram.

5.4.1.1   Amplitude Modulation

Considering that both the carrier and the modulating signals are analog sinusoidal signals, an amplitude modulation can mathematically be described as follows:

$$s_{mod}(t) = [S_C \times S_M \cos(2\pi f_m t + \phi_m)] \cos(2\pi f_c t + \phi_c) \qquad (5.34)$$

In other words, the amplitude of $s_c(t)$ is varied according to the modulating signal, $s_m(t)$. To simplify the analysis, assume that the two signals are in phase ($\phi_m = \phi_c = 0$) and thus, Equation (5.34) reduces to:

$$s_{mod}(t) = [S_C \times S_M \cos(2\pi f_m t)] \cos(2\pi f_c t) \qquad (5.35)$$

Applying Euler's formula ($e^{j\omega t} = \cos(\omega t) + j \sin(\omega t)$), Equation (5.35) reduces to:

$$s_{mod}(t) = \frac{S_C \times S_M}{2} [\cos(2\pi (f_c + f_m) t) + \cos(2\pi (f_c - f_m) t)] \qquad (5.36)$$

In reality, the message signal is not a mere sinusoidal signal. Instead, it is a baseband signal that has a bandwidth of B in which the amplitude and frequency change as functions of time. The Fourier transformation of such a baseband signal resembles that displayed in Figure 5.12. The Fourier transformation of the carrier signal is displayed in Figure 5.13. Hence, the spectrum of the amplitude modulated signal based on Figure 5.12 and Figure 5.13 looks like the one displayed in Figure 5.14.

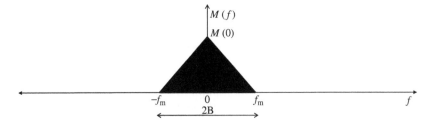

Figure 5.12   The spectrum of a baseband signal having a bandwidth of B.

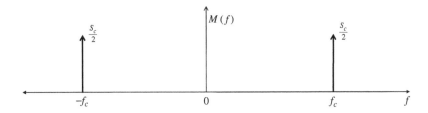

Figure 5.13   The Fourier transformation of a carrier signal having a frequency of f_c.

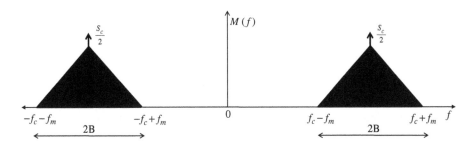

Figure 5.14   The Fourier transformation of an amplitude modulated signal.

Figure 5.15 illustrates the technical aspect of an amplitude modulation. The baseband signal and the carrier signal are mixed (multiplied) by using a mixer, which is, typically, an amplifier having a bandwidth greater than the bandwidth of the baseband signal. Afterwards (not shown here) the signal passes through a number of amplification and filtering processes so that the channel's amplitude and spectrum requirements are fulfilled.

The demodulation process – the extraction of the message signal from the modulated signal – is similar to the modulation process, but involves at least one additional process, that is, lowpass filtering. First, the received modulated signal is mixed (multiplied) with a carrier signal that has the same frequency and, ideally, the same phase as the original carrier signal, $S_C(t)$. Mathematically, this is expressed as:

$$s_{\text{demo}}(t) = S_C \cos(2\pi f_c t) \times s_{\text{mod}}(t) \tag{5.37}$$

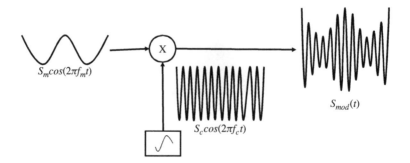

Figure 5.15 Amplitude modulation.

Expanding Equation (5.37) yields:

$$s_{\text{demo}}(t) = S_C \cos\left(2\pi f_c t\right) \times \frac{K S_C \times S_M}{2} \left[\cos\left(2\pi\left(f_c + f_m\right)t\right) + \cos\left(2\pi\left(f_c - f_m\right)t\right)\right]$$

$$(5.38)$$

where $K \ll 1$, which signifies that the modulated signal is attenuated. Applying properties of trigonometry, Equation (5.38) can be simplified into:

$$s_{\text{demo}}(t) = \frac{K S_C^2 S_M}{4} \left[\cos\left(2\pi\left(2 f_c - f_m\right)t\right) + \cos\left(2\pi\left(2 f_c + f_m\right)t\right) + 2\cos\left(2\pi f_m t\right)\right]$$

$$(5.39)$$

As can be seen, Equation (5.39) contains the message signal and a carrier signal whose frequency is much higher than the message signal. The two components can very easily be separated by using a simple envelope detector consisting of a half-wave rectifier and a lowpass filter. Figure 5.16 shows how a modulated signal is mixed with a carrier signal generated by the local oscillator of the receiver. The result passes through a bandpass filter (not shown here) to remove the f_c component. Afterwards, a simple half-wave rectifier and a lowpass filter are used to retrieve the message (baseband) signal.

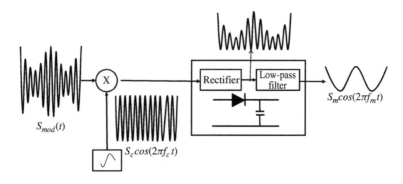

Figure 5.16 Demodulating an AM carrier signal.

5.4.1.2 Frequency and Phase Modulation

In frequency modulation, the amplitude of the carrier signal, $s_c(t)$, remains intact, but its frequency changes according to the message signal, $s_m(t)$. Here, it is essential to restrict the amplitude of the modulating signal such that $|s_m(t)| \leq 1$. Hence, the modulated signal is described as follows:

$$s_{FM}(t) = S_C \cos\left(2\pi \int_0^t f(\tau)\,d\tau\right) \tag{5.40}$$

where $\int_0^t f(\tau)d\tau$ is the instantaneous variation of the local oscillator's frequency. Expressing this frequency variation as a function of the modulating signal yields:

$$s_{FM}(t) = S_C \cos\left(2\pi \int_0^t [f_c + f_\delta s_m(\tau)]\,d\tau\right) \tag{5.41}$$

where f_δ is the maximum frequency deviation of the carrier frequency, f_c. Rearranging the terms in Equation (5.41) yields:

$$s_{FM}(t) = S_C \cos\left(2\pi f_c t + 2\pi f_\delta \int_0^t s_m(\tau)d\tau\right) \tag{5.42}$$

In phase modulation, the phase of the carrier changes in accordance with the message signal.

5.4.1.3 Amplitude Shift Keying

So far, the modulated signal was considered to be analog. In digital communication, the modulated signal is a binary stream.

Amplitude shift keying is a digital modulation technique in which the amplitude of an analog carrier signal is varied in accordance with a binary stream. The frequency and phase of the carrier signal remain unchanged.

There are several possibilities to realize amplitude shift keying. The simplest is to use an on–off modulation system, as shown in Figure 5.17. Accordingly to Figure 5.17, the mixer (multiplier) produces an output that is the multiplication of the two input signals – one of which is the message stream and the other is the output of the local oscillator, namely, the sinusoidal carrier signal having a frequency of f_c.

Direct mixing a square wave (the bit stream) requires a mixer with an excessive bandwidth, which is expensive to afford. Alternatively, an amplitude shift keying can take place by using a pulse-shaping filter (PSF). The PSF removes high-frequency components from

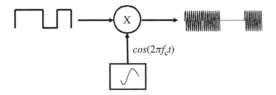

Figure 5.17   Amplitude shift-keying technique using an on-off switch.

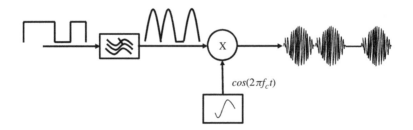

Figure 5.18   An amplitude shift-keying process using a pulse-shaping filter.

the square wave signal and approximates it with a low-frequency signal, which will then modulate the carrier signal. This is displayed in Figure 5.18.

The demodulation process employs a mixer, a local oscillator, a PSF, and a comparator. The mixer and the PSF are used to remove the high-frequency component from the modulated signal. The comparator changes the analog wave form into a stream of bits.

5.4.1.4 Frequency Shift Keying

In frequency shift keying, the frequency of a carrier signal changes in accordance with the message bit stream. Since the message bit stream will have either 0 or 1, the carrier frequency also changes between two values. Figure 5.19 demonstrates how a simple switching amplifier and two local oscillators with carrier frequencies f_1 and f_2 can be used in frequency shift-keying modulation. The switching amplifier is controlled by the message bit stream.

The demodulation process requires two local oscillators (with frequency f_1 and f_2), two PSFs, and a comparator, as illustrated in Figure 5.20.

5.4.1.5 Phase Shift Keying

In phase shift keying the phase of a carrier signal is changed according to the message bit stream. The simplest form of phase shift keying is to make a phase shift of 180° when the bit stream changes from 1 to 0 or vice versa. Figure 5.21 shows a phase shift-keying

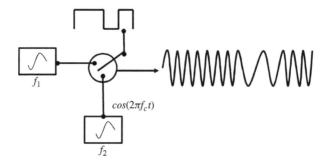

Figure 5.19 A frequency shift-keying modulation.

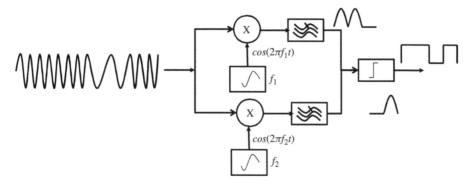

Figure 5.20   Demodulation in a frequency shift-keying process.

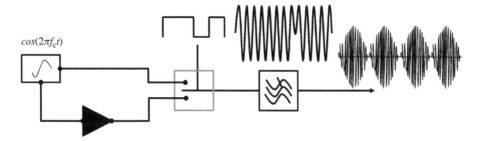

Figure 5.21   A phase shift-keying modulation process.

process in which a transition from 1 to 0 results in a phase shift of 180°. The modulation process requires a local oscillator, an inverter, a switching amplifier, and a PSF. The inverter is responsible for inverting the carrier signal by 180°.

Alternatively, a PSF, a mixer, and a local oscillator can be used as shown in Figure 5.22. The demodulation process uses a local oscillator, a mixer, a PSF, and a comparator, as shown in Figure 5.23.

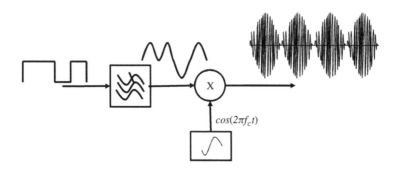

Figure 5.22   A phase shift-keying modulation with a PSF.

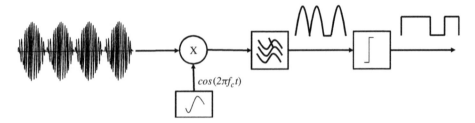

Figure 5.23   A demodulation scheme for a phase shift keying.

5.4.2   Quadratic Amplitude Modulation

So far a single message source is used to modulate a single carrier signal. This, however, is not efficient enough. One can employ orthogonal signals to effectively exploit the channel's bandwidth. In the QAM process (Figure 5.24), two amplitude-modulated, orthogonal carriers are combined as a composite signal, thereby achieving double bandwidth efficiency compared to the normal amplitude modulation. QAM is used with pulse amplitude modulation (PAM) in digital systems, especially in wireless applications. The modulated bit stream is divided into two parallel substreams each of which independently modulates the two orthogonal carrier signals.

The carrier signals have the same frequency, f_c, but they are out of phase by $90°$. Since the signals are orthogonal, they do not interfere with each other. One of the carriers is called the I carrier (in-phase signal) and the other is called the Q signal (quadrature signal).

Recall that:

$$s_Q(t) = S_C \cos(2\pi f t + 90°) = S_C \sin(2\pi f t) \qquad (5.43)$$

At the receiver side, the composite modulated signal bearing the magnitude and phase information of the Q and I signals will be mixed with two demodulating signals which are

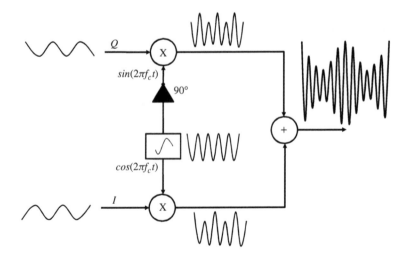

Figure 5.24   A quadratic amplitude modulation process.

identical in frequency but out of phase with each other by 90°. Then the usual detection procedure carries on to extract and aggregate (mix) the message.

Figure 5.25 shows the demodulation process of a QAM signal. The composite signal with magnitude and phase (or I and Q) information arrives at the receiver. The input signal is mixed with the carrier signal from the local oscillator in two forms. One has a reference zero phase while the other has a 90° phase shift. The composite input signal (in terms of magnitude and phase) is thus split into an in-phase, I, and a quadrature, Q, component. These two components of the signal are independent and orthogonal. One can be changed without affecting the other.

Digital modulation is easy to accomplish with I/Q modulators, which map the data to a number of discrete points on the I/Q plane. These are known as constellation points. As the modulated signal moves from one point to another, simultaneous amplitude and phase modulation take place. To accomplish this with an amplitude modulator and a phase modulator is difficult and complex. Simultaneous amplitude and phase modulation can easily be achieved with an I/Q modulator. The I and Q control signals are bounded, but infinite phase wrap is possible by properly phasing the I and Q signals.

5.4.2.1   Modulation Efficiency

Depending on the modulation type for each substream, the relative amplitude, phase, or frequency of the modulated signal carries part of the message bit stream. The modulation efficiency refers to the number of bits of information that can be conveyed in a single symbol. In a QAM, the composite carrier signal contains two orthogonal signals. The amplitude and phase of these signals are modified according to the message bit stream. Inasmuch as a receiver is sensitive enough to detect the relative differences in magnitude and phase between these two signals, much information can be conveyed with a single state of the composite carrier signal. However, there is a tradeoff between the compactness of the modulated technique and the receiver's complexity.

To evaluate the efficiency of a modulation technique, it is important to distinguish between *symbol rate* and *bit rate*. Bit rate refers to the frequency of a system's bit stream. A symbol rate, also referred to as a baud rate, refers to the bit rate divided by the number of bits that can

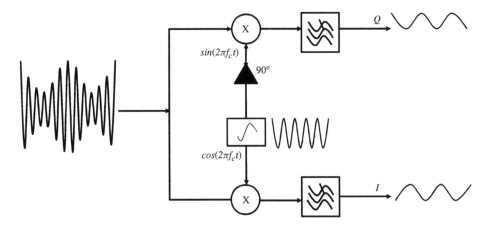

Figure 5.25   Demodulating a QAM signal.

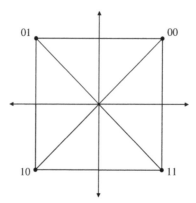

Figure 5.26   Binary phase shift keying: 2 bits per symbol.

be transmitted with each symbol. For example, a 10-bit ADC that samples an accelerometer sensor at a rate of 1 kHz has a bit stream of 10 bits multiplied by 1 kHz samples per second, or 10 kbps.

Now consider a quadrature phase shift keying (QPSK) digital modulation, in which a phase difference of 90° between the I and Q carrier signals indicates a message of 1 or 0. As can be seen in Figure 5.26, four distinct states of the composite carrier signal can be discriminated by the demodulator. Since the message signal is in binary form, the four states can be represented by two bits: 00, 01, 10, 11. Subsequently, the symbol rate is half of the bit rate. For the ADC example, the symbol rate is 5 kbps.

In an eight-state phase shift keying modulation, the phase of the composite carrier signal can have eight distinct states which can be mapped into eight distinct symbols by the demodulator. Since the eight symbols can be represented by 3 bits, the symbol rate is one-third of the bit rate. In other words, the spectrum required by an eight-state phase shift keying is one-third of the spectrum required by a binary phase shift keying modulator. However, the efficiency in spectrum is achieved at the cost of complex system design. Unlike the QPSK modulator, the 8PSK modulator (Figure 5.27) should be able to discriminate eight different transitions in phase of the composite carrier signal.

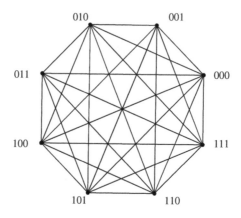

Figure 5.27   8PSK modulation with 3 bits per symbol.

5.4.3 Summary

The choice of a modulation technique depends on the design goals of the communication subsystem. There is a tradeoff between power consumption, spectrum efficiency, and cost. A power efficient modulator enables a communication system to reliably transmit information at the lowest practical power cost. A spectrally efficient modulator enables a communication subsystem to send as many bits of information as possible within a limited bandwidth. Power and spectrum efficiency cannot be achieved at the same time.

For terrestrial links, such as microwave radios, the concern is bandwidth efficiency with low bit-error-rate. Since sufficient power is available, power efficiency is not a concern. Moreover, for such types of links, the receiver's cost or complexity are not usually prior concerns. In wireless sensor networks, power efficiency is a major concern while bandwidth is not, as the nodes produce small-volume data. The cost of the transceivers is also of prime concern in large-scale deployments. Subsequently, the communication subsystems sacrifice bandwidth efficiency to achieve power and cost efficiency.

5.5 Signal Propagation

Wireless sensor networks operate in the license-free ISM spectrum (see Table 5.4) and therefore, they must share the spectrum with and accept interference from devices that operate in the same spectrum – such as cordless phones, WLAN, Bluetooth, Microwave ovens, etc.

A simple channel model (Figure 5.28) ignores the effect of interference and considers the surrounding noise as the predominant factor that affects the transmitted signal. Furthermore, the noise can be modeled as an additive white Gaussian noise (AWGN) that has a constant spectral density over the entire operating spectrum and a normal amplitude distribution. In this model, the noise distorts the amplitude of the transmitted signal.

There are two approaches to deal with noise. First, one can increase the received power so that the signal-to-noise ratio is significantly high and the channel becomes agnostic to noise.

Table 5.4 The Industry, Scientific and Medical (ISM) spectrum as defined by the ITU-R

Spectrum	Center frequency	Availability
6.765–6.795 MHz	6.780 MHz	Subject to local regulations
13.553–13.567 MHz	13.560 MHz	
26.957–27.283 MHz	27.120 MHz	
40.66–40.70 MHz	40.68 MHz	
433.05–434.79 MHz	433.92 MHz	Europe, Africa, the Middle East west of the Persian Gulf including Iraq, the former Soviet Union and Mongolia
902–928 MHz	915 MHz	The Americas, Greenland and some of the eastern Pacific Islands
2.400–2.500 GHz	2.450 GHz	
5.725–5.875 GHz	5.800 GHz	
24–24.25 GHz	24.125 GHz	
61–61.5 GHz	61.25 GHz	Subject to local regulations
122–123 GHz	122.5 GHz	Subject to local regulations
244–246 GHz	245 GHz	Subject to local regulations

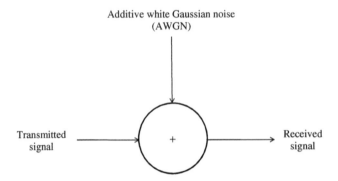

Figure 5.28   An additive white Gaussian noise channel.

Second, one can use a spread spectrum technique to distribute the energy of the transmitted signal so that a wider effective bandwidth can be achieved.

The received power can be improved by adjusting a number of parameters at the side of the transmitter as well as the receiver. The relationship between the received power and the transmitted power can be expressed using Figure 5.29.

Suppose the power amplifier – the final stage in the transmitter before the electric signal is converted into electromagnetic waves – outputs a constant transmission power, P_t, to transmit the signal over a distance of ρ. The relationship between the transmitter's antenna gain, g_t, and the antenna's effective area, A_t, is expressed as:

$$A_t = g_t \frac{\lambda^2}{4\pi} \tag{5.44}$$

where λ is the wavelength of the carrier signal.

At the receiver's side, the transmitted signal will be received and the received power is a function of the distance, the path loss index, and the receiver's antenna gain and effective area. For a line-of-sight (LOS) communication link, the path loss index is 2; for a non-LOS communication link, it lies between 2 and 4. Consequently, the relationship between the received power and the transmitted power for a LOS link is expressed as:

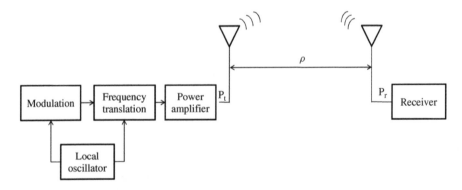

Figure 5.29   Relationship between the transmitted power and the received power.

$$P_r = \frac{P_t}{4\pi\rho^2} g_t \times A_r \tag{5.45}$$

where ρ is the distance that separates the transmitter and the receiver. Since the receiver's antenna gain, g_r, and the effective area, A_r, are related, Equation (5.45) can be reformulated:

$$P_r = \frac{P_t}{4\pi\rho^\gamma} g_t \times g_r \frac{\lambda^2}{4\pi} \tag{5.46}$$

The ratio of the transmitted power to the received power, Pt/Pr is the propagation loss and it is customary to quantify this ratio in decibels (dBs).

$$a(t) = \frac{P_t}{P_r} = \left(\frac{4\pi\rho}{\lambda}\right) \times \frac{1}{g_r g_t} \tag{5.47}$$

Hence, the propagation loss expressed in dBs is:

$$a(t)/\mathrm{dB} = 20\log\left(\frac{4\pi\rho}{\lambda}\right) - 10\log\left(g_r g_t\right) \tag{5.48}$$

The term $20\log(4\pi\rho/\lambda)$ is called the basic transmission loss and is independent of the transmitter and receiver antennas.

Exercises

5.1 How can a single ADC be employed by multiple sensors to convert their analog output into corresponding discrete sequences?

5.2 Suppose a discrete memoryless channel (DMS) source emits symbols from the ternary alphabet $A = \{-1, 0, 1\}$ with a probability, $P(-1) = 0.5$, $P(0) = P(1) = 0.25$. If the source can also be configured such that instead of emitting one symbol at a time, it can emit two symbols (A^2) with a probability that is the multiplication of the probabilities of the individual symbols, show that the entropy of the second configuration is two times greater than the entropy of the first configuration.

5.3 The following codes are given:

$C_1 = \{1, 10, 01\}$
$C_2 = \{0, 00001\}$
$C_3 = \{0, 10, 11\}$
$C_4 = \{01, 11\}$
$C_5 = \{0, 00, 000\}$

(a) Which of the given codes are uniquely decodable?
(b) Which of the given codes are instantaneously decodable?
(c) Which of the given codes can be an optimal prefix-free code for some probability assignment?

5.4 Suppose an information source emits symbols from an alphabet $X = \{x_1, ..., x_8\}$ with corresponding probabilities $\{0.2, 0.35, 0.15, 0.1, 0.09, 0.06, 0.04, 0.01\}$.

   (a) Calculate an upper bound on the average codeword length achievable with a binary Shannon or Huffman code if single symbols are encoded at a time.

   (b) Construct a binary Huffman code for the given source.

   (c) Calculate the average codeword length achieved by the Huffman code and compare it with the calculated bounds.

5.5 Refer to the analog signal shown in Figure 5.30.

   (a) How can the signal be encoded with a 3-bit PCM?

   (b) How can the signal be encoded with a delta encoder?

   (c) Illustrate how a PCM encoder with a codebook of different symbol length can be used to efficiently encode the signal.

   (d) Manchester coding is a line encoding technique that is useful for minimizing the effect of DC voltage during data transmission and for dynamic clock recovery. Illustrate how the PCM stream can be encoded with a Manchester coding.

   (e) Now instead of Manchester coding, encode the PCM stream with a differential Manchester coding.

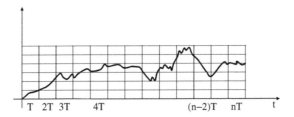

Figure 5.30   Source coding an analog signal.

   (f) Discuss the difference between the bit streams generated by the Manchester and differential Manchester encoding techniques.

5.6 The feedback loop shown in Figure 5.31 is a very useful concept in linear systems. It is the basic principle for designing stable amplifiers and oscillators. Moreover, most receivers employ the feedback loop to set up an automatic gain control (AGC) to ensure that the power of the received message signal remains constant despite changes in the channel's properties. The feedback loop is characterized by the overall loop gain, which is the ratio of the output voltage to the input voltage, $G_{loop} = V_{out}/V_{in}$. Calculate the overall loop gain, G_{loop}.

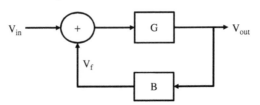

Figure 5.31   A feedback loop.

5.7 The half-wave rectifier shown in Figure 5.32 is one of the two components of an envelope detector which is responsible for extracting the baseband message signal from the
carrier. Sketch the output of the rectifier for the corresponding sinusoidal input signal.

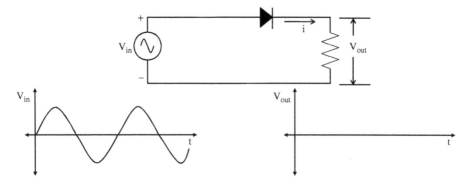

Figure 5.32   A half-wave rectifier.

5.8 Now instead of the half-wave rectifier, the full-wave bridge rectifier shown in
Figure 5.33 is used. How does the output wave form look?

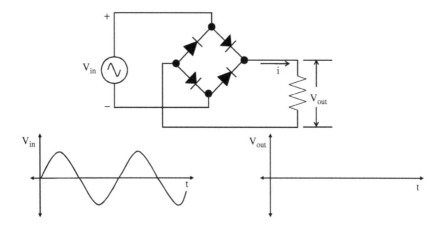

Figure 5.33   A full-wave bridge rectifier.

5.9 A lowpass filter is required to separate the baseband signal from the carrier in an amplitude modulated signal. Explain how the lowpass filter shown in Figure 5.34 can be used
for this purpose.

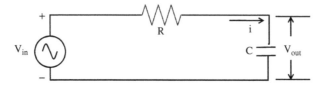

Figure 5.34   An RC lowpass filter.

Based on Figure 5.34, derive an expression for:

(a) The voltage drops across the resistor and the capacitor.
(b) The current that circulates in the filter.
(c) The transfer function, $H_C(s) = V_{out}(s)/V_{in}(s)$, where $s = j\omega$ is the Laplace operator.

5.10 A modulating signal, $m(t) = 5\cos(2\pi\ 1\,\text{kHz}\ t)$, is used to amplitude modulate a carrier signal, $c(t) = 10\cos(2\pi\ 100\,\text{MHz}\ t)$.

(a) Compute the time domain expression of the modulated signal.
(b) Compute the frequency domain expression of the modulated signal.
(c) Suppose the message signal is sampled at a period T using the Dirac delta function as shown in Figure 5.35. What does the spectrum of the sampled signal look like?

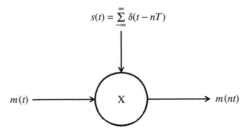

$$s(t) = \sum_{-\infty}^{\infty} \delta(t - nT)$$

$m(t) \longrightarrow \bigcirc\ X \longrightarrow m(nt)$

Figure 5.35 Sampling a modulating signal with Dirac's delta function.

(d) What precondition should be satisfied in order to reconstruct the continuous modulating signal from the sampled sequences?

5.11 The path loss index, γ, describes how an electromagnetic wave is attenuated as it propagates through a space. In free space, where there is no obstacle between the transmitter and the receiver, $\gamma = 2$. That means that the power of a propagating electromagnetic wave falls as the function of the square of a distance – $P_r \approx P_t/4\pi\rho^2$, where ρ is the distance in meters. If, however, there is an obstacle between the transmitter and the receiver, this figure is greater than 2. Figure 5.36 displays a simple model in which the electromagnetic wave reaches the receiver after being reflected once. Derive an expression of the path loss index for the model. Assume that the reflector is an ideal reflector and that there is a line-of-sight link between the transmitter and the reflector and the receiver and the reflector. Assume also that $\rho \gg h$.

5.12 Figure 5.37 displays the block diagram of a part of a receiver. It consists of an omnidirectional antenna, a RF amplifier, a local oscillator, an intermediate-frequency amplifier, and a detector (an envelope detector). While it is possible to detect the modulating signal after mixing the received, modulated signal with the local carrier signal, it is useful to have the intermediate state.

(a) Why is the intermediate frequency amplifier desirable?
(b) Suppose the receiver is used for receiving an amplitude modulated signal. How can the intermediate frequency be obtained?

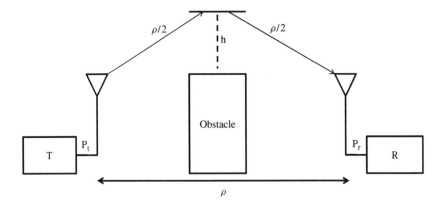

Figure 5.36   Single reflection model for an electromagnetic propagation.

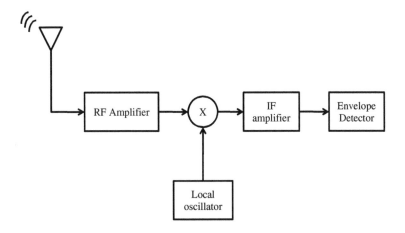

Figure 5.37   Block diagram of a receiver.

5.13 What type of roles do the transmitter's and the receiver's antennas play to enhance signal propagation and reception?

5.14 Why is a considerable amount of power wasted at the power amplifier of a transmitter?

5.15 Explain the tradeoff between modulation efficiency and design complexity in quadrature amplitude modulation.

References

Proakis, J.G. (2000) *Digital Communications* (4th edn). McGraw-Hill Publishing Company.
Wilson, S.G. (1995) *Digital Modulation and Coding*. Prentice Hall.

6

Medium Access Control

In most networks, multiple nodes share a communication medium for transmitting their data packets. The medium access control (MAC) protocol (often referred to as a sublayer of the data link layer of the OSI reference model) is primarily responsible for regulating access to the common medium. Most sensor networks and sensing applications rely on radio transmissions in the unlicensed ISM (Industrial, Scientific, and Medical) band, which may result in communications significantly affected by noise and interferences. The choice of MAC protocol has a direct bearing on the reliability and efficiency of network transmissions due to these errors and interferences in wireless communications and to other challenges such as the hidden-terminal and exposed-terminal problems. Other Further concerns include signal fading, simultaneous medium access by multiple nodes, and asymmetric (unidirectional) links. Since energy efficiency is a primary concern in a wireless sensor network, it also affects the design of the MAC protocol. Energy is not only consumed for transmitting and receiving data, but also for sensing the medium for activity (idle listening). Other reasons for energy consumption include data retransmissions (e.g., due to collisions), packet overheads, control packet transmissions, and transmit power levels that are higher than necessary to reach a receiver. It is common for a MAC protocol in WSNs to trade energy efficiency for increased latency or a reduction in throughput or fairness. This chapter reviews the responsibilities of the MAC layer in general, discusses the characteristics of MAC protocols for WSNs, describes the main classes of MAC protocols for wireless communication, and provides descriptions of a selection of MAC protocols for WSNs.

6.1 Overview

The wireless medium must be shared by multiple network devices, therefore a mechanism is required to control access to the medium. This responsibility is carried out by the second layer of the OSI reference model (Figure 6.1), called the *data link* layer. According to the IEEE 802 reference model (also shown in Figure 6.1), this layer is further divided into the *logical link control* layer and the *medium access control* layer. The MAC layer operates directly on top of the physical layer, thereby assuming full control over the medium. The main function of the MAC layer is to decide when a node accesses a shared medium and to resolve any potential conflicts between competing nodes. It is also responsible for correcting communication errors occurring at the physical layer and performing other activities such as framing, addressing, and flow control.

Fundamentals of Wireless Sensor Networks: Theory and Practice   Waltenegus Dargie and Christian Poellabauer
© 2010 John Wiley & Sons, Ltd

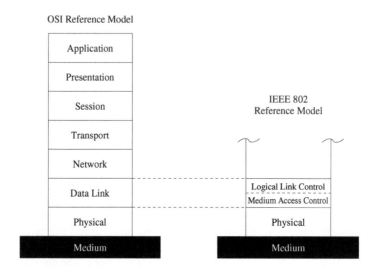

Figure 6.1   The MAC layer in the IEEE 802 reference model.

Existing MAC protocols can be categorized by the way they control access to the medium. Figure 6.2 shows an example of such a categorization. Most MAC protocols fall either into the categories of *contention-free* or *contention-based* protocols. In the first category, MAC protocols provide a medium sharing approach that ensures that only one device accesses the wireless medium at any given time. This category can further be divided into *fixed* and *dynamic* assignment classes, indicating whether the slot reservations are fixed or on-demand. In contrast to contention-free techniques, contention-based protocols allow nodes to access the medium simultaneously, but provide mechanisms to reduce the number of collisions and to recover from such collisions. Finally, some MAC protocols do not easily fit into this classification since they share characteristics of both contention-free and contention-based techniques. These *hybrid* approaches often aim to inherit the advantages of these main categories, while minimizing their weaknesses.

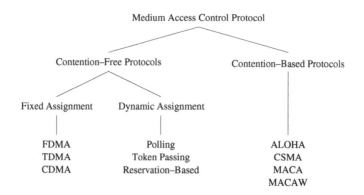

Figure 6.2   Categories and examples of medium access protocols.

6.1.1   Contention-Free Medium Access

Collisions can be avoided by allocating resources to nodes such that each node can use its resources exclusively. For example, the frequency division multiple access (FDMA) protocol is one of the oldest methods of sharing a communication medium. In FDMA, the frequency band is divided into several smaller frequency bands, which can be used for data transfer between a pair of nodes, while all other nodes that could potentially interfere with this transfer use a different frequency band. Similarly, the time division multiple access (TDMA) protocol allows multiple devices to use the same frequency band, but it uses periodic time windows (called *frames*), consisting of a fixed number of transmission slots, to separate the medium accesses of different devices. A time *schedule* indicates which node may transmit data during a certain slot, that is, each slot is assigned to at most one node. The main advantage of TDMA is that nodes do not have to contend to access the medium, thereby avoiding collisions. A downside of TDMA is that changes in the network topology necessitate changes to the slot allocations. Further, TDMA protocols can be inefficient in their bandwidth utilization when slots are of fixed size (and packet sizes can differ) and when slots allocated to a node are not used in every frame iteration. A third class of MAC protocols is based on the concept of code division multiple access (CDMA), where simultaneous accesses of the wireless medium are supported using different *codes*. If these codes are orthogonal, it is possible for multiple communications to share the same frequency band, where forward error correction (FEC) at the receiver is used to recover from interferences among these simultaneous communications.

Fixed assignment strategies can be inefficient in that it is typically not possible to reallocate slots belonging to one device to other devices if not needed in every frame. Also, generating schedules for an entire network (especially in large-scale wireless sensor networks) can be a taunting task and these schedules may require modifications every time the network topology or traffic characteristics in the network change. Therefore, dynamic assignment strategies avoid such rigid allocations by allowing nodes to access the medium on demand. For example, in polling-based protocols, a controller device (e.g., a base station in the case of an infrastructure-based wireless network), issues small polling frames in a round-robin fashion, asking each station if it has data to send. If a station has no data to be sent, the controller polls the next station. A variant of this approach is *token passing*, where stations pass a polling request to each other (again in a round-robin fashion) using a special frame called a token. A station is allowed to transmit data only when it holds the token. Finally, reservation-based protocols use static time slots to allow nodes to reserve future access to the medium based on demand. For example, a node can indicate its desire to transmit data by toggling a reservation bit in a fixed location. These often very complex protocols then ensure that other potentially conflicting nodes take note of such a reservation to avoid collisions.

6.1.2   Contention-Based Medium Access

In contrast to contention-free techniques, contention-based protocols allow nodes to contend to access the medium simultaneously, but provide mechanisms to reduce the number of collisions and to recover from such collisions. For example, the ALOHA (Kuo 1995) protocol uses acknowledgments to confirm the success of a broadcast data transmission.

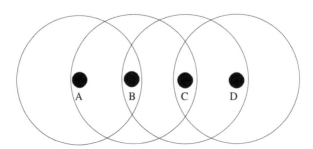

Figure 6.3   Hidden and exposed terminal problem.

ALOHA allows nodes to access the medium immediately, but addresses collisions with approaches such as exponential backoff to increase the likelihood of successful transmissions. The slotted-ALOHA protocol tries to reduce the probability of collisions by requiring that a station may commence transmission only at predefined points in time (the beginning of a time slot). While the slotted-ALOHA increases the efficiency of ALOHA, it also introduces the need for synchronization among nodes.

A popular contention-based MAC scheme is the Carrier Sense Multiple Access (CSMA) approach, including its variations Collision Detection (CSMA/CD) and Collision Avoidance (CSMA/CA). In CSMA/CD-based schemes, the sender first senses the medium to determine whether it is idle or busy. If it is found busy, the sender refrains from transmitting packets. If the medium is idle, the sender can initiate data transmission. In wired systems, the sender continues to listen to the medium to detect collisions of its own data with other transmissions. However, in wireless systems, collisions occur at the receiver, and the sender will therefore be unaware of a collision. The hidden-terminal problem occurs when two sender devices A and C are able to reach a receiver device B, but cannot overhear each other's signals (see Figure 6.3, where circles indicate transmission and interference ranges of nodes). Therefore it is possible for A and C to transmit data to B at the same time, causing a collision at B, without being able to directly detect this collision. A related problem is the exposed-terminal problem, where C wants to transmit data to a fourth node D, but decides to wait because it overhears an ongoing transmission from B to A. However, B's transmission will not interfere with data reception at D since D is outside the transmission range of B. As a consequence, node C's decision to wait delays its transmission unnecessarily. Many MAC protocols for WSNs attempt to address these two challenges.

6.2   Wireless MAC Protocols

A variety of wireless MAC protocols and standards are available today and this section provides an overview of the most common approaches. While these protocols may not be the best choices for wireless sensor networks, they introduce basic concepts, many of which can be found in protocols targeted specifically at sensor networks and their constraints.

6.2.1 Carrier Sense Multiple Access

Many contention-based protocols for wireless sensor networks rely on the concept of CSMA. The main difference between CSMA and ALOHA is that, in CSMA, nodes first sense the medium before they begin a transmission. This reduces the number of collisions. In non-persistent CSMA, a wireless node is allowed to immediately transmit data once it finds the medium idle. If the medium is busy, the node performs a backoff operation, that is, it waits for a certain amount of time before attempting to transmit again. In contrast, in *1*-persistent CSMA, a node wishing to transmit data continuously senses the medium for activity. Once the medium is found idle, the node transmits its data immediately. If a collision occurs, the node waits for a random period of time before attempting to transmit again. In *p*-persistent CSMA, the node also continuously senses the medium, but transmits data with a probability p once the medium becomes idle and delays transmission with a probability $1 - p$. Random backoff values are either continuous values in the case of unslotted CSMA or multiples of a fixed slot size in slotted CSMA.

CSMA/CA (CSMA with Collision Avoidance) is a variation of CSMA that aims to improve the performance by avoiding collisions. In CSMA/CA, nodes sense the medium, but do not immediately access the channel when it is found idle. Instead, a node waits for a time period called DCF interframe space (DIFS) plus the random backoff value, which is a multiple of a slot size (see Figure 6.4). In case there are multiple nodes attempting to access the medium, the one with the shorter backoff period will win. For example, in Figure 6.4, node A waits for DIFS $+ 4 \times s$ (where s represents the slot size), while node B's backoff is DIFS $+ 7 \times s$. Once node A begins with its transmission, node B freezes its own backoff timer and resumes the timer after node A completes its transmission plus another period of DIFS. Finally, once node B's backoff timer expires, it can also begin its transmission.

6.2.2 Multiple Access with Collision Avoidance (MACA) and MACAW

Other collision avoidance schemes rely on a dynamic reservation mechanism such as ready-to-send (RTS) and clear-to-send (CTS) control packets as in MACA (Karn 1990). With RTS, a sender device indicates its desire to transmit a data packet to an intended receiver. If the

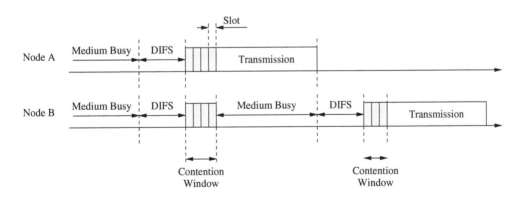

Figure 6.4 Medium access using CSMA/CA.

RTS arrives without collision and the receiver is ready to receive the packet, it responds with a CTS control message. If a sender does not receive a CTS in response to its RTS, it will retry at a later time. However, if the CTS message has been received, the channel reservation has concluded successfully. Other nodes that overhear either the RTS or CTS message know that a data transfer will occur, and they wait before they attempt to reserve the channel. In MACA, this wait time can be based upon the size of the data transmission, which can be indicated as part of the RTS and CTS messages. Using this handshake, MACA addresses the hidden terminal problem and reduces the number of collisions by reserving the medium for data transmissions.

In MACAW (Bharghavan *et al.* 1994) (MACA for Wireless LANs), the receiver responds with an acknowledgment (ACK) control message once the packet has been received correctly, allowing other nodes to learn that the channel is available again and to increase the reliability of transmissions. The consequence is that nodes overhearing an RTS message must remain silent to ensure that the sender of the RTS is able to receive the ACK. Nodes that overheard RTS, but did not acknowledge CTS, do not know whether they did not hear the CTS signal because they are out of reach of the destination or because the CTS message was never sent. In either case, they will also not hear the ACK from the destination, that is, they must stay silent until the expected completion time of the transmission, based on the information carried in the RTS message. However, if no CTS was sent, they remain silent and delay their own transmission, even though no interfering transmissions are occurring. Therefore, the MACAW protocol introduces another control packet, called the data sending (DS) message. The DS message is sent by the node issuing the RTS message after receiving the corresponding CTS message to confirm that a transmission will actually take place. A node overhearing an RTS message, but not the corresponding DS message, may assume that the medium reservation has failed and can attempt to reserve the medium for its own communication.

6.2.3 MACA By Invitation

Another improvement is provided in the MACA By Invitation (MACA-BI) Protocol (Talucci *et al.* 1997), where a destination device initiates data transfers by sending a Ready To Receive (RTR) packet to the source. The source then responds with the data message. Compared to MACA, MACA-BI reduces the overhead (thereby increasing the theoretical maximum throughput), but it depends on the destination knowing when to receive data. Source nodes can use an optional field within the data message to indicate the number of queued messages, thereby providing the destination with an indication that more RTS packets will be required.

6.2.4 IEEE 802.11

In 1999, the Institute of Electrical and Electronics Engineers (IEEE) published the 802.11 wireless LAN standard, specifying the physical and data link layers of the OSI model for wireless connections. This section briefly introduces some characteristics of this set of protocols, due to its pervasiveness and popularity. IEEE 802.11 is also often referred to as "Wireless Fidelity" (Wi-Fi), a certification given by the Wi-Fi Alliance, a group which ensures compatibility between hardware devices that use the 802.11 standard. Wi-Fi combines concepts found in CSMA/CA and MACAW, but also offers features to preserve energy.

IEEE 802.11 can be used in the point coordination function (PCF) or the distributed coordination function (DCF) mode. In the PCF mode, communication among devices goes through a central entity called an access point (AP) or a base station (often referred to as the *managed mode*). In the DCF mode, devices communicate directly with each other (referred to as the *ad hoc mode*). IEEE 802.11 is based on CSMA/CA, that is, before a node transmits, it first senses the medium for activity. If the medium is idle for at least a time period called the DCF interframe space (DIFS), the node is allowed to transmit (see Figure 6.5). Otherwise the device executes a backoff algorithm to defer transmission to a later time. This algorithm randomly selects a number of time slots to wait and stores this value in a backoff counter. For every time slot that passes without activity on the network, the counter is decremented and the device can attempt transmission when this counter reaches zero. If activity is detected before the counter reaches zero, the device waits until the channel has been idle for a period of DIFS before it continues to decrement the counter value.

After a successful transmission, a receiver device responds with an acknowledgment after waiting for a time period called the short interframe space (SIFS). The value of SIFS is smaller than the value of DIFS to ensure that no other device accesses the channel before the receiver can transmit its acknowledgment.

Once a node A makes a reservation using RTS and CTS control messages, another neighboring node B, overhearing the RTS message, must refrain from accessing the medium until node A's transmission has been completed and acknowledged. However, this could mean that node B had to continuously sense the medium to detect when it becomes idle again. Instead, A's RTS message carries the size of the data it will transmit, allowing node B to estimate how long the transmission will take and to decide whether to enter a low-power sleep mode. Some neighboring nodes may only overhear the CTS message the intended receiver sends to node A, but not node A's RTS message. Therefore, the data size information is also carried in the corresponding CTS response. Using the data size information, neighboring nodes set a network allocation vector (NAV) that indicates how long the medium will be unavailable (Figure 6.5). The use of NAV reduces the need for continuously sensing the medium, allowing a node to save power.

In the PCF mode, the access point coordinates channel access to ensure collision-free communication. The AP periodically broadcasts a beacon to its client devices, which includes a list of all devices that have packets pending at the AP. During the contention-free

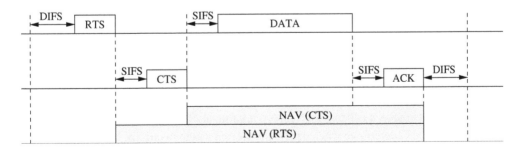

Figure 6.5   IEEE 802.11 medium access control.

period, the AP then transmits these packets to its client devices. Optionally, it can also poll client devices to allow them to initiate data transfer to the AP. In the contention-free period, the AP uses a wait period called the PCF interframe space (PIFS), which is shorter than DIFS, but longer than SIFS. This ensures that PCF traffic has priority over traffic generated by devices operating in the DCF mode, without interfering with control messages in the DCF mode such as CTS and ACK.

The focus of IEEE 802.11 is on providing fair access to the medium with support for high throughput, and mobility. However, since devices spend a large amount of time listening to the medium and collisions occur frequently, this standard incurs large overheads, including significant energy costs. To address the energy consumption problem, IEEE 802.11 offers a power saving mode (PSM) for devices in the PCF mode. Devices can inform the AP that they wish to enter a low-power sleep mode using special control messages. These devices still wake up periodically to receive beacon messages from the AP to determine if they must stay awake to receive incoming messages. While the PS mode improves the problem of energy waste, it only works for the infrastructure mode and it is not specified when or how long devices should sleep (Ye *et al.* 2004).

6.2.5 IEEE 802.15.4 and ZigBee

The IEEE 802.15.4 standard (Gutierrez *et al.* 2001) was created for low-power devices that operate in the 868 MHz, 915 MHz, and 2.45 GHz frequency bands. The data rates supported by this standard are 20, 40, and 250 kbps; rather modest compared to other protocols such as IEEE 802.11 (e.g., IEEE 802.11a offers data rates of up to 54 Mbps). Before this standard was developed, the ZigBee Alliance worked on a low-cost communication technology for low data rates and low power consumption. The IEEE and the ZigBee Alliance ultimately joined forces and ZigBee has become the commercial name for the IEEE 802.15.4 technology.

The standard further offers two topology modes: star and peer-to-peer. In the star topology, similar to Bluetooth, all communication occurs through the Personal Area Network (PAN) coordinator. In the peer-to-peer approach, devices are free to communicate directly with each other. However, they still must associate with the PAN coordinator before they can participate in peer-to-peer communication. In the star topology, there are two types of modes: the synchronized (or beacon-enabled) mode and the unsynchronized mode. In the synchronized mode, the PAN coordinator periodically broadcasts beacon messages for synchronization and management purposes. The synchronization is used to perform slotted channel access, so that a device performs a random backoff before the channel is sensed. If there is no channel activity, the device waits until the next slot and senses the channel again until no activity has been detected for two consecutive slots (after the initial backoff time). If activity has been detected, the backoff procedure is repeated, otherwise the channel can be accessed. The only difference in the unsynchronized mode is that the device can access the channel immediately when no activity has been detected during the first initial backoff time.

Data transfer between the device and its PAN coordinator is always initiated by the device, allowing a device to determine when data is transferred and to maximize its energy savings. When a device wants to send data to the PAN coordinator, it may do so at any time using the channel access method described above. The PAN coordinator transmits data intended for a device only after the device explicitly requested such a transmission. In both cases, optional acknowledgments can be used to let the PAN coordinator or device know that the transmission was successful.

While IEEE 802.15.4 has found widespread use in WSNs, there are a number of problems with this standard. For example, while the message exchange and operation are well defined for the star topology, the standard does not clearly define the operation of the peer-to-peer approach. In large WSNs, it is unlikely that all devices will be able to use the same PAN coordinator. Even though the standard does allow communication among PAN coordinators, this again is not well defined.

6.3 Characteristics of MAC Protocols in Sensor Networks

Most MAC protocols are built for fairness, that is, everybody should get an equal amount of resources (access to the wireless medium) and no one should receive special treatment. In a WSN, all nodes cooperate to achieve a common purpose, therefore fairness is less of a concern. Instead, wireless nodes are mostly concerned with energy consumption and sensing applications may value low latency or high reliability over fairness. This section discusses the main characteristics and design goals for MAC protocols of WSNs.

6.3.1 Energy Efficiency

Sensor nodes must operate using finite energy sources (batteries), therefore MAC protocols must be energy-efficient. Since MAC protocols have full control over the wireless radio, their design can contribute significantly to the overall energy requirements of a sensor node. A common technique to preserve energy is described as dynamic power management (DPM), where a resource can be moved between different operational modes such as *active*, *idle*, and *asleep*. For resources such as the network, the active mode can group together multiple different mode of activity, for example, transmitting and receiving. Without power management, most transceivers switch between transmit, receive, and idle modes, although the receive and idle modes are typically similar in their power consumption. However, dramatic energy savings can be obtained by putting the device into the low-power sleep mode. Periodic traffic models are very common for sensor networks (e.g., environmental monitoring) and many networks can benefit from MAC schemes that do not require nodes to be active at all times. Instead they allow nodes to obtain periodic access to the medium for transmission of data and to put their radios into low-power sleep modes between periodic transmissions. The fraction of time a sensor node spends in active mode is called the *duty cycle*, which is often very small due to the infrequent and brief data transmissions occurring in most sensor networks.

Table 6.1 compares the energy requirements of wireless radios in several widely deployed sensor nodes. The table shows the maximum data rate of each radio and the current consumption for the transmit, receive, idle, and standby operations. The Mica and Mica2 motes employ the Atmel ATmega 128L microcontroller (8-bit RISC processor, 128 KB flash memory, 4 KB SRAM) and use an RFM TR1000/TR3000 transceiver module (Mica) or a Chipcon CC1000 (see CC1000 2004) transceiver module (Mica2). The values shown for the CC1000 radio are for the 868 MHz mode. In addition to the standby mode, the Freescale MC13202 (see MC13202 2008) transceiver module also supports "hibernate" and "sleep"

Table 6.1   Characteristics of typical radios used by state-of-the-art sensor nodes

	RFM TR1000	RFM TR3000	MC13202	CC1000	CC2420
Data rate (kbps)	115.2	115.2	250	76.8	250
Transmit current	12 mA	7.5 mA	35 mA	16.5 mA	17.4 mA
Receive current	3.8 mA	3.8 mA	42 mA	9.6 mA	18.8 mA
Idle current	3.8 mA	3.8 mA	800 μA	9.6 mA	18.8 mA
Standby current	0.7 μA	0.7 μA	102 μA	96 μA	426 μA

modes with 6 and 1 μA respectively. Finally, the CC2420 (see CC2420 2004) transceiver module is used by the XYZ sensor node and Intel's Imote.

In addition to "idle listening" (i.e., a device staying in idle mode unnecessarily), overheads are also caused by inefficient protocol designs (e.g., large packet headers), reliability features (e.g., collisions requiring retransmissions or other error control mechanisms), and control messages to address the hidden-terminal problem. The choice of modulation scheme and transmission rate further affects the resource and energy requirements of a sensor node. Finally, most modern radios can adjust their transmit power, thereby adapting not only the range of communications, but also the energy consumption. "Overemitting", that is, using larger transmit powers than necessary, is another contributor to excessive energy consumption on a sensor node.

6.3.2 Scalability

Many wireless MAC protocols have been designed for use in infrastructure-based networks, where access points or controller nodes arbitrate access to the channel or perform some other centralized coordination and management functions. Most wireless sensor networks (WSNs) rely on multi-hop and peer-to-peer communications without centralized coordinators and they can consist of hundreds or thousands of nodes. As a consequence, MAC protocols must be able to allow for efficient use of resources without incurring unacceptable overheads, particularly in very large networks. For example, centralized protocols would incur significant overheads for the distribution of medium access schedules and are therefore unsuitable for many WSNs. MAC protocols based on CDMA may have to cache a large number of codes, which may be impractical for resource-constrained sensor devices. In general, wireless sensor nodes are not only constrained in their energy resources, but also in their processing and memory capacities. Therefore, protocols must not impose excessive computational burden or require too much memory to save state information.

6.3.3 Adaptability

A key characteristic of a WSN is its ability to self-manage, that is, it can adapt to changes in the network, including changes in topology, network size, density, and traffic characteristics. A MAC protocol for a WSN should be able to gracefully adapt to such changes without significant overheads. This requirement generally favors protocols that are dynamic in nature, that is, protocols that make medium access decisions based on current demand and network state. Protocols with fixed assignments (e.g., TDMA with fixed-size frames and slots) may incur large overheads due to adaptations of such assignments that may affect many or all nodes in the network.

6.3.4 Low Latency and Predictability

Many WSN applications have timeliness requirements, that is, sensor data must be collected, aggregated, and delivered within certain latency constraints or deadlines. For example, in a network that monitors the spreading of a wildfire, sensor data must be delivered to monitoring stations in a timely fashion to ensure accurate information and timely responses. Numerous network activities, protocols, and mechanisms contribute to the delays experienced by such data, including the MAC protocol. For example, a large frame size and a small number of slots allocated to a node in a TDMA-based protocol lead to potential delays before critical data can be transmitted over the wireless medium. In a contention-based protocol, nodes may be able to access the wireless medium sooner, but collisions and the resulting retransmissions incur delays. The choice of MAC protocol can also affect how predictable the experienced delay is, for example, expressed as upper latency bounds. Even if the average latency is large in contention-free protocols with fixed slot assignments it may be easy to determine the maximum latency that a transmission can experience. On the other hand, while average latencies in contention-based protocols could be smaller, it may be more difficult to determine an exact upper latency bound. Some contention-based MAC protocols may even allow the theoretical possibility of starvation, that is, a critical data transmission may continuously be delayed or interfered with by the transmissions of other nodes.

6.3.5 Reliability

Finally, reliability is a common requirement for most communication networks. The design of the MAC protocol can contribute to increased reliability by detecting and recovering from transmission errors and collisions (e.g., using acknowledgments and retransmissions). Particularly in wireless sensor networks, where node failures and channel errors are common, reliability is a key concern for many link-layer protocols.

6.4   Contention-Free MAC Protocols

The idea behind contention-free or schedule-based MAC protocols is to allow only one sensor node to access the channel at any given time, thereby avoiding collisions and message retransmissions. Note, however, that this assumes a perfect medium and environment, where no other competing networks or misbehaving devices exist that could otherwise cause collisions or even jam a channel. This section discusses some of the common characteristics of contention-free MAC protocols for wireless sensor networks and provides an overview of several representative examples.

6.4.1   Characteristics

Contention-free protocols allocate resources to individual nodes to ensure exclusive resource use (e.g., access to the wireless medium) by only one node. This approach eliminates collisions among sensor nodes, exposing a number of desirable characteristics. First, a fixed allocation of slots allows nodes to determine precisely when they have to activate their radio for transmission or reception of data. During all other slots, the radio (or even the entire sensor node) can be switched into a low-power sleep mode. Therefore, typical contention-free protocols are advantageous in terms of energy efficiency. With respect to predictability, fixed slot allocations also impose upper bounds on the delay that data may experience on a node, thereby facilitating the provision of delay-bounded data delivery.

While these advantages make contention-free protocols desirable choices for energy-conscious networks, they also have some disadvantages. Even though scalability of a sensor network depends on a variety of factors, the design of the MAC protocol affects how well resources are utilized in large-scale networks. Contention-free protocols with fixed slot assignments can pose significant design challenges, that is, it may be difficult to design schedules for all nodes that effectively utilize the available bandwidth when frame and slot sizes are the same for all nodes. This becomes even more pronounced when the network experiences changes in topology, density, size, or traffic characteristics, which may require the reallocation of slots or even the resizing of frames and slots. In networks with frequent changes, these disadvantages proscribe the use of protocols with fixed schedules.

6.4.2   Traffic-Adaptive Medium Access

The Traffic-Adaptive Medium Access (TRAMA) protocol (Rajendran *et al.* 2003) is a contention-free MAC protocol that aims to increase the network throughput and energy-efficiency, compared to traditional TDMA and contention-based solutions. It uses a distributed election scheme based on information about the traffic at each node to determine when nodes are allowed to transmit. This helps to avoid assigning slots to nodes with no traffic to send (leading to increased throughput) and allows nodes to determine when they can become idle and do not have to listen to the channel (increased energy efficiency).

TRAMA assumes a time-slotted channel, where time is divided into periodic random-access intervals (signaling slots) and scheduled-access intervals (transmission slots). During random-access intervals, the Neighbor Protocol (NP) is used to propagate one-hop neighbor information among neighboring nodes, allowing them to obtain consistent two-hop topology information. During the random-access interval, nodes join a network by transmitting during a randomly selected slot. The packets transmitted during these slots are used to gather neighborhood information by carrying a set of added and deleted neighbors. In case no changes have occurred, these packets serve as "keep-alive" beacons. By collecting such updates, a node knows the one-hop neighbors of its own one-hop neighbors, thereby obtaining information about its two-hop neighborhood.

A second protocol, called the Schedule Exchange Protocol (SEP), is used to establish and broadcast actual schedules, that is, allocations of slots to a node. Each node computes a duration SCHEDULE_INTERVAL, which represents the number of slots for which the node can announce its schedule to its neighbors. This duration depends on the rate at which the node's applications can produce packets. At time t, the node then computes the number of slots within $[t, t+$ SCHEDULE_INTERVAL$]$ for which it has the highest priority among its two-hop neighbors. The node announces the selected slots and the intended receivers using a *schedule packet*. The last slot in this schedule is used to announce the next schedule for the next interval. For example, if a node's SCHEDULE_INTERVAL is 100 slots and the current time (slot number) is 1000, a possible slot selection for interval [1000, 1100] for this node could be 1011, 1021, 1049, 1050, and 1093. During slot 1093, the node broadcasts its new schedule for interval [1093, 1193].

The list of intended receivers in the schedule packet is implemented as a bitmap, whose length is equal to the number of one-hop neighbors. Each bit in the bitmap corresponds to one particular receiver ordered by its identity. Since every node knows the topology within its two-hop neighborhood, it can determine the receiver address based on the bitmap and its list of neighbors.

Slot selection is based on the node's priority at time t using a pseudo-random hash of the concatenation of the node's identity i and t:

$$\text{prio}(i, t) = \text{hash}(i \oplus t) \tag{6.1}$$

If the node does not require all its slots, it can indicate which of the slots it gives up (using the bitmap in the schedule packet), allowing other nodes to claim these unused slots. Based on a node's two-hop neighborhood information and the announced schedules, a node can determine its state for any given time slot t. A node i is in the transmit (TX) state if i has the highest priority and if it has data to send. Node i is in the receive (RX) state when it is the intended receiver of the transmitter during slot t. Otherwise, the node can be switched into the sleep (SL) state.

In summary, TRAMA reduces the probability of collisions and increases the sleep time (and energy savings) compared to CSMA-based protocols. Unlike standard TDMA approaches, TRAMA divides time into random-access and scheduled-access intervals. During the random-access intervals, nodes are awake to either transmit or receive topology information, that is, the length of the random-access interval (relative to the scheduled-access interval) affects the overall duty cycle and achievable energy savings of a node.

6.4.3 Y-MAC

Another protocol that uses TDMA-based medium access, however for multiple channels, is Y-MAC (Kim *et al.* 2008). Similar to TDMA, Y-MAC divides time into frames and slots, where each frame contains a broadcast period and a unicast period. Every node must wake up at the beginning of a broadcast period and nodes contend to access the medium during this period. If there are no incoming broadcast messages, each node turns off its radio awaiting its first assigned slot in the unicast period. Each slot in the unicast period is assigned to only one node for receiving data. This receiver-driven model can be more energy-efficient under light traffic conditions, because each node samples the medium only in its own receive time slots. This is particularly important for radio transceivers where the energy costs for receiving are greater than for transmitting (e.g., due to sophisticated despreading and error correction techniques).

Medium access in Y-MAC is based on synchronous low power listening. Contention between multiple senders is resolved in the contention window, which is at the beginning of each slot. A node wishing to send data sets a random wait time (backoff value) within the contention window. After this wait time, the node wakes up and senses the medium for activity for a specific amount of time. If the medium is free, the node sends a preamble until the end of the contention window to suppress competing transmissions. The receiver wakes up at the end of the contention window to wait for packets in its assigned slot. If it receives no signal from any of its neighboring nodes, it turns off the radio and returns to the sleep mode.

During the unicast period, messages are initially exchanged on the base channel. At the beginning of its receive slot, a receiver switches its frequency to the base channel. The node that won the medium also uses the base channel to transmit its packet. The receiver acknowledges this packet if the acknowledgment request flag was set in the packet. Similarly, in the broadcast period, every node tunes to the base channel and potential senders take part in the contention process described above.

Every node polls the medium only during broadcast time slots and its own unicast receive time slots, making this approach energy-efficient. However, under heavy traffic conditions, many unicast messages may have to wait in the message queue or are dropped due to the limited bandwidth reserved for the receiving node. As a consequence, Y-MAC uses a channel-hopping mechanism to reduce packet delivery latency. Figure 6.6 shows an example with four channels. After receiving a packet during its time slot on the base channel, the receiving node hops to the next channel and sends a notification that it can continue to receive packets on the second channel. Contention for the medium in the second channel is resolved as before. At the end of this slot, the receiving node can decide to hop again to another channel until reaching the last channel or until no more data is being received. The actual hopping sequence among the available channels is determined by the hopping sequence generation algorithm, which should guarantee that there is only one receiver among one-hop neighbors on any particular channel.

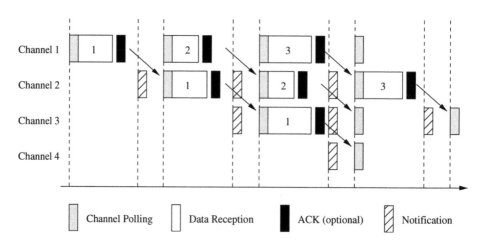

Figure 6.6   Example of channel hopping in Y-MAC (using four channels).

In summary, Y-MAC uses slot assignments such as TDMA, but communication is receiver-driven to ensure low-energy consumption (i.e., a receiver briefly samples the medium during its slot and returns to the sleep mode if no packets arrive). It further uses multiple channels to increase the achievable throughput and reduce delivery latency. The main drawbacks of the Y-MAC approach are that it has the same flexibility and scalability issues as TDMA (i.e., fixed slot allocations) and that it requires sensor nodes with multiple radio channels.

6.4.4 DESYNC-TDMA

DESYNC (Degesys *et al.* 2007) is a self-organizing *desynchronization* algorithm used to implement a collision-free MAC protocol (called DESYNC-TDMA) based on TDMA. This MAC protocol focuses on two shortcomings of traditional TDMA: it does not require a global clock and it automatically adjusts to the number of participating nodes to ensure that the available bandwidth is always fully utilized. Desynchronization is a useful primitive for periodic resource sharing in a variety of sensor applications. For example, sensors sampling a common geographic region can desynchronize their sampling schedule such that the requirements of the monitoring task are equally distributed among the sensors. In DESYNC, desynchronization is used to implement TDMA-style medium access.

Consider a network of n nodes that communicate with each other and each node performs a periodic task with a period T. Each node i can be modeled as an oscillator with a frequency $\omega = 1/T$ and a phase $\phi_i(t) \in [0, 1]$. For example, a phase of 0.75 indicates that the node is 75% through its cycle. Once a node reaches phase 1, it "fires" and resets its phase to 0. One can imagine the nodes as beads moving along a ring with period T, as shown in Figure 6.7, where a node fires once it reaches the top. The only information nodes can observe about the current state of the ring is the firing of events and they can use this to jump forward or backward in phase. The goal is then to have nodes adjust their phase independently such that eventually the network is desynchronized (i.e., nodes are equally spaced around the ring). Specifically, a node i keeps track of the firings of its immediate neighbors, that is, nodes $i + 1 \pmod{n}$ and $i - 1 \pmod{n}$. Assume that Δ_i represents the distance between oscillators on the ring, that is, $\Delta_i(t) = \phi_i(t) - \phi_{i-1}(t)$. Then, node i records its neighbors' firing times as $\bar{\Delta}_{i+1}$ and $\bar{\Delta}_i$, respectively. Node i can then approximate the phases of its neighbors as $\phi_{i+1}(t) = \phi_i(t) + \bar{\Delta}_{i+1} \pmod{1}$ and $\phi_{i-1} = \phi_i - \bar{\Delta}_i \pmod{1}$. The midpoint between these

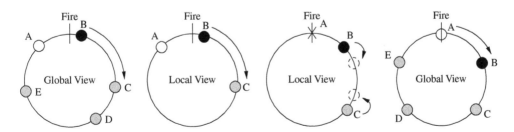

Figure 6.7 Concept of the DESYNC algorithm.

neighbors can then be determined as:

$$\phi_{\text{mid}}(t) = \frac{1}{2}[\phi_{i+1}(t) + \phi_{i-1}(t)](\text{mod } 1) \tag{6.2}$$

$$= \phi_i(t) + \frac{1}{2}(\bar{\Delta}_{i+1} - \bar{\Delta}_i)(\text{mod } 1) \tag{6.3}$$

Once the midpoint has been determined, node i can jump to it. This concept is illustrated in Figure 6.7, where the first ring shows the global view of 5 nodes that are not desynchronized. The second and third rings show B's local view. When A fires, the node that fired immediately before it (node B), knows the positions of both of its neighbors; that is, it overheard the firings of both A and C. Node B can now compute its ideal position for a desynchronized network and jumps to this position. However, C may have changed its own position in the meantime, unknown to B. If each node can fire closer to the midpoint of its neighbors, this process will eventually bring the system to a state where all nodes are exactly at the midpoint of their neighbors. The last ring in Figure 6.7 shows the global view for the desynchronized state where the distances between any two neighboring nodes are identical. Since the nodes are equally distributed, the system is stable and no more position changes are necessary.

In a wireless sensor network, the firing corresponds to a node broadcasting a "firing message". Node i keeps track of the times of the firings occurring immediately before and after its own firing. The senders of these firing messages are then *phase neighbors* of node i. Applied to TDMA, node i's TDMA slot begins at the previously computed midpoint between node i and its previous phase neighbor and ends at the previously computed midpoint between itself and its next phase neighbor. This way, a node will never fire outside its own slot. This algorithm defines a set of non-overlapping slots over the period T and nodes can transmit data without collisions, even during desynchronization. Once desynchronization has completed, the slots have converged to be of equal size.

DESYNC-TDMA ensures that the bandwidth is always fully used. When a node leaves the network, the desynchronization process ensures that slot boundaries are adjusted over time such that their sizes are equalized again. When a node joins the network, it first sends a series of short interrupt messages before sending its initial firing message. These interrupt messages notify the owner of the current slot that a new node wants to join and that the slot owner should temporarily pause its transmissions to avoid message collisions.

In summary, DESYNC-TDMA is an adaptive TDMA-based protocol that does not require explicit scheduling or time synchronization. It provides collision-free communication even during desynchronization. It further can provide high throughput, while guaranteeing predictable message latencies and fairness. DESYNC-TDMA adjusts the schedule autonomically to accommodate new nodes or to recapture slots given up by leaving nodes. However, fairness is often not a key concern in wireless sensor networks, and ensuring equal slot sizes can lead to inefficient bandwidth usage, that is, unused slot portions are therefore wasted. Similarly, if a node has more data to transmit than fits into its slot, the queuing latencies can be high.

6.4.5 Low-Energy Adaptive Clustering Hierarchy

The Low-Energy Adaptive Clustering Hierarchy (LEACH) protocol (Heinzelman *et al.* 2002) combines TDMA-style contention-free communication with a clustering algorithm

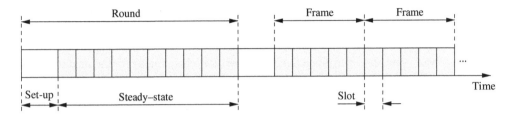

Figure 6.8 Operation and communication structure of LEACH.

for wireless sensor networks. A cluster consists of a single cluster head and any number of cluster members, which only communicate with their cluster head. Clustering is a popular approach for sensor networks since it facilitates data aggregation and in-network processing at the cluster head to reduce the amount of data that needs to be transmitted to the base station. LEACH operates in rounds consisting of two phases: a *setup phase* and a *steady-state phase* (Figure 6.8), both of which are described below.

6.4.5.1   Setup Phase

During the setup phase, cluster heads are determined and communication schedules within each cluster are established. Since the cluster head is responsible for coordinating cluster activity and forwarding data to the base station, its energy requirements will be significantly large compared to other sensor nodes. Therefore, LEACH rotates the cluster head responsibility among sensor nodes to evenly distribute the energy load. Specifically, at the beginning of a round, every sensor i elects itself to be a cluster head with a certain probability $P_i(t)$. In a network with N nodes and a desired number of cluster heads of k, the probabilities can be chosen to satisfy:

$$\sum_{i=1}^{N} P_i(t) = k \tag{6.4}$$

There are various approaches to choose $P_i(t)$, for example:

$$P_i(t) = \begin{cases} \dfrac{k}{N - k * (r \bmod N/k)}, & C_i(t) = 1; \\ 0, & C_i(t) = 0 \end{cases} \tag{6.5}$$

This approach uses an indicator function $C_i(t)$ to determine whether node i has been a cluster head in the $r\bmod(N/k)$ previous rounds. Only nodes that have not been cluster heads recently are candidates for the cluster head role. This approach to selecting cluster heads aims to evenly distribute the cluster head responsibility, and, therefore, the energy overhead, among all sensor nodes. However, this does not consider the actual amount of energy available to each node. Therefore, an alternative approach to determining the probability of becoming a cluster head can be used:

$$P_i(t) = \min \left\{ \frac{E_i(t)}{E_{\text{total}}(t)} k, 1 \right\} \tag{6.6}$$

where $E_i(t)$ is node i's actual current energy and $E_{total}(t)$ is the sum of the energy levels of all nodes. One disadvantage of this approach is that every node must know (or estimate) $E_{total}(t)$.

Once a sensor node has determined that it will serve as cluster head for the next round, it informs other sensor nodes of its new role by broadcasting an advertisement message (ADV) using a non-persistent CSMA protocol. Every sensor node joins a cluster by selecting the cluster head that can be reached with the smallest amount of transmit energy (based on *received signal strength* of the ADV messages from the cluster heads) and by transmitting a join-request (Join-REQ) message to the chosen cluster head (again, using CSMA). The cluster head establishes a transmission schedule for its cluster and transmits this schedule to each node in its cluster.

6.4.5.2 Steady-State Phase

A sensor node communicates only with the cluster head and is allowed to transmit data only during its allocated slots indicated by the schedule received from the cluster head. It is then the responsibility of the cluster head to forward sensor data originating at one of its sensor nodes to the base station. To preserve energy, each cluster member uses the minimum required transmit power to reach the cluster head and turns off the wireless radio between its designated slots. On the other hand, the cluster head must be awake at all times to receive sensor data from its cluster members and to communicate with the base station.

While intra-cluster communication is contention-free using TDMA-style frames and slots, communication occurring in one cluster can still interfere with communication in another cluster. Therefore, sensor nodes use the direct sequence spread spectrum (DSSS) technique to limit the interference among clusters, that is, each cluster uses a spreading sequence that is different from the spreading sequence used in neighboring clusters. Another reserved sequence is used for communication between cluster heads and the base station. Communication between cluster heads and the base station is based on this fixed spreading code and CSMA. Before a cluster head transmits data, it first senses the channel to see if there is an ongoing transmission using the same spreading code.

A variation of this protocol, called LEACH-C, relies on the base station to determine the cluster heads. This occurs during the setup phase, where each sensor node transmits its location and energy levels to the base station. Based on this information, the base station determines the cluster heads and informs the cluster heads of their new role. Other sensor nodes can then join clusters using join messages as described in the original LEACH protocol.

In summary, LEACH utilizes a variety of techniques to reduce energy consumption (minimum transmit energy, avoiding idle listening of cluster members) and to obtain contention-free communication (schedule-based communication, DSSS). While intra-cluster communication is contention-free and interferences among clusters are avoided, communication between the cluster heads and the base station is still based on CSMA. Furthermore, LEACH assumes that all nodes are able to reach the base station, which affects the scalability of this protocol. However, this can be addressed by either adding multi-hop routing support between the base station and all cluster heads or by implementing a hierarchical clustering approach, where some cluster heads have the responsibility of collecting data from other cluster heads.

6.4.6 Lightweight Medium Access Control

The Lightweight Medium Access Control (LMAC) protocol (Van Hoesel and Havinga 2004) is based on TDMA, that is, time is again divided into frames and slots, where each slot is owned by exactly one node. However, instead of relying on a central manager to assign slots to nodes, nodes execute a distributed algorithm to allocate slots.

Each node uses its slot to transmit a message consisting of two parts: a control message and a data unit. The fixed-size control message carries information such as the identity of the time slot controller, the distance (in hops) of the node to the gateway (base station), the address of the intended receiver, and the length of the data unit (Table 6.2 summarizes the contents of the LMAC control message). Upon receiving a control message, a node determines if it is the intended receiver and decides whether to stay awake or to turn off the radio until the next slot. The *Occupied Slots* field of the control message is a bitmask of slots, where an unoccupied slot is represented by 0 and an occupied slot is represented by 1. By combining control messages from all neighbors, a node is able to determine unoccupied slots. The process of claiming slots starts at the gateway device, which determines its own slots. After one frame, all direct neighbors of the gateway know the gateway's slots and choose their own slots. This process continues throughout the network and during each frame, a new set of nodes with a higher hop distance from the gateway determine their slots. Each node must select slots that are not in use within a two-hop neighborhood. Slots are selected randomly, therefore, it is possible for multiple nodes to select the same slot. This will result in a collision of control messages during a slot, which can be observed by the competing nodes, which, in turn, results in a restart of the selection process.

6.4.6.1 Mobile LMAC

The slot allocations in LMAC are computed only once, therefore, this protocol is not suitable for mobile sensor networks, in which nodes frequently join and leave other nodes' radio

Table 6.2 Content of control message in LMAC

Description	Size (bytes)
Identification of time slot controller	2
Current slot number	1
Occupied slots	4
Distance to gateway in hops	1
Collision in slot	1
Destination address	2
Data size (bytes)	1
Total	12

ranges. The Mobile LMAC (MLMAC) (Mank *et al.* 2007) protocol uses a distributed slot allocation mechanism, but it is able to adapt to changes in the network topology. When a node X leaves the radio range of node Y, both nodes will realize that they no longer receive control messages from each other and will remove each other from their neighbor lists. Now assume that node X moves into the radio range of node Z and that another node in Z's range, node W, uses the same slot as X. In this case, the control messages from X and W will collide at Z. Node Z will no longer receive any correct control messages during this slot and will therefore mark this slot as unused. Nodes X and W will receive Z's control message, indicating that their slot is unused, meaning that there must have been a collision. As a consequence, they give up their current slot and restart the slot selection mechanism.

Both LMAC and MLMAC have the same advantages as TDMA (collision-free communication, energy efficiency), but additionally they are able to establish transmission schedules in a distributed fashion. However, in both protocols, the slot size is fixed and slot allocations are also fixed (except when a node has to restart the slot selection mechanism), which can lead to bandwidth inefficiency.

6.5 Contention-Based MAC Protocols

Contention-based MAC protocols do not rely on transmission schedules, but instead on other mechanisms to resolve contention when it occurs. The main advantage of contention-based techniques is their simplicity compared to most schedule-based techniques. For example, where schedule-based MAC protocols must save and maintain schedules or tables indicating the transmission order, most contention-based protocols do not require to save, maintain, or share state information. This also allows contention-based protocols to adapt quickly to changes in network topologies or traffic characteristics. However, contention-based MAC protocols typically result in higher collision rates and energy costs due to idle listening and overhearing. Contention-based techniques may also suffer from fairness issues, that is, some nodes may be able to obtain more frequent channel accesses than others.

6.5.1 Power Aware Multi-Access with Signaling

The focus of the Power Aware Multi-Access with Signaling (PAMAS) protocol (Singh and Raghavendra 1998) is to avoid unnecessary energy expenditure caused by overhearing. For example, in Figure 6.9, node B's transmission to node A is overheard by node C since it is an immediate neighbor of node B. Therefore, node C consumes energy for receiving a frame intended for another node. Further, since C is in B's interference range, C cannot receive a frame from another node during B's transmission. Therefore, to conserve energy, C can turn its radio into a low-power sleep mode for the duration of B's transmission. This is particularly useful in dense networks where a node can be in the interference ranges of many other nodes.

PAMAS uses two separate channels, one for data frames and one for control frames, to prevent collisions among data transmissions. The control messages exchanged in PAMAS are ready-to-send (RTS) and clear-to-send (CTS) messages, similar to the MACA protocol. This separate signaling channel allows nodes to determine when and how long to power down their wireless transceivers. In addition to RTS/CTS, devices transmit *busy tones* on

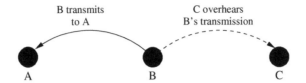

Figure 6.9   Power consumption caused by unnecessary overhearing.

the control channel to ensure that devices that did not overhear either RTS or CTS will not access the data channel for transmissions.

To initiate a data transfer, a PAMAS device sends an RTS message over the control channel to the receiver. If the receiver does not detect activity on the data channel and has not overheard other recent RTS or CTS messages, it will respond with a CTS message. If the source does not receive a CTS within a specific timeout interval, it will attempt to transmit again after a backoff time (determined by an exponential backoff algorithm). Otherwise, it begins data transmission and the receiver node issues a busy tone over the control channel (whose length is greater than twice the length of a CTS). The receiver device also issues a busy tone over the control channel whenever it receives an RTS message or detects noise on the control channel while it receives a frame. This is done to corrupt possible CTS message replies to the detected RTS, thereby blocking any data transmission of the receiver's neighbors.

Every node in a PAMAS network independently decides when to power off its transceiver. Specifically, a node decides to turn off its transceiver whenever one of two conditions holds:

- a neighbor begins a transmission and the node has no frames to transmit; and
- a neighbor transmits a frame to another neighbor, even if the node has frames to transmit.

A node can easily detect either condition by overhearing its neighbor's transmissions (condition 1) or its neighbor's busy tone (condition 2). Embedding the size or expected transmission duration into messages allows a node to identify how long to power down its transceiver. However, when a transmission begins while a node is still asleep, the node does not know how long this transmission will last and how long the node should continue to sleep. As a consequence, the node issues a *probe frame* over the control channel to all transmitting nodes in its neighborhood. The probe frame contains a time interval and all transmitters that will complete during this interval respond with their predicted completion time. If such a response is received by the awakening node without collision, the node can return to the sleep mode until the completion time indicated by the transmitting node. If multiple transmitters respond and their responses collide, the node reissues the probe frame with a shorter time interval. Similarly, if the node did not receive a response, it can reissue the probe with a different time interval. In effect, the node chooses time intervals to perform a binary search to identify when the last ongoing transmission will end.

In summary, PAMAS attempts to reduce the significant amounts of energy waste by those nodes which remain active during time periods when they can neither transmit nor receive data. However, PAMAS relies on the presence of two radios, which in itself can greatly increase the energy consumption and implementation cost.

6.5.2 Sensor MAC

The goal of the sensor MAC (S-MAC) protocol (Ye *et al.* 2002) is to reduce unnecessary energy consumption, while providing good scalability and collision avoidance. S-MAC adopts a duty-cycle approach, that is, nodes periodically transition between a *listen state* and a *sleep state*. Each node chooses its own schedule, though it is preferred when nodes synchronize their schedules such that they listen or sleep at the same time. In this case, nodes using the same schedule are considered to belong to the same *virtual cluster*, but no real clustering takes place and all nodes are free to communicate with nodes outside their clusters. Nodes periodically exchange their schedules with their neighbors using SYNC messages, that is, every node knows when any of its neighbors will be awake. If node A wants to communicate with a neighbor B that uses a different schedule, A waits until B is listening and then initiates the data transfer. Contention for the medium is resolved using the RTS/CTS scheme.

In order to choose a schedule, a node initially listens to the medium for a certain amount of time. If this node receives a schedule from a neighbor, it chooses this schedule as its own and this node becomes a *follower*. The node broadcasts its new schedule after a random delay t_d (to minimize the possibility for collisions from multiple new followers). Nodes can adopt multiple schedules, that is, if a node receives a different schedule after it has broadcast its own schedule, it adopts both schedules. Further, if a node does not hear a schedule from another node, it determines its own schedule and broadcasts it to any potential neighbors. This node becomes a *synchronizer* in that other nodes will begin to synchronize themselves with it.

S-MAC divides a node's listen interval further into a part for receiving SYNC packets and a part for receiving RTS messages (top graph in Figure 6.10). Each part is further divided into small slots to facilitate carrier sensing. A node trying to send a SYNC or RTS message randomly selects a time slot (within the SYNC or RTS part of the interval, respectively) and senses the carrier for activity from when the receiver begins listening to the selected slot. If no activity has been detected, it wins the medium and begins transmission. Figure 6.10 shows the timing relationship between a receiver and different senders: one sending a SYNC (middle graph) and another one sending data (bottom graph).

S-MAC adopts a contention-based approach, where contention for the medium is addressed using collision avoidance based on RTS/CTS handshakes. When a node hears an RTS or CTS and concludes that it cannot transmit or receive at the same time, it can go to sleep to avoid energy waste through overhearing (a node may only overhear brief control messages, but not the typically longer data messages).

In summary, S-MAC is a contention-based protocol that utilizes the sleep mode of wireless radios to trade energy for throughput and latency. Collision avoidance is based on RTS/CTS, which is not used by broadcast packets, thereby increasing the collision probability. Finally, duty cycle parameters (sleep and listen periods) are decided beforehand and may be inefficient for the actual traffic characteristics in the network.

6.5.3 Timeout MAC

The listening period of S-MAC is of fixed duration, that is, if there is only little traffic, this can actually waste energy. On the other hand, if traffic is heavy, the fixed duration may not be large enough. Therefore, the Timeout MAC (T-MAC) protocol (Van Dam and

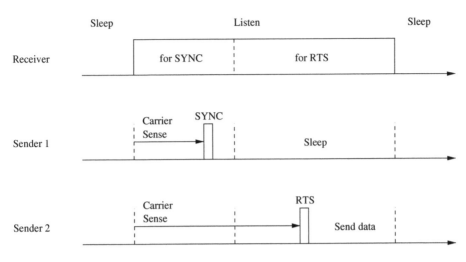

Figure 6.10   S-MAC timing and messaging relationships.

Langendoen 2003) is a variation of S-MAC that uses an active period that adapts to traffic density. Nodes wake up during the beginning of a slot to listen very briefly for activity and return to the sleep mode when no communication has been observed. When a node transmits, receives, or overhears a message, it remains awake for a brief period of time after completion of the message transfer to see if more traffic can be observed. This brief *timeout* interval allows a node to return to the sleep mode as quickly as possible. The end effect is that a node's awake times will increase with the heavier traffic and will be very brief if traffic is light.

Figure 6.11 shows the basic operation and data exchange of T-MAC. To reduce potential collisions, each node waits for a random period of time within a fixed contention interval before the medium is accessed. For example, in Figure 6.11(a), nodes A and C are trying to send data to node B, but node A wins the medium and transmits its data to node B. The minimum time a node remains active to listen for activity is expressed as *TA* and it must be long enough to hear a potential CTS from one of its neighbors. Once a node hears a CTS, it knows that another node won the medium. This node then stays awake until the end of the transmission, which can be observed by overhearing the acknowledgment (ACK) sent by node B. This event initiates the beginning of the next contention interval and node C will have an opportunity to transmit its data if it wins the medium.

Figure 6.11(a) also shows a potential problem occurring in T-MAC. Assume that messages flow from top to bottom, that is, node A sends only to node B, node B sends to node C, etc. Every time node C wants to send a message to node D, it must contend for the medium and may lose to either node B (which may transmit an RTS before C does) or to node A (node C overhears a CTS transmitted by node B). While node C stays awake after overhearing node B's CTS message, its intended receiver (node D) is not aware of C's intention to transmit data and therefore returns to the sleep mode after *TA* has expired. This problem is referred to as the *early sleeping* problem, and one possible solution to this problem is shown in Figure 6.11(b). In the *future-request-to-send* technique, a node with pending data can inform its intended receiver by transmitting a future-request-to-send (FRTS) packet immediately after overhearing a CTS message. Node D, upon receiving the FRTS message, knows

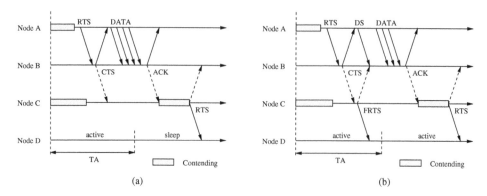

Figure 6.11 Data exchange in T-MAC, showing the early sleeping problem in (a) and the future-request-to-send technique in (b).

that node C will attempt to send data to it and will therefore remain active. However, sending the FRTS message immediately after CTS could interfere with node B's reception of node A's data, therefore, node A first sends a *dummy* message called Data-Send (DS) to delay the transmission of the actual data. DS has the same size as FRTS and can collide with FRTS at node B, which is of no consequence since it does not contain any useful information.

In summary, T-MAC's adaptive approach allows it to adjust a node's sleep and awake intervals based on the traffic load. In T-MAC, nodes send messages as bursts of variable length and sleep between such bursts to conserve energy. Both S-MAC and T-MAC concentrate message exchanges to small periods of time, which results in inefficiencies under high traffic loads. Finally, intended receivers are kept awake using messages that indicate future transmissions, which can significantly increase the idle listening times (and therefore energy consumption) of nodes.

6.5.4 Pattern MAC

The Pattern MAC (PMAC) protocol (Zheng *et al.* 2005) is another example of a TDMA-style protocol that uses frames and slots, but it adapts its sleep schedules on the basis of its own traffic and the traffic patterns of its neighbors. Compared to S-MAC and T-MAC, PMAC further reduces energy costs of idle listening by allowing devices to turn off their radios for long durations during periods of inactivity. Nodes use *patterns* to describe their *tentative* sleep and awake times, where a pattern is a string of bits, each bit representing a time slot and indicating whether a node plans to sleep (bit is 0) or be awake (bit is 1). While patterns are only tentative, *schedules* represent the *actual* sequence of sleep and awake times. The format of a pattern is always 0^m1, where $m = 0, 1, \ldots, N - 1$ and N time slots are considered to be a period. For example, a pattern of 001 and $N = 6$ indicate a node's plan to be awake during the third and sixth slot of the period (i.e., the pattern is repeated whenever its length is less than N). The value of m (i.e., the number of leading zeros) is an indicator of traffic load around the node – that is, a small value indicates heavy traffic and a large value indicates light traffic.

At network activation, every node's pattern during the first period is 1, that is, $m = 0$, and every node assumes a heavy traffic load and that it should be awake at all times. If a node does not have any data to send during the first slot, then it uses this as an indicator that the traffic around it is potentially light and it updates its own pattern to 01. The node continues to double the sleep interval (doubling the number of zeros) every time it has no data to send, allowing it to sleep longer. This process (which mimics the slow-start behavior of TCP) is continued until a predefined threshold is reached, after which the number of zeros is increased linearly. That is, if there is no data for node i to send, the following sequence of patterns will be generated: $1, 01, 0^21, 0^41, \ldots, 0^81, 0^801, 0^80^21, 0^80^31, \ldots, 0^{N-1}1$. Whenever a node has data to send, the pattern is immediately reset to 1, allowing the node to wake up quickly to handle the traffic load.

While a pattern is only a tentative sleep plan, patterns are used to derive actual sleep schedules. A node broadcasts its own pattern at the end of a period, during a time interval called the Pattern Exchange Time Frame (PETF). The PETF is divided into a sequence of brief slots, where the number of slots is set to the maximum number of neighbors a node could have. These slots are accessed using CSMA and collisions can occur. If a node does not receive a pattern update from one of its neighbors (most likely due to a collision), the node simply assumes that the neighbor's pattern remains unchanged. Once a node has received the patterns from its neighbors, it determines its own schedule, where each slot can be used for one of three possible operations. A node wakes up and transmits a message to a neighbor if the neighbor has advertised a 1 for that slot. If a node has advertised a 1, but has no data to send, the slot is used to listen. If none of these two conditions holds, the node sleeps.

In summary, PMAC provides a simple mechanism to build schedules that adapt to the amount of traffic in a neighborhood. When traffic loads are light, a node is able to spend considerable amounts of time in the sleep mode, thereby preserving energy. However, collisions during the PETF may prevent nodes from receiving pattern updates from all neighbors, while other nodes may have received these updates. This leads to inconsistent schedules among nodes in a neighborhood, which can cause further collisions, wasted transmissions, and unnecessary idle listening.

6.5.5 Routing-Enhanced MAC

The Routing-Enhanced MAC (RMAC) protocol (Du *et al.* 2007) is another example of a protocol that exploits duty cycles to preserve energy. Compared to S-MAC, it attempts to improve upon end-to-end latency and contention avoidance. The key idea behind RMAC is to align the sleep/wake periods of nodes along the path of sensor data such that a packet can be forwarded to the destination within a single *operational cycle*. It achieves this by sending a control frame along the route to inform nodes of the upcoming packet, allowing them to learn when to be awake to receive and forward this packet.

RMAC partitions an operational cycle into three components: the SYNC period, the DATA period, and the SLEEP period (Figure 6.12). During the SYNC phase, nodes synchronize their clocks to ensure that they maintain sufficient precision. The DATA period is used to announce and initiate the packet transmission process along the packet's route to the destination. The DATA period is contention-based and the sender waits for a randomly chosen period of time plus an additional DIFS period, during which it senses the medium. If no activity is detected, the sender transmits a Pioneer Control Frame (PION), containing the addresses of the sender, destination, and next hop; the duration of the transmission, and the number of hops the PION has travelled so far (which is set to zero at the sender). The

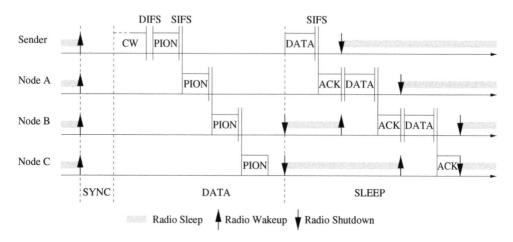

Figure 6.12 Duty cycle and communication pattern in RMAC.

next hop along the route (node A in Figure 6.12) looks up the next hop for this route (from the network layer) and forwards the PION to the next hop after waiting for a SIFS period. This process continues until the PION reaches the destination.

Actual data transmission takes place during the SLEEP period of the protocol. In Figure 6.12, node A stays awake to receive the data packet from the sender and after successful transmission, A returns an acknowledgment (ACK). Similar to the PION schedule during the DATA period, all data and ACK packets are separated by a SIFS period. After receiving the ACK from node A, the sender has completed its part and can return its radio into the sleep mode. Node A relays the received packet to the next hop, node B, and also returns its radio to sleep mode once B has acknowledged the data packet. This process continues until the data packet has been received and acknowledged by the destination.

In this example, the sender and node A stay awake after the DATA period to immediately begin the transmission of the data packet over the first hop. All other nodes along the route can turn off their radios after the DATA period has completed to further preserve energy. Each node wakes up at the right time to receive the data packet from the upstream node. This time to wake up can be computed by node i as:

$$T_{\text{wakeup}}(i) = (i - 1) \times (\text{size(DATA)} + \text{size(ACK)} + 2 \times \text{SIFS}) \qquad (6.7)$$

where size(DATA) and size(ACK) are the times required to send a single data and ACK frame, respectively.

In summary, RMAC addresses the large latencies often experienced in MAC protocols that use duty cycles. It is able to perform end-to-end packet delivery within a single operational cycle. It also alleviates contention by separating medium contention and data transfer into two separate periods. However, collisions can still occur, even on data packets during the SLEEP period. A source always commences transmission at the beginning of the SLEEP period. Therefore, it is possible that data packets coming from two different sources (which succeeded in the PION scheduling process during the DATA period, but cannot see each other) may still collide. This problem has been addressed by a similar protocol called Demand Wakeup (DW-MAC) (Sun *et al.* 2008a). In DW-MAC, the schedule is a one-to-one mapping between a DATA period and the following SLEEP period:

$$T_i^S = T_i^D \times \frac{T_{\text{sleep}}}{T_{\text{data}}} \tag{6.8}$$

where T_i^S is the start time of the *scheduling frame* (SCH) (the equivalent of the pioneer frame in RMAC) measured from the start of the DATA period and T_i^D is the start time of the data transmission in the SLEEP period (measured from the start of the SLEEP period). T_{sleep} and T_{data} are the durations of the SLEEP and DATA periods, respectively. This means that data packet transmissions do not necessarily coincide with the start of the SLEEP period, but instead depend on the contention window during the DATA period (the delay in data transmission corresponds to the delay in the transmission of the SCH frame). This approach, compared to RMAC, reduces the risk of collisions during the SLEEP period.

6.5.6 Data-Gathering MAC

The Data-Gathering MAC (DMAC) (Lu *et al.* 2004) protocol exploits the fact that many wireless sensor networks rely on *convergecast* as communication pattern, that is, data from sensor nodes are collected at a central node (the "sink") in a *data-gathering tree*. The goal of DMAC is to deliver data along the data gathering tree with low latency and high energy efficiency.

In DMAC, the duty cycles of nodes along the multi-hop path to the sink are "staggered"; nodes wake up sequentially like a chain reaction. Figure 6.13 illustrates this concept, showing an example of a data-gathering tree and the staggered wakeup scheme. Nodes switch between sending, receiving, and sleep states. During the sending state, a node sends a packet to the next hop node on the route and awaits an acknowledgment (ACK). At the same time, the next hop node is in the receiving state, immediately followed by a sending state (unless the node is the destination of the packet) to forward the packet to the next hop. Between these intervals of receiving and sending of packets, a node enters the sleep state, where it can power down its radio to preserve energy.

The sending and receiving intervals are large enough for exactly one packet. Since there are no queuing delays, a node at depth d in the tree can then deliver a packet to the sink within

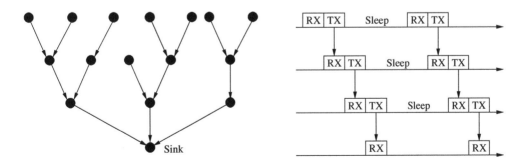

Figure 6.13   Data-gathering tree and convergecast communication in DMAC.

d intervals. While limiting a node's activity to brief intervals for sending and receiving reduces the contention, collisions still do occur. Particularly, nodes with the same depth in the tree will have synchronized schedules. In DMAC, if a sender does not receive an ACK, it queues the packet until the next sending interval. After three failed retransmissions, the packet will be dropped. To reduce collisions, nodes do not transmit immediately at the beginning of the sending slot, but instead have a backoff period (BP) plus a random time within a contention window.

When a node has multiple packets to send during a sending slot, it can increase its own duty cycle and request other nodes along the route to the sink to do the same. This is implemented through a slot-by-slot renewal mechanism using a *more data* flag in the MAC header. A receiver checks for this flag and, if set, it returns an acknowledgment that has also the more data flag set. It then stays awake to receive and forward one additional packet.

In summary, DMAC's staggering technique achieves very low latency and nodes only stay awake for brief receive and send intervals. Since many nodes in the data-gathering tree share the same schedule, collisions will occur and DMAC only employs limited collision avoidance methods. DMAC works best for networks in which transmission paths and rates are well known and do not change over time.

6.5.7 Preamble Sampling and WiseMAC

WiseMAC (El-Hoiydi and Decotignie 2004) is a MAC protocol that is concerned with the energy consumption of *downlink* communication in infrastructure-based sensor networks, that is, communication from a base station to the sensor nodes. To avoid energy consumption due to idle listening, WiseMAC relies on the *preamble sampling* technique (El-Hoiydi 2002). In this technique, the base station transmits a preamble preceding the actual data transmission to alert the receiving node (see left graph in Figure 6.14). All sensor nodes sample the medium (listen to the channel) with a fixed period T_w, but with independent and constant relative sampling schedule offsets. If the medium is busy, a sensor node continues to listen until the medium becomes idle or a data frame is received. The preamble's size is equal to the sampling period, which ensures that the receiver will be awake to receive the data portion of the packet. This approach allows the energy-constrained sensor nodes to turn off the radio when the channel is idle, without the risk of missing a packet. A disadvantage of this approach is that the size of the preamble affects the achievable throughput and that a device must stay awake when a preamble is detected, even if it is not the intended receiver.

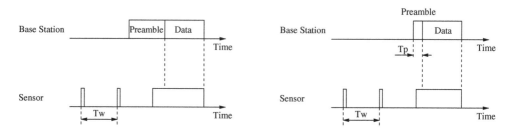

Figure 6.14 Preamble sampling.

WiseMAC improves upon this by adding a technique that lets a base station learn the sampling schedules of the destination, thereby allowing it to start the transmission of the preamble immediately before the receiver wakes up. This allows the base station to reduce the size of the preamble as shown in the right graph of Figure 6.14. Further, since the data portion of the packet will start shortly after the receiver's radio has turned on, the awake time of the receiver is also shortened. A node's schedule offset is embedded into the acknowledgment (ACK) message, allowing the base station to learn the sampling schedules. The duration of the preamble (T_p) is then determined as the minimum of the destination's sampling period (T_w) and a multiple of the clock drift between the clock at the base station and on the receiver, which grows over time. Therefore, the preamble length depends on the traffic load: the preamble is shorter when traffic is high (brief intervals between two consecutive communications) and larger under low load.

In summary, WiseMAC implements energy-efficient wake/sleep schedules for sensor nodes, while ensuring that all data transmissions from a base station to the sensors will be received (the receiver will be awake). However, the approach is inefficient for broadcast messages since the preamble is likely to be very large, that is, it must span over the sampling points of all receiver devices. Finally, WiseMAC is also affected by the hidden terminal problem, that is, a sender's preamble can interfere with ongoing transmissions when the sender is not aware of this other transmission.

6.5.8 Receiver-Initiated MAC

Another contention-based solution is the Receiver-Initiated MAC (RI-MAC) protocol (Sun *et al.* 2008b), where a transmission is always initiated by the receiver of the data. Each node wakes up periodically to check whether there is an incoming data packet. That is, immediately after turning on its radio, a node checks if the medium is idle and, if so, broadcasts a beacon message, announcing that it is awake and ready to receive data. A node with pending data to transmit stays awake and listens for a beacon from its intended receiver. Once this beacon has been received, the sender immediately transmits the data, which will be acknowledged by the receiver with another beacon (see left graph in Figure 6.15). That is, the beacon serves two purposes: it invites new data transmissions and it acknowledges previous transmissions. If there is no incoming data packet for a certain amount of time after the beacon broadcast, the node goes back to sleep after waiting a certain time.

If there are multiple contending senders, a receiver uses its beacon frames to coordinate transmissions. A field in the beacon, called the backoff window size (BW), specifies the window over which to select a backoff value. If the beacon does not contain a BW (the first beacon sent out after waking up does not contain a BW), senders immediately commence transmission. Otherwise, each sender randomly selects a backoff value within BW and the receiver increases the BW value in the next beacon when it detects a collision. The right graph in Figure 6.15 shows an example with two senders immediately transmitting data packets after receiving the receiver's beacon. The receiver notices the collision and sends another beacon, this time containing a BW. Should the transmission collide multiple times and the receiver was not able to receive a packet for several beacon intervals, it simply goes to sleep without further attempts.

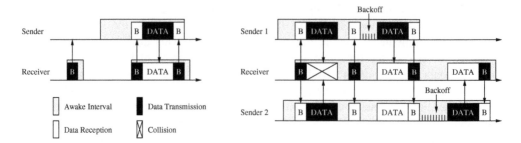

Figure 6.15 Data transmission from a single sender (left) and two contending senders (right).

In RI-MAC, the receiver is in control of when to receive data and it is responsible for detecting collisions and recovering lost data. Since transmissions are triggered by beacons, the receiver will have very little overhead due to overhearing. On the other hand, senders must wait for the receiver's beacon before they can transmit, potentially leading to large overhearing costs. Finally, when packets collide, the senders will retry until the receiver gives up, potentially leading to more collisions in the network and to increased data delivery latencies.

6.6 Hybrid MAC Protocols

Some MAC protocols do not easily fit into schedule-based or contention-based categories, but instead display characteristics of both categories, for example, they may attempt to reduce the number of collisions by relying on features present in periodic contention-free medium access protocols, while taking advantage of the flexibility and low complexity of contention-based protocols. This section describes representative examples of hybrid protocols.

6.6.1 Zebra MAC

The Zebra MAC (Z-MAC) protocol (Rhee *et al.* 2005) uses frames and slots, similar to TDMA-based protocols, to provide contention-free access to the wireless medium. However, Z-MAC also allows nodes to utilize slots they do not own using CSMA with prioritized backoff times. As a consequence, Z-MAC emulates a CSMA-based approach in low-traffic scenarios and a TDMA-based approach when traffic loads are high.

When a node starts up, it enters a setup phase to allow it to discover its neighbors and to obtain its slot in the TDMA frame. Every node periodically broadcasts a message containing a list of its neighbors. Through this process, a node learns about its 1-hop and 2-hop neighbors. This information is used as input to a distributed slot assignment protocol (Rhee *et al.* 2006), which provides each node with time slots, ensuring a schedule where no two nodes within a 2-hop neighborhood will be assigned the same slot. Further, Z-MAC allows nodes to select the periodicity of their assigned slots, where different nodes can have different periods, which are called *time frames* (TF). The advantage of this approach is that it is not necessary to propagate a maximum slot number (MSN) to the entire network and that the protocol can adapt slot allocations locally. Specifically, if node i is assigned slot s_i

and F_i represents the MSN within the node's 2-hop neighborhood, then i's TF is set to be 2^a, where a is a positive integer that satisfies $2^{a-1} \leq F_i < 2^a - 1$. Node i then uses the s_ith slot in every 2^a time frame.

Figure 6.16 shows an example with eight nodes, where the number indicates the assigned slot for each node and the number in parenthesis is F_i. The bottom part of the figure shows the corresponding schedule for all nodes, where light-shaded slots are the ones used for transmissions and dark-shaded slots are the empty slots that are not used by any 1-hop or 2-hop neighbors. If a global time frame is used, the chosen time frame size will be 6 and nodes A and B will be allowed to use their slots only once every 6 slots (even though their frame sizes are 2 each). However, in Z-MAC, they can use frame size 4, which increases the concurrency in the channel usage and reduces message delays. The resulting schedule shows that some slots (specifically slots 6 and 7) are not assigned to any node. In a global time frame, a frame size could have been chosen that reduces the number of empty slots. However, Z-MAC allows nodes to compete for these "extra" slots using CSMA.

After the schedule has been determined, every node forwards its frame size and slot number to its 1-hop and 2-hop neighbors. Even though slots are owned by nodes, Z-MAC uses CSMA to determine who may transmit. However, slot owners are given preference by using a random backoff value chosen from the range $[0, T_o]$, whereas other nodes choose their

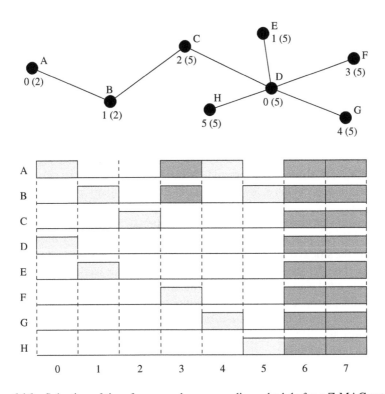

Figure 6.16   Selection of time frames and corresponding schedule for a Z-MAC network.

backoff values from the range $[T_o, T_{no}]$. Z-MAC also uses explicit contention notification (ECN), where, based on its local estimate of the contention level (e.g., determined using the packet loss rate or channel noise level), each node decides whether to send an ECN message to a neighbor for which it has a message. This neighbor then broadcasts the ECN to its own neighbors, which then enter a *high contention level* (HCL) mode. A node in the HCL mode only transmits data in its own slots or slots belonging to its 1-hop neighbors, reducing the contention between 2-hop neighbors. Nodes in HCL mode return to a *low contention level* (LCL) mode if they have not received any ECN messages for a certain amount of time.

In summary, Z-MAC adopts characteristics found in both TDMA and CSMA protocols, allowing it to quickly adapt to changing traffic conditions. Under light traffic loads, Z-MAC behaves more like CSMA, while under heavy traffic loads, contention for slots is reduced. Z-MAC requires an explicit setup phase, which consumes both time and energy. While ECN messages can be used to reduce the contention locally, these messages add more traffic to an already busy network and take time to propagate, causing delays in the adaptation to a more TDMA-like behavior.

6.6.2   Mobility Adaptive Hybrid MAC

In many sensor networks, some or all nodes can be mobile, which can bear significant challenges for the design of a MAC protocol. The Mobility Adaptive Hybrid MAC (MH-MAC) protocol (Raja and Su 2008) proposes a hybrid solution, where a schedule-based approach is used for static nodes and a contention-based approach is used for mobile nodes. While it is straightforward to determine a TDMA-style schedule for static nodes, this is not the case for mobile nodes. Therefore, MH-MAC allows mobile nodes entering a neighborhood to use a contention-based approach to avoid the delays often needed to be inserted into the schedule.

In MH-MAC, the slots of a frame belong to one of two categories: static slots or mobile slots. Each node uses a mobility estimation algorithm to determine its mobility and which type of slots the node should use. The mobility estimation is based on periodic *hello messages* and received signal strength. The hello messages are always transmitted with the same transmit power and the receiving nodes compare consecutive message signal strengths to estimate the relative position displacement between itself and each of its neighbors. A *mobility beacon interval* is provided at the beginning of a frame to distribute mobility information to neighbors.

Static slots use an approach similar to the LMAC approach described in Section 6.4.6, that is, a static slot has two portions: a control section and a data section. The control section is used to indicate the slot assignment information in a neighborhood and all static nodes must listen to this part of the static slot. However, during the data section, only the transmitter and receiver stay awake and all other nodes can turn their radios off.

For mobile slots, nodes contend for the medium in a two-phase contention period. First, a *wakeup tone* is sent during the first phase, and then the data is sent during the second phase. In order to reduce the effective contention, LMAC uses a priority ordering among mobile nodes based on node addresses.

Since the ratio between static and mobile nodes in a network can vary, MH-MAC provides a mechanism to dynamically adjust the ratio between static and mobile slots based on the observed mobility. Each node estimates its own mobility and broadcasts this information in the previously mentioned beacon time slot at the beginning of a frame. Using this mobility information, each node calculates a mobility parameter for the network, which determines the ratio of static and mobile slots.

In summary, MH-MAC combines characteristics of the LMAC protocol for static nodes and features of contention-based protocols for mobile nodes. Therefore, mobile nodes can quickly join a network without long setup or adaptation delays. Compared to LMAC, MH-MAC allows nodes to own more than one slot in a frame, which increases bandwidth utilization and decreasing latencies.

6.7 Summary

The choice of a medium access protocol has a substantial impact on the performance and energy-efficiency of a WSN. MAC protocols should also be designed to accommodate changes in network topology and traffic characteristics. Further, latency, throughput, and fairness among competing nodes are also determined or affected by the characteristics of the MAC layer. This chapter discussed the major categories of medium access control protocols and illustrated each category with several examples. A major distinction can be made between protocols that are based on transmission schedules (e.g., TDMA-style protocols) and protocols that do not use such schedules, but instead let nodes compete for the medium. The main advantage of schedule-based protocols is that communication can occur collision-free. However, such protocols can be resource-inefficient, may require well-synchronized nodes throughout the network, and are often difficult to adapt to changing topologies. On the other hand, contention-based solutions are more flexible in that they easily accommodate changing network topologies. They typically also require less overhead. However, such protocols are not collision-free, so they must possess features that allow them to recover from collisions, and the network utilization may suffer when collisions occur frequently.

Exercises

6.1 What is the main purpose of the MAC layer and why is this challenging in networks with shared media?

6.2 What are the advantages and disadvantages of contention-free and contention-based medium access strategies? Can you think of scenarios where one would be preferable over the other?

6.3 The key idea behind CSMA/CD is that the sender detects collisions, allowing it to react correspondingly. Why is this approach not practical in wireless networks?

6.4 What are "hidden terminals" and how do they affect the performance of wireless sensor networks?

6.5 Consider the network topology in Figure 6.17, where circles indicate the communication and interference range of each node, that is, each node can hear the immediate neighbors to the left and right. Assume that RTS/CTS is not being used.

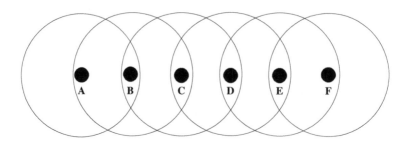

Figure 6.17   Hidden-terminal problem (Exercise 6.5).

(a) Node B currently sends to node A and node C wants to send to node D. Is node C allowed to do so (i.e., can it do so without causing a collision) and will it decide to do so?

(b) Node C sends to node B and node E wants to send to node D. Is E allowed to do so and will it do so?

(c) Node A sends to node B and node D sends to node C. Which other nodes are allowed to send at the same time?

(d) Node A sends to node B and node E sends to node F. Which other nodes are allowed to send at the same time?

6.6 Describe the problems in using CSMA as a medium access control mechanism in a WSN.

6.7 In a CSMA/CA network, nodes use a random delay before accessing the medium. Why is this being done?

6.8 Assume that the RTS and CTS frames were as long as the DATA and ACK frames. Would there be any advantage to using the RTS/CTS approach? Explain why or why not.

6.9 How does MACAW extend MACA and what is the purpose of the additional control messages?

6.10 What are the specific features of the IEEE 802.11 PSM (Power Saving Mode) and what are the main difficulties of using it in wireless sensor networks?

6.11 Does the NAV field in IEEE 802.11 networks prevent the hidden terminal problem?

6.12 Explain why the IEEE 802.11 standard uses three different "interframe spaces".

6.13 Consider the network topology in Figure 6.18, where the lines indicate which nodes can communicate and interfere with each other. Assume a TDMA protocol with a frame size of 5 slots and that each node can only be sender or receiver during any time slot.

(a) Generate a schedule such that every node has an opportunity to communicate to all if its neighbors.

(b) For your schedule, how many slots in a frame could each node sleep to preserve energy? What is your insight with respect to node density and energy preservation?

(c) Assume that node A sends a message to node E; how long (in number of time slots) does it take for E to receive the message using your schedule? (Explain your answer.)

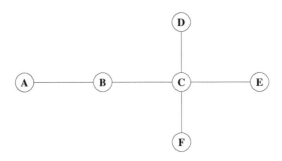

Figure 6.18   TDMA protocol (Exercise 6.13).

6.14 Why is the IEEE 802.15.4 standard preferable over the IEEE 802.11 standard for most wireless sensor networks?

6.15 Describe how the design of the MAC protocol affects the energy efficiency of a sensor node.

6.16 This chapter described five requirements of MAC protocols for wireless sensor networks: energy efficiency, scalability, adaptability, low latency, and reliability. Can you describe a concrete WSN application for each of these five requirements, where the requirement would be more important than the others?

6.17 The TRAMA protocol is an example of a contention-free MAC scheme. Answer the following questions about TRAMA.

(a) What are the advantages and disadvantages of the TRAMA protocol (compared to contention-based protocols)?

(b) What is the difference between transmission slots and signaling slots?

(c) What is the purpose of the NP component?

6.18 What is the advantage of a receiver-initiated MAC scheme such as Y-MAC? What is the main disadvantage of Y-MAC that makes it unsuitable for most low-power and low-cost sensor nodes?

6.19 Demonstrate the concept of DESYNC using the example ring shown in Figure 6.19. The ring has 16 positions [0 ... 15], with node A currently in position 0 (the firing position), B in position 14, etc. Every unit of time, each node moves one position clockwise along the ring. The table indicates the positions of the four nodes, including the new distance information that is learned at each firing. Assume that node A has received D's last firing, indicating a distance of 10 between nodes A and D. In this table, at time 0, node A fires, allowing node B to learn its distance to A (i.e., 2). At time 1, no node is in the firing position. At time 2, node B fires, allowing node C to learn its distance to node B (i.e., 2). At the same time, node A now knows its distance to node B and its distance to node D. According to the description of the DESYNC algorithm in this chapter, node A can now find the midpoint between nodes B and D and jump to this new location (i.e., 6), which is indicated in the table at time 2. At time 3, again each node moves ahead one position. Continue this table using the DESYNC algorithm until time 19. Compare the average distance between neighboring nodes at time 19 with that of time 0.

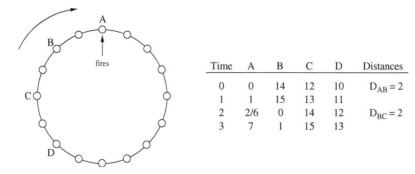

Time	A	B	C	D	Distances
0	0	14	12	10	$D_{AB} = 2$
1	1	15	13	11	
2	2/6	0	14	12	$D_{BC} = 2$
3	7	1	15	13	

Figure 6.19   DESYNC ring (Exercise 6.19).

6.20 Discuss the cluster head election policy in the LEACH protocol and explain how LEACH can consider the available energy on each node in this election process. What is the problem with this energy-aware election policy? Further, LEACH uses TDMA within a cluster; explain the advantages and disadvantages of this approach.

6.21 Why does LEACH use the direct sequence spread spectrum technique?

6.22 How does the Mobile LMAC protocol handle changes in network topology?

6.23 Discuss why overhearing is a problem in a wireless sensor network and explain how PAMAS addresses this problem.

6.24 Explain how the busy-tone scheme of PAMAS helps to avoid the hidden-terminal problem.

6.25 How does the S-MAC protocol reduce the duty cycles of sensor nodes? How does the S-MAC protocol attempt to reduce collisions? How does it address the hidden-terminal problem? Name at least three disadvantages of the S-MAC protocol.

6.26 Which shortcoming of S-MAC does T-MAC address? Explain briefly T-MAC's ability to adapt to traffic density.

6.27 What is the "early sleeping problem" and how does T-MAC address this problem?

6.28 Describe the concept behind PMAC's approach to adapting a node's sleep durations to observed traffic.

6.29 The use of duty cycles allows nodes to alternate periods of activity and low-power sleep intervals. However, this often introduces large communication latencies. In Figure 6.20, node A wishes to send a packet to node F using the route A–B–C–D–E–F. Assume that the interval between two dashed vertical lines is one unit of time and that each transmission requires exactly one unit of time of overlapping periods of activity of two neighboring nodes. The first transmission between A and its neighbor is already shown in the figure. Complete this graph to determine the end-to-end latency experienced by the packet. Further, explain how RMAC reduces these end-to-end latencies and what would this latency be in an RMAC network? Finally, explain how the RMAC protocol reduces collisions.

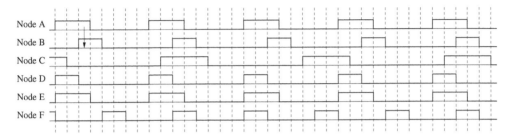

Figure 6.20   RMAC duty cycle pattern (Exercise 6.29).

6.30 For what type of WSN applications would you use the DMAC protocol?

6.31 Explain the problem of "idle listing" and describe how preamble sampling addresses this problem. How does WiseMAC improve upon "standard" preamble sampling?

6.32 What is the advantage of having the receiver (instead of the sender) in control over the timing of transmissions (e.g., as in the RI-MAC protocol)? How does the RI-MAC protocol handle multiple contending senders?

6.33 Explain how the Z-MAC protocol allows nodes to determine their own local time frames instead of using a single global time frame. What are the disadvantages of Z-MAC?

References

Bharghavan, V., Demers, A., Shenker, S., and Zhang, L. (1994) MACAW: A medium access protocol for wireless LANs. *Proc. of the Conference on Communication Architectures, Protocols and Applications*.

CC1000 (2004) Single chip very low power RF transceiver. http://focus.ti.com/lit/ds/symlink/cc1000.pdf.

CC2420 (2004) 2.4-GHz IEEE 802.15.4/ZigBee-ready RF transceiver datasheet. http://enaweb.eng.yale.edu/drupal/system/files/CC2420_Data_Sheet_1_4.pdf.

Degesys, J., Rose, I., Patel, A., and Nagpal, R. (2007) DESYNC: Self-organizing desynchronization and TDMA on wireless sensor networks. *Proc. of the 6th International Conference on Information Processing in Sensor Networks (IPSN)*.

Du, S., Kumar, A., David, S., and Johnson, B. (2007) RMAC: A routing-enhanced duty-cycle MAC protocol for wireless sensor networks. *Proc. of the 26th IEEE International Conference on Computer Communications (INFOCOM)*.

El-Hoiydi, A. (2002) Spatial TDMA and CSMA with preamble sampling for low power ad hoc wireless sensor networks. *Proc. of the 7th IEEE Symposium on Computers and Communications*.

El-Hoiydi, A., and Decotignie, J.D. (2004) WiseMAC: An ultra low power MAC protocol for the downlink of infrastructure wireless sensor networks. *Proc. of the 9th IEEE Symposium on Computers and Communication*.

Gutierrez, J.A., Naeve, M., Callaway, E., Bourgeois, M., Mitter, V., and Heile, B. (2001) IEEE 802.15.4: A developing standard for low-power low-cost wireless personal area networks. *IEEE Network* **15** (5), 12–19.

Heinzelman, W.B., Chandrakasan, A.P., and Balakrishnan, H. (2002) An application specific protocol architecture for wireless microsensor networks. *IEEE Transactions on Wireless Communications*.

Karn, P. (1990) A new channel access method for packet radio. *Proc. of the ARRL 9th Computer Networking Conference*.

Kim, Y., Shin, H., and Cha, H. (2008) YMAC: An energy-efficient multi-channel MAC protocol for dense wireless sensor networks. *Proc. of the International Conference on Information Processing in Sensor Networks (IPSN)*.

Kuo, F.F. (1995) The ALOHA system. *ACM SIGCOMM Computer Communication Review* **25** (1), 41–44.

Lu, G., Krishnamachari, B., and Raghavendra, C.S. (2004) An adaptive energy-efficient and low-latency MAC for data gathering in wireless sensor networks. *Proc. of the 18th International Parallel and Distributed Processing Symposium*.

Mank, S., Karnapke, R., and Nolte, J. (2007) An adaptive TDMA based MAC protocol for mobile wireless sensor networks. *Proc. of the International Conference on Sensor Technologies and Applications (SENSORCOMM)*.

MC13202 (2008) 2.4-GHz low power transceiver for the IEEE 802.15.4 standard. www.freescale.com /files/rf_if/doc/data_sheet/MC13202.pdf.

Raja, A., and Su, X. (2008) A mobility adaptive hybrid protocol for wireless sensor networks. *Proc. of the Consumer Communications and Networking Conference*.

Rajendran, V., Obraczka, K., and Garcia Luna Aceves, J.J. (2003) Energy efficient, collision free medium access control for wireless sensor networks. *Proc. of the 1st International Conference on Embedded Networked Sensor Systems (SenSys)*.

Rhee, I., Warrier, A., Aia, M., and Min, J. (2005) ZMAC: A hybrid MAC for wireless sensor networks. *Proc. of the 3rd ACM Conference on Embedded Networked Sensor Systems (SenSys)*.

Rhee, I., Warrier, A., Min, J., and Xu, L. (2006) DRAND: Distributed randomized TDMA scheduling for wireless ad hoc networks. *Proc. of the 7th ACM International Symposium on Mobile Ad Hoc Networking and Computing (MobiHoc)*.

Singh, S., and Raghavendra, C. (1998) PAMAS: Power aware multi-access protocol with signaling for ad hoc networks. *SIGCOMM Computer Communications Review* **28** (3), 5–26.

Sun Y., Du, S., Gurewitz, O., and Johnson, D.B. (2008a) DWMAC: Low latency, energy efficient demand wakeup MAC protocol for wireless sensor networks. *Proc. of the 9th International Symposium on Mobile Ad Hoc Networking and Computing (MobiHoc)*.

Sun, Y., Gurewitz, O., and Johnson, D.B. (2008b) RIMAC: A receiver initiated asynchronous duty cycle MAC protocol for dynamic traffic loads in wireless sensor networks. *Proc. of the 6th ACM Conference on Embedded Networked Sensor Systems (SenSys)*.

Talucci, F., Gerla, M., and Fratta, L. (1997) MACABI (MACA By Invitation): A receiver oriented access protocol for wireless multi-hop networks. *Proc. of the 8th IEEE International Symposium on Personal, Indoor and Mobile Radio Communications*.

Van Dam, T., and Langendoen, K. (2003) An adaptive energy-efficient MAC protocol for wireless sensor networks. *Proc. of the 1st ACM Conference on Embedded Networked Sensor Systems (SenSys)*.

Van Hoesel, L., and Havinga, P. (2004) A lightweight medium access protocol (LMAC) for wireless sensor networks: Reducing preamble transmissions and transceiver state switches. *Proc. of the 1st International Conference on Networked Sensing Systems (INSS)*.

Ye, W., Heidemann, J., and Estrin, D. (2002) An energy-efficient MAC protocol for wireless sensor networks. *Proc. of the 21st Annual Joint Conference of the IEEE Computer and Communications Societies (INFOCOM)*.

Ye, W., Heidemann, J., and Estrin, D. (2004) Medium access control with coordinated active sleeping for wireless sensor networks. *IEEE/ACM Transactions on Networking* **12** (3), 493–506.

Zheng, T., Radhakrishnan, S., and Sarangan, V. (2005) PMAC: An adaptive energy-efficient MAC protocol for wireless sensor networks. *Proc. of the 19th IEEE International Parallel and Distributed Processing Symposium (IPDPS)*.

7

Network Layer

Data collected by sensor nodes in a WSN is typically propagated toward a base station (gateway) that links the WSN with other networks where the data can be visualized, analyzed, and acted upon. In small sensor networks where sensor nodes and a gateway are in close proximity, direct (single-hop) communication between all sensor nodes and the gateway may be feasible. However, most WSN applications require large numbers of sensor nodes that cover large areas, necessitating an indirect (multi-hop) communication approach. That is, sensor nodes must not only generate and disseminate their own information, but also serve as relays or forwarding nodes for other sensor nodes. The process of establishing paths from a source to a sink (e.g., a gateway device) across one or more relays is called *routing* and is a key responsibility of the network layer of the communication protocol stack. When the nodes of a WSN are deployed in a deterministic manner (i.e., they are placed at certain predetermined locations), communication between them and the gateway can occur using predetermined routes. However, when the nodes are deployed in a randomized fashion (i.e., they are scattered into an environment randomly), the resulting topologies are nonuniform and unpredictable. In this case, it is essential for these nodes to self-organize, that is, they must cooperate to determine their positions, identify their neighbors, and discover paths to the gateway device. This chapter introduces the main categories of routing protocols and data dissemination strategies and discusses state-of-the-art solutions for each category.

7.1 Overview

The key responsibility of the network layer is to find paths from data sources to sink devices (e.g., gateways). In the single-hop routing model (left graph in Figure 7.1), all sensor nodes are able to communicate directly with the sink device. This direct communication model is the simplest approach, where all data travels a single hop to reach the destination. However, in practical settings, this single-hop approach is unrealistic and a multi-hop communication model (right graph in Figure 7.1) must be used. In this case, the critical task of the network layer of all sensor nodes is to identify a path from the sensor to the sink across multiple other sensor nodes acting as relays. This design of a routing protocol is challenging due to the unique characteristics of WSNs, including resource scarcity or the unreliability of the wireless medium. For example, the limited processing, storage, bandwidth, and energy capacities require routing solutions that are lightweight, while the frequent dynamic changes in a WSN (e.g., topology changes due to node failures) require routing solutions

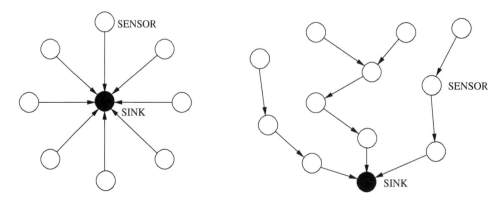

Figure 7.1 Single-hop routing model (left) versus multi-hop routing model (right).

that are adaptive and flexible. Further, unlike traditional routing protocols for wired networks, protocols for sensor networks may not be able to rely on global addressing schemes (e.g., IP addresses on the Internet).

There are various ways to classify routing protocols. Figure 7.2 presents three different classifications based on the network structure or organization, the route discovery process, and the protocol operation (Al-Karaki and Kamal 2004). With respect to network organization, most routing protocols fit into one of three classes. Flat-based routing protocols consider all nodes of equal functionality or role. In contrast, in hierarchical-based routing protocols, different nodes may assume different roles in the routing process, that is, some nodes may forward data on behalf of others, while other nodes only generate and propagate their own sensor data. Location-based routing protocols rely on the location information from nodes to make routing decisions. Routing protocols are responsible for identifying or discovering routes from a source or sender to the intended receiver. This route discovery process can also be used to distinguish between different types of routing protocols. Reactive protocols discover routes *on-demand*, that is, whenever a source wants to send data to a receiver and does not already have a route established. While reactive route discovery

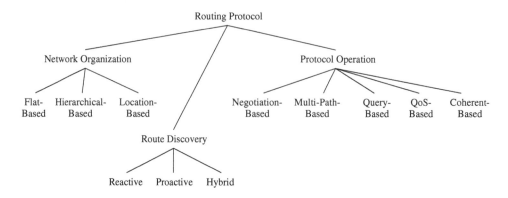

Figure 7.2 Categories of routing protocols.

incurs delays before actual data transmission can occur, proactive routing protocols establish routes before they are actually needed. This category of protocols is also often described as *table-driven*, because local forwarding decisions are based on the contents of a *routing table* that contains a list of destinations, combined with one or more next-hop neighbors that lead toward these destinations and costs associated with each next hop option. While table-driven protocols eliminate the route discovery delays, they may be overly aggressive in that routes are established that may never be needed. Further, the time interval between route discovery and actual use of the route can be very large, potentially leading to outdated routes (e.g., a link along the route may have broken in the meantime). Finally, the cost of establishing a routing table can be significant, for example, in some protocols it involves propagating a node's local information (such as its list of neighbors) to all other nodes in the network. Some protocols exhibit characteristics of both reactive and proactive protocols and belong to the category of hybrid routing protocols. Finally, routing protocols also differ in their operation, for example, negotiation-based protocols aim to reduce redundant data transmissions by relying on the exchange of negotiation messages between neighboring sensor nodes before actual data transfers occur. The SPIN family of protocols (Section 7.5) belongs to this category. Multipath-based protocols use multiple routes simultaneously to achieve higher performance or fault tolerance. Query-based routing protocols are receiver-initiated, that is, sensor nodes send data in response to queries issued by the destination node. The goal of QoS-based routing protocols is to satisfy a certain Quality-of-Service (QoS) metric (or a combination of multiple metrics), such as low latency, low energy consumption, or low packet loss. Finally, routing protocols also differ in the way they support in-network data processing. Coherent-based protocols perform only a minimum amount of processing (e.g., eliminating duplicates, time-stamping) before sensor data is sent to receivers and data aggregators. However, in noncoherent-based protocols, nodes may perform significant local processing of the raw data before it is sent to other nodes for further processing.

Further, when sensor data is explicitly sent to one or more receivers, routing is considered *node-centric*. Most routing protocols focus on unicast routing, that is, forwarding of sensor data to exactly one receiver. Multicast and broadcast routing approaches, on the other hand, disseminate data to multiple or all nodes, respectively. *Data-centric* routing is used when nodes are not explicitly addressed, but instead receivers are implicitly described by certain attributes. For example, a query issued by the gateway device may request temperature readings and only sensors that can collect such information respond to the query.

7.2 Routing Metrics

Wireless sensor networks and their applications vary widely in their constraints and characteristics, which must be taken into consideration in the design of a routing protocol. For example, most WSNs will be constrained in terms of their available energy, processing power, and storage capacities. Sensor networks can vary widely in scale, the geographic areas they cover, and their position-awareness. Global addressing schemes (such as IP addresses used on the Internet) may be unavailable and even nonfeasible, particularly in networks with heterogeneous nodes and node mobility. Finally, from the application's perspective, sensor data may be collected in various different approaches. In *time-driven* schemes (e.g., environmental monitoring), nodes propagate their collected sensor data periodically to a sink or gateway device. In *event-driven* schemes (e.g., wildfire detection), nodes only report their collected information when events of interest occur. Finally, in *query-driven* schemes, it is the responsibility of the sink to request data from sensors when needed. Regardless of the scheme used in the network, the design of a routing protocol is driven by the resources available in the network and the needs of the applications. Toward this end, routing metrics are used to express a variety of objectives of a routing protocol with respect to the consumption of these resources or the performance an application perceives. This section provides a brief overview of commonly used routing metrics in WSNs.

7.2.1   Commonly Used Metrics

7.2.1.1   Minimum Hop

The most common metric used in routing protocols is *minimum hop* (or *shortest hop*), that is, the routing protocol attempts to find the path from the sender to the destination that requires the smallest number of relay nodes (hops). In this simple technique, every link has the same cost and a minimum-hop routing protocol selects the path that minimizes the total cost of data propagation from source to destination. The basic idea behind this metric is that using the shortest path will result in low end-to-end delays and low resource consumptions, because the smallest possible number of forwarding nodes will be involved. However, since the minimum-hop approach does not consider the actual resource availability on each node, the resulting route is probably nonoptimal in terms of delay, energy, and congestion avoidance. Nevertheless, the minimum-hop metric is being used in many routing protocols due to its simplicity and its isotonicity, that is, its ability to ensure that the order of weights of two paths is preserved even when they are appended or prefixed by a common third path.

7.2.1.2   Energy

Undoubtedly the most crucial aspect of routing in WSNs is energy efficiency. However, there is not one unique energy metric that can be applied to the routing problem; instead, there are various different interpretations of energy efficiency, including (Singh *et al.* 1998):

1. *Minimum energy consumed per packet:* This is the most natural concept of energy efficiency, that is, the goal is to minimize the total amount of energy expended for the propagation of a single packet from the source to the destination. The total energy is then the sum of the energy consumed by each node along a route for receiving and transmitting the packet. Figure 7.3 shows an example of a small sensor network, where a source node wishes to transmit a packet to a destination node using a route that minimizes the packet's energy overheads. The number on each link indicates the cost of propagating the packet over this link. As a consequence, the packet will travel via nodes A–D–G (with a total cost of 5). Note that this is different from the minimum-hop route (B–G).
2. *Maximum time to network partition:* A network partitions into several smaller subnetworks when the last node that links two parts of the network expires or fails. As a consequence, a subnetwork may not be reachable, rendering the sensor nodes within the subnetwork useless. Therefore, the challenge is to reduce the energy consumption on nodes that are crucial to maintaining a network where every sensor node can be reached

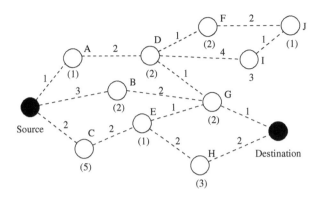

Figure 7.3 Comparison of routing choices using different energy metrics.

via at least one route. For example, a minimal set of nodes, whose removal will cause a network to partition, can be found using the max-flow–min-cut theorem. Once a routing protocol has identified these critical nodes, it can attempt to balance the traffic load such that premature expiration of these nodes is prevented. In Figure 7.3, node D is such a node, for example, if node D's battery becomes depleted, nodes F, I, and J would not be reachable from any other node in the network.

3. *Minimum variance in node power levels:* In this scenario, all nodes within the network are considered equally important and the challenge is to distribute the energy consumption across all nodes in the network as equally as possible. The goal of such an approach could be to maximize the lifetime of the entire network, for example, instead of some nodes expiring sooner than others and thereby continuously decreasing the network size, one could aim at keeping as many nodes alive as long as possible. In the ideal (but practically impossible) case, all nodes would expire at exactly the same time.

4. *Maximum (average) energy capacity:* In this approach, the focus is less on the energy cost of packet propagation, but instead on the energy capacity (i.e., the current battery charge level) of the nodes. A routing protocol that uses this metric would then favor routes that have the largest total energy capacity from source to destination. In Figure 7.3, the numbers in parentheses below the nodes indicate the nodes' remaining energy capacity. In this example, a routing protocol could select path C–E–G, which has the largest total capacity (i.e., 8). A routing protocol that uses this metric must be carefully designed to avoid the pitfall of choosing unnecessarily long routes in order to maximize the total energy capacity. A variation of this metric is to maximize the average energy capacity, which can avoid this problem.

5. *Maximum minimum energy capacity:* Finally, instead of maximizing the energy capacities of the entire path, the primary routing goal could be to select the path with the largest minimum energy capacity. This technique also favors routes with larger energy reserves, but also protects low-capacity nodes from premature expiration. In Figure 7.3, a protocol using this metric would choose B–G, since the minimum capacity along this route is 2, which is larger than the minimum capacities of all other possible routes.

These different formulations of energy awareness lead to very different protocol implementations that differ in their results (i.e., the selected routes) and their overheads. For example, to determine the minimum energy consumed per packet, the cost for receiving and transmitting a packet may be based on a cost function with the packet size as input. On the other hand, energy capacities change over time and therefore a routing protocol using a capacity-based metric must obtain these capacities from other nodes from time to time.

7.2.1.3 Quality-of-Service

The term *Quality-of-Service* (QoS) refers to defined measures of performance in networks, including end-to-end latency (or delay) and throughput, but also jitter (variation in latency) and packet loss (or error rate). The choice of a QoS metric depends on the type of application. Sensor networks performing target detection and tracking will require low end-to-end transmission delays for their time-sensitive sensor data, while data-intensive networks (e.g., multimedia sensor networks) may require high throughput. The Expected Transmission Time (ETT) is a common metric to express latency and is defined as (Draves *et al.* 2004):

$$\text{ETT} = \text{ETT} \times \frac{S}{B} \tag{7.1}$$

where S is the average size of a packet and B is a link's bandwidth. It expresses the expected time necessary to successfully transmit a packet at the MAC layer. To capture packet loss as a routing metric, the Expected Transmission Count (ETX) can be used, which is defined as the number of transmissions necessary to successfully deliver a packet over a wireless link (Couto *et al.* 2003). Very often multiple QoS metrics (e.g., end-to-end latency and packet loss rate) are combined, for example, the bandwidth-delay product refers to the product of a link's bandwidth and its end-to-end delay. Which metrics are chosen affects the design of the network at different levels, including the network (routing) and MAC layers. Most WSNs must strike a balance between satisfying the application-specific QoS requirements and the goal of energy efficiency in the network as whole.

7.2.1.4 Robustness

Many sensor applications may wish to use routes that stay stable and reliable for long periods of time. Toward this end, a node can measure or estimate the *link quality* to each of its neighbors and then select a next hop neighbor that increases the probability of a successful transmission. However, this metric is rarely used alone. A routing protocol could identify several minimum-hop paths and then select the one with the highest total or average link quality along these paths. In networks with mobile nodes, a routing protocol could also use the *link stability* metric, which measures how likely it is that a link will be available in the future. These metrics can be used to bias route selection toward more robust paths and stationary nodes.

7.3 Flooding and Gossiping

An old and simple strategy to disseminate information into a network or to reach a node at an unknown location is to *flood* the entire network. A sender node broadcasts packets to its immediate neighbors, which will repeat this process by rebroadcasting the packets to their own neighbors until all nodes have received the packets or the packets have traveled for a maximum number of hops. With flooding, if there exists a path to the destination (and assuming lossless communication), the destination is guaranteed to receive the data. The main advantage of flooding is its simplicity, while its main disadvantage is that it causes heavy traffic. Therefore, measures must be taken to ensure that packets do not travel through the network indefinitely. For example, *maximum-hop counts* are used to limit the number of times a packet is forwarded. It should be set large enough so that every intended receiver can be reached, but also small enough to ensure that packets do not travel too long in the network. Further, *sequence numbers* in packets (combined with the address of the source) can be used to uniquely identify a packet. When a node receives a packet that it has already forwarded (i.e., with the same destination–source pair and the same sequence number), it simply discards this duplicate. However, even with these mechanisms, flooding faces a number of additional challenges (Heinzelman *et al.* 1999):

1. ***Implosion:*** A node receiving a packet relays this packet to all its neighbors using broadcasting, regardless of whether these neighbors have already received this packet from other neighbors. This leads to resource waste due to unnecessary transmit-and-receive operations. The left graph in Figure 7.4 illustrates this problem. Node A broadcasts packet P1 to both of its neighbors, B and C. B forwards this packet to its own neighbor D and, finally, C also forwards this packet to node D. Even if D discards the duplicate packet, energy has been wasted on the transmission of the packet from C to D.
2. ***Overlap:*** Sensors are often used to monitor overlapping geographic areas, as illustrated in the right graph in Figure 7.4. Here, sensors A and B share the region marked as Y. Therefore, these sensors gather overlapping data and both forward their collected information to their neighbor C. Similar to the implosion problem, this also leads to resource waste since the same information is sent twice to the same receiver. Unlike the implosion problem, the overlap problem is more difficult to address, because a solution to this problem must not only consider the topology of the sensor network, but also the mapping of the collected sensor information to sensor nodes.
3. ***Resource Blindness:*** The simplicity of the protocol also means that flooding is incognizant of the resource constraints of individual nodes. As a consequence, flooding is not able to adapt its behavior based on the amount of energy available to a specific node.

A variation of flooding is gossiping (Hedetniemi *et al.* 1988), where a node does not necessarily broadcast data. Instead, it uses a probabilistic approach, where it decides to forward the data to its own neighbors with a probability p and to discard the data with a probability

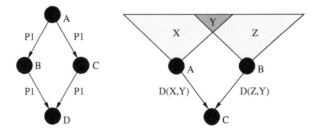

Figure 7.4   The implosion problem (left) and the overlap problem (right).

$1 - p$. It therefore can reduce the amount of traffic and achieve energy conservation by randomization. However, this only addresses the implosion problem of flooding, not the overlapping and resource blindness problems. With gossiping, it is also possible that delivery of sensor data fails when, for example, a node's only neighbor decides not to forward the data to this node. If a high probability for forwarding is chosen, the consequence is that the message volume will be high (a probability of 1 corresponds to classic flooding), thereby limiting the benefits of gossiping. On the other hand, if the probability is low, the overhead can be significantly smaller, but the probability of unsuccessful data delivery will increase.

7.4 Data-Centric Routing

In most sensor networks, the sensor nodes themselves are less important than the information they generate. Therefore, in data-centric routing techniques, the focus is on the retrieval and dissemination of information of a particular type or described by certain attributes, as opposed to the data collection from particular sensors. This section provides an overview of data-centric routing and dissemination protocols in flat-based networks, where all nodes play the same role concerning routing and all nodes collaborate to perform the routing task (i.e., no topology management is necessary).

7.4.1 Sensor Protocols for Information via Negotiation

Sensor Protocols for Information via Negotiation (SPIN) (Kulik *et al.* 2002) is a family of negotiation-based, data-centric, and time-driven flooding protocols. However, compared to classic flooding, SPIN nodes rely on two key techniques to overcome the deficiencies of flooding. To address the problems of implosion and overlap, SPIN nodes negotiate with their neighbors before they transmit data, allowing them to avoid unnecessary communications. To address the problem of resource blindness, each SPIN node uses a resource manager to keep track of actual resource consumption, allowing them to adapt routing and communication behavior based on resource availability.

SPIN uses meta-data to succinctly and completely describe the data collected by sensor nodes. To ensure that the meta-data is useful for SPIN, a key requirement is that if x describes the meta-data for some sensor data X, the size of x (in bytes) must be less than the size of X. Further, two identical pieces of sensor data should have the same meta-data representation. Similarly, if two pieces of sensor data differ, their meta-data representations should differ too. The actual translation of sensor data to meta-data is application-specific and SPIN relies on each application to interpret and synthesize its own meta-data. For example, a camera sensor may use (x, y, ϕ) as meta-data, where (x, y) is a geographic coordinate and ϕ is an orientation.

7.4.1.1 SPIN-PP

The first member of the SPIN family, SPIN-PP, is optimized for networks using point-to-point transmission media, where two nodes can communicate exclusively with each other without interference from other nodes. In SPIN-PP, data is flooded in three steps via a 3-way handshake protocol (Figure 7.5). First, when new data arrives, a node advertises this event

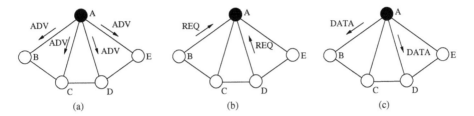

Figure 7.5 The SPIN-PP protocol: (a) advertisement phase, (b) request phase, and (c) data transmission.

using an advertisement message (ADV) to its neighbors via the data's meta-data. Upon receiving an advertisement, a node checks whether it has already received the described sensor data. If not, the node responds with a request for data (REQ) message, indicating its desire to receive the advertised data. Finally, the sender node responds to the REQ message with a DATA message, containing the advertised data.

As shown in Figure 7.5(b), only nodes that do not yet possess a copy of the advertised data respond to an ADV message. Further, upon receiving the DATA message from A, nodes B and D can aggregate this data with their own data and advertise the aggregate to their neighbors. A key strength of this protocol is its simplicity and nodes only need to know their single-hop neighbors to run this protocol. While this protocol has been designed for lossless environments with symmetric communication links, nodes could compensate for lost ADV messages by periodically readvertising their data and for lost REQ and DATA messages by rerequesting data of interest if such data does not arrive within certain timeout intervals. As an alternative, the protocol could be modified to use explicit acknowledgments. For example, REQ messages could contain explicit lists of data that a node wants or does not want to receive. Based on the list, a node can identify if previous advertisements were successfully received by the neighbor.

7.4.1.2 SPIN-EC

A variation of this protocol, called SPIN-EC, adds a simple heuristic to add energy conservation to the SPIN-PP protocol. As long as all nodes have sufficient energy, they participate in the three-way handshake of the SPIN-PP protocol. However, once the energy of a node approaches a specific *low-energy threshold*, it becomes more selective in its participation in the protocol. That is, a node should only participate in the three-way handshake if it believes that it can complete all stages of the protocol without falling below the energy threshold. Therefore, a node replies to an advertisement only if it has sufficient energy to transmit the request and receive the requested data. Similarly, a node initiates the three-way handshake with its neighbors only if it believes that it can complete the protocol even if all neighbors request a copy of the data.

7.4.1.3 SPIN-BC

While the SPIN-PP and SPIN-EC protocols assume point-to-point communication networks, the SPIN-BC protocol improves upon these protocols by exploiting the characteristics of broadcast transmission. In broadcast transmission, every message sent by a sender will be received by all nodes within the sender's radio range. SPIN-BC uses cheap, one-to-many communications and nodes can coordinate their resource-conserving efforts more effectively because they are able to overhear all transactions within their radio ranges.

The SPIN-BC uses a three-way handshake consisting of ADV, REQ, and DATA messages, but with three central differences compared to the approaches taken by SPIN-PP:

1. All messages are directed to the broadcast message, that is, all nodes within a sender's transmission range will receive a copy of the message.
2. Upon receiving an ADV message, a node checks whether it wishes to receive a copy of the advertised data, and if so, sets a random timer, uniformly chosen from a predetermined interval. Only after the timer expires, the node issues the REQ message, again to the broadcast address (specifying the identity of the sender of the ADV message in the message header). When a node overhears the REQ message before its own timer expires, the node cancels its timer and does not send its own REQ message.
3. The advertiser transmits the advertised data to the broadcast address only once, that is, it will ignore duplicate REQ messages for the same data.

The random timer is necessary to avoid collisions of REQ messages from different neighbors and to allow nodes to avoid REQ transmissions when another node has already transmitted an REQ. The consequence of this approach is that the number of transmissions can be significantly reduced, for example, for each transmission, a node needs only to transmit a single ADV message and a single copy of the DATA message.

7.4.1.4 SPIN-RL

The final variant of the SPIN protocol is SPIN-RL, a reliable version of SPIN-BC, addressing packet loss and asymmetric communications. First, each node keeps track of overheard REQ messages and if it does not receive a corresponding DATA message within a certain timeout interval, it assumes that either the REQ message or the DATA message did not arrive. In this case, the node rerequests the data by broadcasting an REQ message, specifying the identity of a randomly selected node among the nodes that previously advertised this data in the message header. In addition, SPIN-RL limits the frequency with which DATA messages are sent out. That is, once a node sends a DATA message, it will wait a predetermined time before responding to any other requests for the same data.

7.4.2 Directed Diffusion

Directed diffusion (Intanagonwiwat *et al.* 2000) is another data-centric data dissemination protocol that is also application-aware in that data generated by sensor nodes is named by attribute-value pairs. The main idea of directed diffusion is that nodes request data by sending *interests* for named data. This interest dissemination sets up *gradients* within the network that are used to direct sensor data toward the recipient, and intermediate nodes along the data paths can combine data from different sources to eliminate redundancy and reduce the number of transmissions.

Directed diffusion does not rely on globally valid node identifiers, but instead uses attribute-value pairs to describe a sensing task and to steer the routing process. For example, a description for a simple vehicle-tracking application could be:

```
type = vehicle                  // detect vehicle location
interval = 20 ms                // send data every 20 ms
duration = 10 s                 // perform task for 10 s
rect = [-100,-100,200,200]      // from sensors within rectangle
```

That is, a task description expresses a node's desire (or interest) to receive data matching the provided attributes. The data sent in response to such interests is also named in the same manner, that is, using attribute-value pairs.

Once an application has been described using this naming approach, the interest must be diffused through the sensor network. This process is shown in Figure 7.6. A sink node periodically broadcasts an interest message to its neighbors, which continue to broadcast the message throughout the network (Figure 7.6(a)). Each node establishes a gradient toward the sink node, where a gradient is a reply link toward the neighbor from which the interest was received. As a consequence, using interests and gradients, paths between event sources and sinks can be established (Figure 7.6(b)). Once a source begins to transmit data, it can use multiple paths for transmission toward the sink. The sink can then *reinforce* one particular neighbor based on some data-driven local rule. For example, a sink could reinforce a neighbor from which the sink has received a previously unseen event. Toward this end, the sink resends the original interest message to the neighbor, which in turn reinforces one or more of its neighbors based on its own local rule (Figure 7.6(c)).

Directed diffusion differs from SPIN in that queries (interests) are issued on demand by the sinks and not advertised by the sources as in SPIN. Based on the process of establishing gradients, all communication is neighbor-to-neighbor, removing the need for addressing schemes and allowing each node to perform aggregation and caching of sensor data, both of which can contribute to reduced energy consumption. Finally, directed diffusion

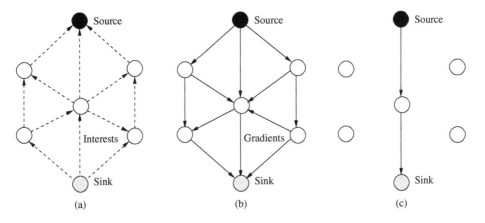

(a)   (b)   (c)

Figure 7.6 Directed diffusion: (a) interest propagation, (b) initial gradients set up, and (c) data delivery.

is a query-based protocol, which may not be a good choice for certain sensor network applications, particularly where continuous data transfers are required (e.g., environmental monitoring applications).

7.4.3 Rumor Routing

While the classic flooding approach can be described as *event flooding* (i.e., a source propagates its sensor data via an event throughout the network), *query flooding* describes the process used to propagate queries to all nodes in the network when no localization information is available to steer the query toward the appropriate sensors. Rumor routing (Braginsky and Estrin 2002) is a variation of directed diffusion that attempts to combine the characteristics of both techniques.

In rumor routing, each node maintains a list of its neighbors and an event table that contains forwarding information to all known events. Once a node witnesses an event (i.e., a phenomenon in the physical world), the event is added to this table (including a distance of zero) and an agent is generated with a certain probability (i.e., not all events may result in an agent generation). This agent is a long-lived packet that travels the network to propagate information about this event and other events encountered along its route to remote nodes. Once an agent arrives at a node, the node can use the agent's content to update its own table. For example, in Figure 7.7(a), the table for node A pointing toward events E1 and E2 is shown, before the arrival of an agent originating at node E. When the agent arrives at node A (via node G), A sees that E1 can be reached via neighbor G using a shorter route than currently stored in its table. It therefore updates its table with the newly obtained information from E's agent (Figure 7.7(b)).

When a node wants to issue a query targeted at a specific event, it first checks whether it has a route toward the target event. If so, it forwards the query to the neighbor indicated by the event's table entry. If there is no route, a random neighbor is selected and the query is passed to this neighbor. This process is continued on each node, where the query message

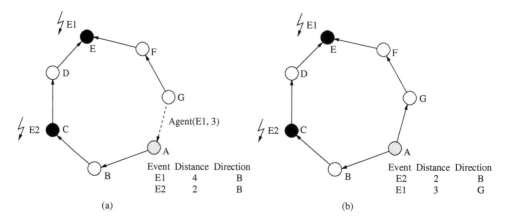

Figure 7.7   Rumor routing: (a) before the arrival of the agent notifying B of a shorter path to E1, (b) after the arrival of the agent.

collects a list of recently seen nodes to avoid revisiting them. Both agent and query messages use a *time-to-live* (TTL) counter value that is decremented at each hop and a message is only forwarded if this counter value is greater than zero. Note that even though a node may not know the direction of the target event, by forwarding it to a random neighbor, the query may be able to come to a node that has an entry for the desired event. Since it is possible that the query may not reach the target event, the querying node can use another technique, for example, flooding the query, when no response is received after a certain timeout value (which depends on the TTL value).

In summary, rumor routing attempts to strike a compromise between query and event flooding. Query flooding can be expensive when the query to event ratio is very high, while event flooding is more beneficial when the number of total events generated in the network is low. Neither protocol works well for moderate query to event ratios. While rumor routing uses a query-based approach, it attempts to route queries to only the nodes that have observed a particular event rather than flooding the entire network. Rumor routing is not concerned with latency, instead a nonoptimal route is satisfactory. Further, a node's table size grows with the number of events it is aware of, therefore, in networks with many events, the cost of storing and maintaining such tables may pose a problem.

7.4.4 Gradient-Based Routing

Another variant of the concept of directed diffusion is Gradient-Based Routing (GBR) (Schurgers and Srivastava 2001), where a gradient is determined on the basis of the number of hops to the sink. Similar to directed diffusion, GBR uses interests to capture a sink's desire to receive certain types of information and during the flooding of these interests, gradients are established on each node. Each interest announcement message records the number of hops it has traveled since leaving the sink. This allows nodes in the network to determine their distance (in terms of number of hops) to the sink, which is called a node's *height* in GBR. Then, the difference between a node's height and the height of its neighbor is considered to be the gradient on the link between these two nodes. A data packet is then forwarded on the link with the largest gradient.

When multiple routes pass through a node, their data may be combined. In GBR, nodes can establish a Data Combining Entity (DCE), which is responsible for data compaction to increase the resource efficiency of the network. Further, GBR uses traffic spreading techniques to more uniformly balance the traffic over the network. In the *stochastic* spreading scheme, each node selects the next hop randomly when there are two or more next hops with the same gradient. In the *energy-based* scheme, a node increases its height when it detects that its energy level has dropped below a certain threshold. This discourages other nodes from sending traffic via this node. Changing a node's height can also affect the heights of the node's neighbors since a node calculates its height to be one more than that of its lowest-height neighbor. Finally, in the *stream-based* scheme, new streams are diverted away from nodes that already serve other streams. To achieve this, a node that receives packets informs all its neighbors (except the one from where the packets originate) that its height has increased. As a consequence, the original stream remains unaffected, but new streams may choose a different route since the height of the nodes of the original stream's route appears to have increased.

7.5 Proactive Routing

Proactive (or table-driven) routing protocols establish paths before they are actually needed. The main advantage of this approach is that routes are available whenever they are needed and no delays are incurred to search for routes such as in on-demand routing protocols. The main disadvantages are the overheads involved in building and maintaining potentially very large routing tables and that stale information in these tables may lead to routing errors.

7.5.1 Destination-Sequenced Distance Vector

The Destination-Sequenced Distance Vector (DSDV) routing protocol (Perkins and Bhagwat 1994) is a modified version of the classic Distributed Bellman-Ford algorithm. In distance-vector algorithms, every node i maintains a list of distances $\{d_{ij}^x\}$ for each destination x via each neighbor j. Then, node i selects node k as the next hop for packet forwarding if $d_{ik}^x = \min\{d_{ij}^x\}$. This information is stored in a routing table, along with a sequence number for each entry, where this number is assigned by the destination node. The purpose of the sequence numbers is to allow nodes to distinguish stale routes from new ones in order to prevent routing loops. Each node broadcasts updates to the routing table periodically, but also immediately whenever significant new information becomes available. DSDV uses two types of packets to share its routing table content. A *full dump* contains all available routing information, whereas an *incremental* packet contains only information that has changed since the last full dump. Incremental packets are typically much smaller than full dumps, therefore reducing the control overhead of DSDV. When a node receives an incremental packet, the received information is compared with the node's current knowledge and a route indicated in the packet replaces the corresponding route in the node's table if the packet's route has a more recent sequence number. A packet's route also replaces the node's route in its table if the sequence numbers are identical, but the packet's route has a shorter distance.

Figure 7.8 shows a possible network topology, indicating node locations and connectivity, including node D's routing table (first table). Suppose that node C moves from its current location to a new location in the vicinity of nodes H and G, which become node C's new neighbors. Update packets from D's neighbors will ultimately inform D that the route to C via B is invalid and that a new route to C via node E exists. Therefore, node D will replace C's information in its routing table to show E as the next hop neighbor and to reflect the new distance of three (second table in Figure 7.8).

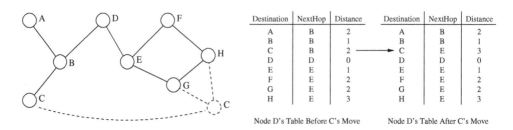

Destination	NextHop	Distance
A	B	2
B	B	1
C	B	2
D	D	0
E	E	1
F	E	2
G	E	2
H	E	3

Destination	NextHop	Distance
A	B	2
B	B	1
C	E	3
D	D	0
E	E	1
F	E	2
G	E	2
H	E	3

Node D's Table Before C's Move   Node D's Table After C's Move

Figure 7.8   Example of a network with a moving node.

7.5.2 Optimized Link State Routing

Another example of a proactive protocol is the Optimized Link State Routing (OLSR) protocol (Clausen *et al.* 2001), which is based on the link state algorithm. In this approach, nodes periodically broadcast topological information updates to all other nodes in the network, allowing them to obtain a complete topological map of the network and to immediately determine paths to any destination in the network.

Every node in OLSR uses *neighbor sensing* to identify its neighbors and to detect changes to the node's neighborhood. Toward this end, a node periodically broadcasts a HELLO message, which contains the node's identity (address) and a list of all known neighbors of this node. For each neighbor, this list also indicates whether the link between the node and the neighbor is symmetric (both can receive each other's messages) or asymmetric. By collecting the neighbors' HELLO messages, a node can determine information about its two-hop neighborhood. To obtain network-wide information, topological information must be flooded throughout the network. Compared to the classic flooding approach, OLSR relies on *multipoint relays* (MPRs) to provide a more efficient way of disseminating such control information. That is, a node selects a set of symmetric neighbor nodes as MPRs, called the MPR selector set. Only MPRs forward messages to other nodes, which may significantly reduce duplicate transmissions. This concept is shown in Figure 7.9, compared to the classic flooding approach.

All nodes select their MPRs independently, possibly using different algorithms and heuristics. For example, a node can determine its two-hop neighbors via the received HELLO messages and can then calculate the minimum set of one-hop neighbors necessary to reach all two-hop neighbors. These one-hop neighbors are then selected as MPRs and informed of their new role using the HELLO messages.

OLSR does not inform all other nodes of all its neighbors; instead, its control messages (which will be relayed by its MPRs) contain the addresses of its MPRs. Effectively, a node announces reachability to all its MPRs and since all nodes have selected an MPR

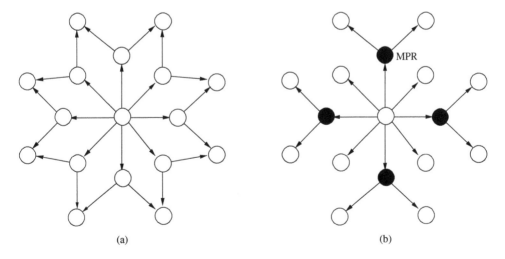

(a)   (b)

Figure 7.9 Comparison of (a) classic flooding and (b) MPR-based OLSR.

set, reachability to all nodes will be announced through the network. Therefore, each node will obtain a partial topographical map of the network, which can be used to determine optimal routes (e.g., using a shortest path algorithm) to any reachable destination in the network.

7.6 On-Demand Routing

Compared to proactive routing protocols, reactive protocols do not discover and maintain routes until they are explicitly requested and used. A source node, knowing the identity or address of the destination node, initiates a route discovery process within the network, which completes when at least one route is found or when all possible routes have been examined. A route is then maintained until it either breaks or is no longer needed by the source.

7.6.1 Ad Hoc On-Demand Distance Vector

An example of an *on-demand* or reactive protocol is the Ad Hoc On-Demand Distance Vector (AODV) protocol (Perkins and Royer 1999). Unlike OLSR, nodes neither maintain any routing information nor participate in periodic routing table exchanges. AODV relies on a broadcast route discovery mechanism, which is used to dynamically establish route table entries at intermediate nodes.

The *path discovery* process of AODV is initiated whenever a source node needs to transmit data to another node, but for which the source node does not have routing information in its table. Toward this end, the source node broadcasts a *route request* (RREQ) packet to its neighbors, which contains the addresses of the source and the destination, a hop count value, a broadcast ID, and two sequence numbers. The broadcast ID is incremented whenever the source issues a new RREQ packet and is combined with the source's address to uniquely identify an RREQ. Upon receiving an RREQ packet, a node that possesses a current route to the specified destination responds by sending a unicast *route reply* (RREP) message directly back to the neighbor from which the RREQ was received. Otherwise the RREQ is rebroadcast to the intermediate node's neighbors. A duplicate RREQ (identified by its source address and broadcast ID) is discarded.

Each node in the network maintains its own sequence number. A source issuing an RREQ packet also includes its own sequence number and the most recent sequence number it has for the destination. Therefore, intermediate nodes reply to an RREQ only if the sequence number of their route to the destination is greater than or equal to the destination sequence number specified in the RREQ packet. When an RREQ is rebroadcast, the intermediate node records the address of the neighbor from which the RREQ was received, thereby establishing a *reverse path* from the destination to the source. As the RREP travels back to the source, each intermediate node sets up a forward pointer to the node from which the RREP came and records the latest destination sequence number for the requested destination. RREP packets contain the addresses of the source and the destination nodes, the destination sequence number, and a hop count. An intermediate node receiving an RREP propagates this packet toward the source only if (1) this is the first copy of this RREP, (2) the RREP contains a greater destination sequence number than the previous RREP, or (3) the destination sequence number is the same as in the previous RREP, but the hop count is smaller. This decreases the number of RREPs traveling toward the source and ensures that the routing information for the shortest

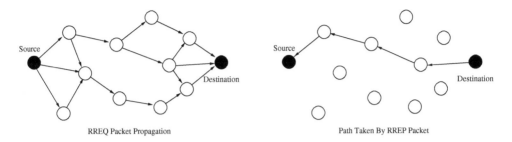

Figure 7.10   Path discovery process of AODV.

route (in terms of number of hops) reaches the source. An example of this path discovery process is shown in Figure 7.10.

A timer for each entry in each node's routing table limits the lifetime of unused routes. Neighboring nodes also exchange periodic HELLO messages to monitor the state of their links. When a link along the route breaks (e.g., due to a moving node), the intermediate node closer to the source noticing the broken link issues a *route error* (RERR) packet upstream toward the source. Upon receiving an RERR packet, a source node can reinitiate the path discovery process.

Using AODV, routes are only established when needed, avoiding costly route table updates and exchanges for routes that may never be used. However, AODV does require periodic exchanges of HELLO messages among neighbors. Further, since a source node may not have a valid route in its table when it wishes to transmit data, there can be an initial delay (for completion of the path discovery process) before the route can be used. Finally, since the route established from the source to the destination is the reverse path of RREP packets that traveled from the destination to the source, AODV assumes that all links are symmetric.

7.6.2   *Dynamic Source Routing*

The Dynamic Source Routing (DSR) protocol (Johnson 1994) employs route discovery and route maintenance procedures similar to AODV. In DSR, each node maintains a *route cache* with entries that are continuously updated as a node learns new routes. Similar to AODV, a node wishing to send a packet will first inspect its route cache to see whether it already has a route to the destination. If there is no valid route in the cache, the sender initiates a route discovery procedure by broadcasting a route request packet, which contains the address of the destination, the address of the source, and a unique request ID. As this request propagates through the network, each node inserts its own address into the request packet before re-broadcasting it. As a consequence, a request packet records a route consisting of all nodes it has visited. When a node receives a request packet and finds its own address recorded in the packet, it discards this packet and does not rebroadcast it further. A node keeps a cache of recently forwarded request packets, recording their sender addresses and request IDs, and discards any duplicate request packets.

Once a request packet arrives at the destination, it will have recorded the entire path from the source to the destination. In symmetric networks, the destination node can unicast a response packet, containing the collected route information, back to the source using the exact same path as taken by the request packet. In networks with asymmetric links, the destination can itself initiate a route discovery procedure to the source, where the request packet also contains the path from the source to the destination. Once the response packet (or the destination's request packet) arrives at the source, the source can add the new route into its cache and begin transmitting packets to the destination. Similar to AODV, DSR also employs a route maintenance procedure based on error messages, which are generated whenever the link layer detects a transmission failure due to a broken link.

Compared to proactive routing protocols, DSR shares similar advantages and disadvantages as AODV. Unlike AODV, each packet in DSR carries route information, which allows intermediate nodes to add new routes proactively to their own caches. Also, DSR's support of asymmetric links is another advantage compared to AODV.

7.7 Hierarchical Routing

Hierarchical routing protocols are based on the grouping of nodes into clusters to address some weaknesses of flat routing protocols, most notably scalability and efficiency. The main idea behind hierarchical routing is that sensor nodes communicate only directly with a leader node in their own cluster, typically referred to as *cluster head*. These cluster heads, which may be more powerful and less energy-constrained devices than "regular" sensor nodes, are then responsible for propagating the sensor data to the sink. This approach can significantly reduce the communication and energy burdens on sensor nodes, while cluster heads will experience significantly more traffic than regular sensor nodes.

Challenges in the design and operation of hierarchical routing protocols include the selection of cluster heads, the formation of clusters, and adaptations to network dynamics such as mobility or cluster head failures. Compared to flat routing approaches, hierarchical solutions may reduce collisions in the wireless medium and facilitate the duty cycling of sensor nodes for increased energy efficiency. Clustering may also facilitate the routing process, but may lead to longer routes than many flat routing protocols. Clustering also facilitates in-network aggregation of sensor data, because data coming from colocated sensors (which may monitor overlapping regions of the environment) are likely to pass through the same cluster head. Figure 7.11 illustrates two variations of the clustering approach. When all cluster heads communicate directly with the sink node (left graph), the routing challenge is reduced to the cluster formation problem. When cluster heads do not directly communicate with the sink (right graph), a cluster-based routing protocol must also establish multi-hop routes from all cluster heads to the sink.

In the landmark routing technique (Tsuchiya 1988), nodes self-organize into hierarchies, where a *landmark* is a node to which its neighbors within a certain number of hops know a route to that node. For example, in the left graph of Figure 7.12, assume that nodes 2–6 have routing information for node 1. On the other hand, nodes 7–11 do not have routes to node 1, therefore node 1 is a landmark "visible" to all nodes within a 2-hop distance. Node 1 is then a landmark of radius 2. In general terms, a node i for which all other nodes within n hops have routing information toward i is a landmark of radius n. Using this definition, a hierarchy of landmarks can be constructed, where a packet can be forwarded toward a destination by choosing an appropriate sequence of landmarks. The right graph in Figure 7.12 shows an example of such a hierarchy, where the dotted lines and the circles

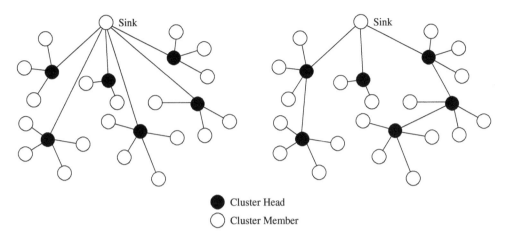

Figure 7.11   Clustering with single-hop connections to the sink (left) and clustering with multi-hop connections to the sink (right).

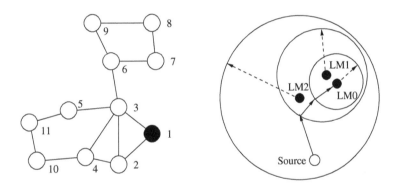

Figure 7.12   Definition of a landmark (left) and routing using a hierarchy of landmarks (right).

indicate the radius of each landmark. Global landmarks are those whose radii are larger than the diameter of the network, that is, routes to these nodes are known by all other nodes in the network. In landmark routing, the address of a node consists of a sequence of identifiers of the nearest landmarks, for example, the landmark address of node LM0 in Figure 7.12 is LM2.LM1.LM0. In this concrete example, the source node will inspect its routing table and it will find an entry for LM2, but not for LM1 or LM0. Therefore, the source will choose a path toward LM2. Each node along the path will make the same decision until the packet reaches a node within the range of LM1. This node will find entries for both LM2 and LM1, but since LM1 has the finer resolution, a path toward LM1 will be chosen. Finally, once the packet reaches a node within the range of LM0, a direct path to the destination LM0 will be chosen.

The LANMAR protocol (Gerla *et al.* 2000) extends the concept of landmark routing by combining it with Fisheye State Routing (Pei *et al.* 2000). It uses landmarks to establish a two-tiered logical hierarchy, where each landmark is a cluster head of a logical subnet. Fisheye State Routing is a link state routing protocol, where the frequency of route updates depends on the distance, that is, routes within a fisheye scope (a certain predefined distance) are more accurate than routes to more distant nodes. In LANMAR, routing updates are only exchanged with nodes in the immediate neighborhood and with landmark nodes; the update frequency of all other nodes is zero. When a node needs to relay a packet, the packet will be forwarded directly to the destination if the destination node is within the node's fisheye scope. Otherwise, the packet will be forwarded toward the landmark corresponding to the destination's logical subnet. Again, once a packet enters the scope of the destination, it is routed directly to it.

Another hierarchical clustering algorithm is the Low-Energy Adaptive Clustering Hierarchy (LEACH) protocol (see Section 6.4.5 for protocol details), which combines a clustering approach with MAC-layer techniques. LEACH assumes that every cluster head can directly communicate with the base station. With LEACH, cluster heads are responsible for all communication between their cluster members and a base station and the aggregation of data coming from its cluster members in order to eliminate redundancies. LEACH can achieve significant energy savings (depending on how much redundancy can be removed) and sensor nodes (apart from the cluster heads) are not responsible for forwarding other nodes' data.

The main idea of the Power-Efficient Gathering in Sensor Information Systems (PEGASIS) protocol (Lindsey and Raghavendra 2002) is for each node to exchange packets with close neighbors and take turns in being responsible for relaying packets to the base station. Toward this end, nodes organize into a chain, for example, using a greedy algorithm initiated by a specific node or computed by the base station, which then broadcasts the chain information to all other nodes. While data travels along a chain, it can be aggregated with other data until relayed to the base station. PEGASIS achieves high-energy efficiency since each node only communicates with its closest neighbors (thereby also allowing a node to reduce its transmission power to the lowest required to reach its neighbors in the chain) and occasionally serves as relay toward the base station. Similar to LEACH, the protocol assumes that all nodes can communicate with the base station. A disadvantage in PEGASIS is that packets may experience significant delays, particularly if they originate from distant nodes in the chain. Finally, the node serving as relay to the base station can become a bottleneck.

The Safari architecture (Du et al. 2008) also provides a self-organizing network hierarchy and a routing protocol with similarity to the landmark approaches. Such landmarks, called *drums* in Safari, use a self-election algorithm (a distributed algorithm with no centralized coordination) to form subnets, called *cells* and *supercells*. At the lowest level of the cell hierarchy (level 0), each individual node forms its own cell. At level 1, Safari defines *fundamental cells*, that is, cells that contain multiple individual nodes, but no other cells. Higher-level cells are then composed of multiple smaller cells at lower levels. Each drum periodically broadcasts beacon messages within well-defined limited scopes in the network. These beacons aid the hierarchy formation, give nodes an indication of their position in the network topology, and provide routing information toward the drum's cell. Safari uses a hybrid approach for routing, that is, routing within cells is based on the reactive DSR protocols, while a proactive routing approach is used to compute routes to more distant nodes. The inter-cell communication relies on a destination node's hierarchical address and on the beacon records stored on each node. Basically, inter-cell communication follows the reverse path of the beacons issued by the cells' drums. The main advantage of Safari over other approaches is its scalability, due to the hybrid routing approach and the hierarchical addressing scheme.

7.8   Location-Based Routing

Location-based or geographic routing can be used in networks where sensor nodes are able to determine their position using a variety of localization systems and algorithms (see Chapter 10 for examples of localization techniques). Instead of topological connectivity information, sensors use geographic information to make forwarding decisions. In unicast location-based routing, packets are sent directly to a single destination, which is identified by its location. That is, a sender must be aware not only of its own location, but also the location of the destination. This location can be obtained either via querying (e.g., flooding a query to request a response from the destination containing its location) or a *location broker*, that is, a service that maps node identities to locations. In broadcast or multicast location-based routing approaches, the same packet must be disseminated to multiple destinations. Multicast protocols take advantage of the known destination locations to minimize resource consumption by reducing redundant links.

The identity of a sensor node is typically less important than its location, that is, data may be disseminated to all nodes that lie within a certain geographic region. This approach is called *geocasting* and can, for example, be used to diffuse queries to specific regions of interest instead of flooding the entire network, significantly reducing both bandwidth and energy requirements. Once a packet reaches the desired region, it must be either disseminated (multicast) to all nodes within this region or transmitted to at least one node within this region (*anycast*).

Typically, location-based routing protocols require that every node in the network knows its own geographic location and the identities and locations of its one-hop neighbors (e.g., obtained via periodic beacon messages). The destination is expressed either as the location of a node (instead of a unique address) or a geographic region. Compared to other routing solutions, an advantage of location-based routing is that only geographic information is needed for forwarding decisions and it is not necessary to maintain routing tables or to establish end-to-end paths between sources and destinations, eliminating the need for control packets (apart from the beacon messages among neighbors).

7.8.1   Unicast Location-Based Routing

In unicast location-based routing, the goal is to propagate a packet to a specific node located at a position known to the sender. The routing protocol's responsibility on each node is to make a local forwarding decision to ensure that a packet moves closer to the destination with each hop. An example of this simple procedure is shown in Figure 7.13, where broken circles indicate the transmission ranges of the forwarding nodes and the arrows indicate the resulting path taken by the packet. In this *greedy forwarding* approach, it is only required that each node knows its own location and the location of its neighbors, and the source must know the location of the destination. However, this approach carries certain risks. Most importantly, a packet may arrive at a node that does not have any neighbors that could serve as next hops to bring the packet closer to the destination. To identify and circumvent

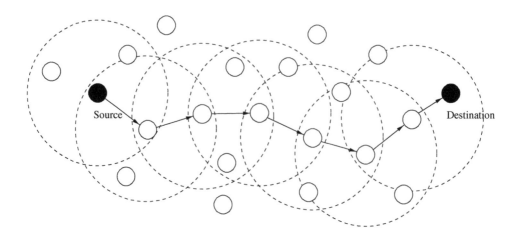

Figure 7.13 Unicast location-based routing.

such *voids* or *holes* is a main challenge being addressed by most location-based routing protocols. Solutions to this problem are discussed in the remainder of this section with the help of several representative protocols for unicast location-based routing.

7.8.1.1 Greedy Perimeter Stateless Routing

An example of a routing protocol that makes forwarding decisions based on the positions of nodes and a packet's destination is Greedy Perimeter Stateless Routing (GPSR) (Karp and Kung 2000). GPRS nodes only require information about their immediate neighbors to decide where to relay a packet. The source of a packet marks the packet with the location of the destination node. If a node knows all its neighbors' positions (e.g., obtained via periodic HELLO or beacon messages), an intermediate node can make a locally optimal forwarding decision by selecting the neighbor that is geographically closest to the destination. Continuing this process node by node, the packet will move closer and closer to the destination with every hop, until the destination is reached.

Since every intermediate node makes a forwarding decision based only on its knowledge of its neighbors' locations, it can happen that a packet must move temporarily farther in geometric distance from the destination to ultimately continue its path toward the destination. Figure 7.14 shows an example of such a situation. Node x is closer to the destination than both its neighbors y and w. Here, the dashed arc around the destination has a radius equal to the distance between the destination and x. Based on the greedy forwarding protocol, x would not select any of the two paths that would lead to the destination.

In this example, the intersection of x's radio range and the circle centered at the destination with a radius equal to the distance between the destination and x is called a void since x does not have any neighbors in this region. Therefore, GPRS provides a mechanism that allows it to route around this void, allowing the packet to continue on its route toward the destination. Toward this end, GPRS relies on the well-known *right-hand rule* for traversing a graph (depicted in Figure 7.14). The rule states that when a packet arrives at node x

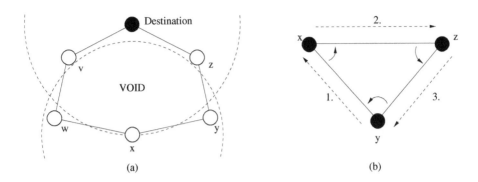

Figure 7.14   GPRS: (a) node x's void with respect to the destination and (b) the right-hand rule.

from node y, the next edge traversed is the next one sequentially counterclockwise about x from edge (x, y). The right-hand rule traverses the interior of a polygon (in this case a triangle) in clockwise edge order and the exterior region (i.e., the region outside the triangle) in counterclockwise edge order.

GPRS exploits this rule to route packets around voids, for example, in Figure 7.14(a), traversing the cycle $(x, w, v, \text{destination}, z, y, x)$ using the right-hand rule amounts to navigating around the void to nodes closer to the destination than x. This sequence of edges traversed according to this rule is called a *perimeter*. Unfortunately, the right-hand rule does not always result in a traversal of the perimeter of a closed polygon. In nonplanar graphs, that is, graphs with crossing edges, it is possible that the right-hand rule may take a degenerate tour of edges that does not trace the boundary of a closed polygon. In GPRS, multiple techniques can be applied to obtain planar graphs (i.e., all crossing edges will be removed), for example, by reducing a graph to a Relative Neighborhood Graph (RNG) or a Gabriel Graph (GG), as long as the removal of edges does not partition the network. For example, to obtain a RNG, we consider the intersection of the radio ranges of two nodes u and v. This region, called the *lune* between u and v, must be empty of any witness node w such that the edge (u, v) can be included in the RNG. That is, if this region is nonempty, the link (u, v) will be removed.

To summarize, GPRS operates in two different modes. Upon receiving a packet, a node searches its neighbor table for the neighbor geographically closest to the destination. If this neighbor is closer to the destination, the packet is relayed to this neighbor. Otherwise, the node enters the perimeter routing mode and records in the packet the location where greedy forwarding failed. Upon receiving a packet in perimeter mode, this location is compared to the forwarding node's location and the packet is returned to the greedy mode if the distance from the forwarding node to the destination is less than that from the recorded location to the destination.

7.8.1.2   Forwarding Strategies

The goal of greedy forwarding is to move a packet closer to the destination with each hop. Each node makes such a forwarding decision based on local information only. However, a variety of forwarding strategies have been explored that all meet this requirement,

but potentially lead to different resource requirements and paths. Common strategies for
forwarding in unicast location-based protocols include:

1. *Greedy:* This common technique chooses a neighbor that minimizes the distance to the
   destination in each hop. In Figure 7.15, this would be node E. The goal of the greedy
   approach is to minimize the number of hops required to reach the destination.
2. *Nearest with Forwarding Progress (NFP):* This strategy chooses the nearest neighbor
   of all neighbors that make a positive progress (in terms of geographic distance to the
   destination) toward the destination (Hou and Li 1986). Sensor nodes that can adapt their
   transmission powers can choose the smallest transmission power necessary to reach this
   neighbor (e.g., node A in Figure 7.15), thereby contributing to reduced packet collisions
   in their neighborhood.
3. *Most Forwarding Progress within Radius (MFR):* The MFR strategy (Takagi and Kleinrock 1984) selects the neighbor that makes the greatest positive progress towards the
   destination, where progress is defined as the distance between the source and its neighbor node projected onto a line drawn from the source to the destination (e.g., node B
   in Figure 7.15). This technique attempts to minimize the number of hops a packet must
   travel.
4. *Compass Routing:* This strategy chooses the neighbor with the smallest angle between
   a line drawn from the source to the neighbor and the line connecting the source and
   the destination (Kranakis *et al.* 1999). This approach (which would select node C in
   Figure 7.15) attempts to minimize the spatial distance that a packet has to travel.

In addition to these strategies that only rely on geometry to determine the next-hop neighbor, forwarding strategies can also combine geometry with additional criteria. For example,
the transmission power needed to reach a neighbor can be used to reduce a node's energy
overhead for the packet forwarding process, the residual energy of the neighbors can be
considered to prolong the lifetime of the network, or the link quality (e.g., signal-to-noise

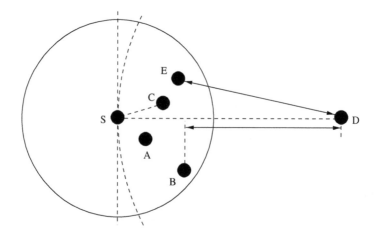

Figure 7.15   Forwarding strategies in location-based routing.

ratio) can be used to maximize the number of successful packet deliveries and to minimize the number of required retransmissions.

7.8.1.3 Geographic Adaptive Fidelity

The Geographic Adaptive Fidelity (GAF) protocol (Xu *et al.* 2001) is another example of an energy-aware unicast location-based routing protocol, but is primarily designed for networks with mobile nodes. The network region is divided into a virtual grid, where only a single device in each grid cell serves as the forwarding node at any given time. This node is then responsible for relaying data to the base station, while all other nodes can go to sleep. Further, GAF assumes that for two adjacent cells A and B, all nodes in A can communicate with all nodes in B and vice versa (Figure 7.16). The grid and cell sizes can be predetermined, allowing each node (assuming that it knows its own location) to determine the cell to which it belongs. That means that most nodes in the network will have neighbors in all four directions (except nodes in border cells).

Nodes in GAF transit between three different states. Initially, each node enters the *discovery* state, where it listens for messages from other nodes within its cell. It also sets a timer for a certain timeout duration and once the timer fires, the node broadcasts a discovery message and enters the *active* state. The node uses another timer to reenter the discovery state once that second timer fires. While in the active state, the node periodically rebroadcasts its discovery message. Further, while in either discovery or active state, a node can enter the *sleep* state whenever it determines that some other node will handle the forwarding of packets. This is achieved using an application-dependent negotiation procedure, for example, based on the expected lifetime of a node. Nodes in the active state win the negotiation process over nodes in the discovery mode. In the case of ties, node IDs can be used to decide which node will serve as forwarder. In general, the goal of this approach is to quickly reach a state where a cell has only one active node. Nodes entering the sleep state periodically reenter the discovery state to repeat the process of negotiating the forwarding roles.

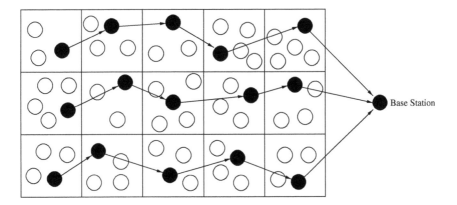

Figure 7.16 Virtual grid approach of GAF.

7.8.2 *Multicast Location-Based Routing*

Multicast is used to deliver the same packet to multiple receivers. A simple approach could deliver the packet to each receiver separately via unicast routing. However, this approach is resource-inefficient in that it does not exploit the fact that routes to different receivers may share paths. Another technique is to simply flood the entire network, which ensures that all receivers will obtain a copy of the packet, but is also very resource costly. Multicast routing is concerned with the efficient delivery of the same packet to all receivers by minimizing the number of links the packet has to travel to reach all destinations. A common technique is to establish a *multicast tree* rooted at the packet source with the destinations as leaf nodes. This section describes representative protocols for multicasts in sensor networks that take advantage of geographic information.

The Scalable Position-Based Multicast (SPBM) protocol (Transier *et al.* 2007) relies on a *group management scheme* to maintain a list of all destinations for a particular packet. However, instead of putting all destinations into the packet header, SPBM uses hierarchical group membership management to ensure that the approach is efficient even when the number of destinations is large. Toward this end, the network is represented as quad-tree with a pre-defined number of levels L, for example, the left graph in Figure 7.17 illustrates an example with L = 4 (levels 0 ... L−1). Squares are identified by concatenating their level numbers, for example, square 442 is a level-0 square, located in the level-3 square that encompasses the entire network, the level-2 square 4, and the level-1 square 44. All nodes in a level-0 square are in radio range of each other.

Based on this hierarchical addressing scheme, each node maintains two tables: a *global member table* containing entries for the three neighboring squares for each level and a *local member table* containing all members of the node's level-0 neighbors. Each entry in the global member table contains the square's identifier and a list of nodes located in the square. Each entry in the local member table contains a node ID and the membership information of that node. This membership information indicates the multicast groups to which a node

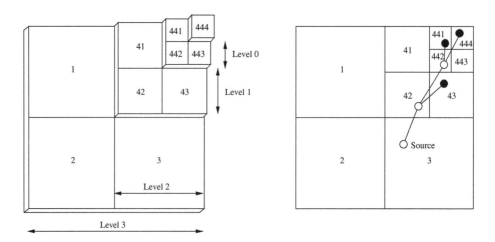

Figure 7.17 Quad-tree network representation in SPBM (left) and routing using the quad-tree (right).

belongs and is encoded as vector where each bit represents a multicast group. For example, an entry of 10100010 for square 41 in the global membership table would indicate that there exist nodes in square 41 that belong to multicast groups 2, 6, and 8. An entry of 00000001 for node 14 in the local membership table indicates that node 14 is a member of multicast group 1 only. The contents of a node's local membership table are periodically broadcast within the node's level-0 square. A randomly chosen node in each level-0 square periodically disseminates its global membership table to all nodes in its level-1 square, and this process is repeated for higher-level squares.

Using these tables, a source can now disseminate a packet to all nodes subscribed to a multicast group. For example, in the right graph of Figure 7.17, the source wishes to transmit a packet to its multicast group members located in squares 441, 444, and 43 (indicated by the black circles). Using its global membership table, it knows that there are multicast members in level-2 square 4 and therefore forwards the packet toward that direction. Similar to GPSR, SPBM uses a greedy forwarding approach by choosing next-hop neighbors that make the largest progress toward a destination. Once the packet arrives at a node in square 42, that node knows that there are multicast members in level-1 squares 43 and 44 and therefore forwards the packets toward each square. The rule for *splitting* a multicast packet is based on a heuristic that provides a tradeoff between the total number of nodes forwarding the packet and the optimality of the individual routes toward the destinations. Once a forwarding node finds a multicast member in its local membership table, it forwards the packet directly to that member node. Similar to GPSR, whenever greedy forwarding fails, the protocol switches into perimeter routing mode.

Other location-based multicast protocols are the Geographic Multicast Routing (GMR) protocol (Sanchez *et al.* 2006) and the Receiver Based Multicast (RBMulticast) protocol (Feng and Heinzelman 2009). GMR uses a heuristic neighbor selection scheme that requires low computational overheads, resulting in efficient routes based on a *cost over progress* metric. This metric is a ratio of the number of selected forwarding nodes over the progress made toward all destinations (i.e., the total remaining distance from the neighbors to the destinations minus the total distance from the forwarding node to all destinations). RBMulticast is a receiver-based multicast approach, that is, a sender can transmit packets without specifying the next-hop node. Similar to SPBM, RBMulticast divides the network into *multicast regions*, splitting packets depending on the locations of the destinations. However, RBMulticast is a completely stateless protocol, eliminating the need for membership tables. This is achieved by representing each multicast region with a *virtual node* and each forwarding node replicates a packet for each region that contains at least one multicast member. The destination of a packet is then the virtual node for a particular multicast region. In RBMulticast, it is up to the MAC layer to ensure that the neighbor closest to the location of the virtual node takes responsibility for forwarding the packet. That is, RBMulticast assumes an underlying MAC protocol where receivers contend for channel access and nodes that make the most forward progress to the destination will contend earlier and have a better chance to become the next-hop node.

7.8.3 Geocasting

In many wireless sensor networks, it is preferred to propagate information to all or some nodes within a specific geographic region. This is a very natural model for many sensor network applications, specifically when the exact location of individual sensors is unknown. For example, in query-based networks, the same query can be propagated to multiple sensors monitoring a specific geographic area, instead of sending the same query repeatedly to different individual sensors. The routing problem then consists of two separate challenges: (1) propagating a packet *near* the target region and (2) distributing a packet within the target region. The first challenge can be addressed using approaches similar to unicast location-based routing as previously described, although no exact location of a sensor node near or within the target region may be known. If it is sufficient for a packet to reach only a single node within the target region, the protocol has succeeded once the packet arrives on at least one node within the region. However, if all nodes within the region must obtain a copy of the packet, the second challenge can be addressed using approaches similar to the broadcast techniques described previously. Therefore, geocasting to multiple receivers is a combination of both unicast and broadcast geographic routing.

7.8.3.1 Geographic and Energy Aware Routing

The Geographic and Energy Aware Routing (GEAR) protocol (Yu *et al.* 2001) is an example of a geocasting protocol, where packets are forwarded to all nodes within a specific target region. GEAR consists of the two phases described above: (1) packets are forwarded toward the target region using a geographical and energy-aware neighbor selection algorithm and (2) packets are disseminated to nodes within the target region using a recursive geographic forwarding algorithm.

Each node in the network maintains two types of costs of reaching a destination via its neighbors. The *estimated cost* $c(N_i, R)$ for each neighbor N_i and a target region R is defined as:

$$c(N_i, R) = \alpha d(N_i, R) + (1 - \alpha)e(N_i) \qquad (7.2)$$

where α is a tunable weight, $d(N_i, R)$ is the distance from neighbor N_i to the centroid D of region R normalized by the largest such distance among all neighbors, and $e(N_i)$ is the consumed energy at node N_i, normalized by the largest consumed energy among all neighbors. That is, the estimated cost is a combination of both residual energy and distance to the target region. The *learned cost* $h(N, R)$ of a node N is then a refinement of the estimated cost that allows nodes to circumvent voids or holes in the network (if there are no holes, the learned cost and the estimated cost are identical). Similar to GPSR, GEAR makes locally greedy forwarding decisions, that is, whenever a node receives a packet, it will pick the next hop among the neighbors that are closer to the destination.

When a node N receives a packet, and if there are no neighbors that are closer to the destination, N knows that it is in a hole. In this case, the learned cost function is used to select one of N's neighbors as the next hop, that is, the packet is forwarded to the node with the minimum learned cost (ties are broken using some predefined ordering). After a node picks the next-hop neighbor N_{min}, it sets its own learned cost $h(N, R)$ to $h(N_{min}, R) + C(N, N_{min})$, where $C(x, y)$ is the cost of transmitting a packet from node x to node y. Therefore, the learned cost will increase, which allows upstream nodes to avoid forwarding packets toward the node in the hole. Figure 7.18(a) shows an example of this procedure, where T represents the centroid of the target region. Node S wishes to forward a packet toward the destination and it has three neighbors that are closer to the destination: B, A, and

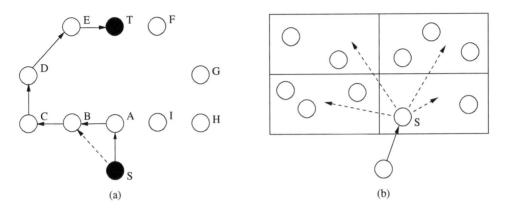

Figure 7.18 GEAR: (a) learning routes around holes and (b) recursive geographic forwarding.

I. B's and I's learned and estimated costs are $\sqrt{5}$ and A's learned and estimated costs are both 2. S will forward the packet to the lowest cost neighbor, which is A. Node A will find itself in a hole and it will forward the packet to the neighbor with the minimum learned cost, for example, node B. Additionally, it will update its own cost $h(A, T) = h(B, T) + C(A, B)$, that is, assuming a cost $(A, B) = 1$, the new learned cost of A will be $\sqrt{5} + 1$. The next time a packet for T arrives at node S, S will forward the packet directly to B instead of A to circumvent the hole.

Once a packet reaches the target region R, a simple flooding with duplicate suppression scheme could be used to disseminate the packet to all nodes within R. However, due to the cost of flooding, GEAR relies on a process called *Recursive Geographic Forwarding*, shown in Figure 7.18(b). Assume that the target region R is the large rectangle and node S received a packet for R and finds itself within R. Then, S creates four new copies of the packet bound to four smaller subregions (shown as the smaller rectangles) of region R. For each subregion, GEAR repeats the forwarding and splitting process until a packet reaches a node that is the only one within the current subregion.

7.8.3.2   Geographic-Forwarding-Perimeter-Geocast

Another protocol that combines geographic routing with region flooding is the Geographic-Forwarding-Perimeter-Geocast (GFPG) protocol (Seada and Helmy 2004). Similar to GPRS, it uses greedy forwarding to propagate a packet toward its geocast region, where the destination is the center of the geocast region. When greedy forwarding fails, perimeter routing is used to circumvent voids. Once the packet enters the geocast region, simple flooding could be used to deliver it to all nodes with the region. However, this assumes that there are no obstacles and gaps, that is, all nodes within the region must be able to reach each other without going out of the region. If this assumption does not hold, delivery cannot be guaranteed. Therefore, GFPG uses a combination of geocast and perimeter routing to guarantee delivery to all nodes. For example, the gray geocast region in Figure 7.19 has two clusters of nodes that cannot directly reach each other within the geocast region (i.e., there is a gap between two nodes in the lower left and the upper right corners).

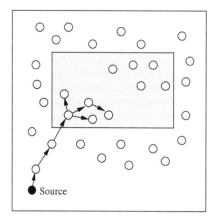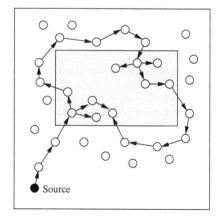

Figure 7.19   An example of a geocast region with a gap (left) and region flooding and perimeter routing used by GFPG to reach all nodes in the geocast region (right).

Once a packet reaches a geocast region, it is flooded to all nodes, but in addition, *region border nodes*, that is, nodes that have at least one neighbor outside the region, also send the packet to their neighbors outside the region in the planar graph. Nodes outside the region forward the packet using the right-hand rule to neighbors in the planar graph and, as a consequence, the packet travels around the face until it enters the region again (Figure 7.19). The first node inside the region to receive the perimeter packet floods this packet to its neighbors if it has not seen this packet before. Perimeter routing is therefore able to link disconnected clusters of a geocast region together.

7.9   QoS-Based Routing Protocols

Although most routing and data dissemination protocols aim for some kind of Quality-of-Service (QoS), for example, minimum hop routing protocols try to achieve low latencies by using "short" paths, some protocols proposed for sensor networks explicitly address one or more QoS routing metrics. The goal of these protocols is to find feasible paths between sender and destination, while satisfying one or more QoS metrics (latency, energy, bandwidth, reliability), but also optimizing the use of the scarce network resources. Wireless sensor networks pose numerous challenges to providing satisfactory QoS, including dynamic topologies, resource scarcity (including power limitations), varying quality of the radio channels, the lack of centralized control, and the heterogeneity of network devices. This section introduces several representative QoS-based routing protocols for ad hoc and sensor networks.

7.9.1   Sequential Assignment Routing

One of the first routing protocols to explicitly consider Quality-of-Service is the Sequential Assignment Routing (SAR) protocol (Sohrabi *et al.* 2000), which is also an example of a *multipath* routing approach. SAR creates multiple trees, each rooted at a 1-hop neighbor of the sink, to establish multiple paths from each node to the sink. These trees grow outward from the sink, while avoiding nodes with low QoS (e.g., high delay). The QoS associated with a path is expressed as an additive QoS metric where higher values imply lower QoS. After the tree-building procedure has completed, it is likely that a node is part of multiple trees, that is, it can choose from multiple routes toward the sink. SAR selects a route for a packet based on the QoS metric, energy (in terms of number of packets that can be transmitted without energy depletion, assuming exclusive use of the path), and the priority level of the packet. The goal of SAR is to minimize the average weighted QoS metric over the lifetime of the network. The availability of multiple routes ensures fault-tolerance and quick recovery from broken paths. However, establishing and maintaining the trees (i.e., routing tables) are expensive tasks, particularly in large sensor networks.

7.9.2 SPEED

Numerous wireless sensor applications require sensor data collection within certain time constraints to ensure that the collected information is useful and can be acted upon in a timely fashion. For example, events of interest such as the detection of moving objects in surveillance systems or the impending failure of a bridge require rapid responses.

For applications with *soft* real-time requirements, SPEED (He *et al.* 2003) is an example of a protocol that provides real-time communication services, including real-time unicast, real-time area-multicast, and real-time area-anycast. SPEED is also an example of a location-based routing protocol, that is, a node relies on position information from its neighbors instead of routing tables. Position information is obtained through periodic HELLO (or beacon) messages that contain a node's ID, position, and an average receive delay. Each node also maintains a neighbor table containing the node ID and position for each of its neighbors, but also an expiration time (ExpireTime) and two delays called ReceiveFromDelay and SendToDelay. The SendToDelay is the delay received from the beacon message coming from the neighbor, while the ReceiveFromDelay is estimated by measuring the delay experienced by a packet in the MAC layer of the sender plus a propagation delay. The ReceiveFromDelay values of all neighbors are averaged periodically to obtain a single receive delay.

The routing component of the SPEED protocol is called Stateless Nondeterministic Geographic Forwarding (SGNF). The neighbor set of a node i is defined as the set of neighbors of i (i.e., all nodes within i's radio range) that are at least a distance of K away from i. The forwarding candidate set (FS_i^{Dest}) of a node i for destination Dest consists of all nodes from the node's neighbor set that are at least a distance of K closer to the destination. That is, if L is the distance of node i from the destination and L_{next} is the distance from i's neighbor j to the destination, $L - L_{\text{next}}$ has to be greater than or equal to K in order to add j to i's forwarding candidate set. Packets are only forwarded to nodes belonging to FS_i^{Dest} and if this set is empty, packets are dropped. SPEED further divides the forwarding candidate set into two subsets: one contains nodes that have a SendToDelay less than a certain single hop delay D, and the other contains the remaining nodes. The forwarding candidate is then selected from the first group where nodes with higher relay speed have a greater chance of being chosen. The relay speed considers both distance and delays and is calculated as:

$$\text{RelaySpeed} = \frac{|L - L_{\text{next}}|}{\text{SendToDelay}} \tag{7.3}$$

where a discrete exponential distribution can be used to trade off between load balancing and optimal path length. If there are no nodes in the first subset of forwarding candidates, a relay ratio is calculated, based on another component of the SPEED protocol, the neighborhood feedback loop. This component is responsible for determining the relay ratio by looking at the miss ratios of the neighbors of a node (i.e., the nodes which could not provide the required RelaySpeed). If this relay ratio is less than a randomly generated number between 0 and 1, the packet is dropped. The goal of the neighborhood feedback loop is to keep the system performance at a desired value, that is, it attempts to maintain a single hop delay below a certain value D.

The final component of SPEED is the back-pressure rerouting protocol, which is responsible for (i) preventing voids that occur when a node fails to find a next hop node and (ii) reducing congestion using a feedback approach. Figure 7.20 depicts two examples showing the operation of this technique. In both examples, the shaded regions are areas where traffic is high, causing congestion. In the first case, node 3 will be notified of the delays experienced by nodes 6 and 7 through the beacon exchange process. The SGNF component of SPEED reduces the probability of nodes 6 and 7 being selected as forwarding nodes, therefore reducing the congestion around these nodes. In the second case, all forwarding nodes of 3 are congested and, in this case, both the neighborhood feedback loop and SGNF work together to address the congestion. For example, node 3 may drop a certain number of packets, where these dropped packets count as packet with delay D in terms of computing the delay at this node. The average delay of 3 will increase, which will be detected by 3's upstream nodes (i.e., node 2). Should node 2 be in the same situation as node 3, further back-pressure will be imposed on node 1, that is, back-pressure rerouting may continue to proceed upstream until it reaches the source, which can then suppress further packets.

7.9.3 Multipath Multi-SPEED

The goal of the Multipath Multi-SPEED (MMSPEED) protocol (Felemban *et al.* 2006) is to provide QoS differentiation in terms of *timeliness* and *reliability*, while at the same time

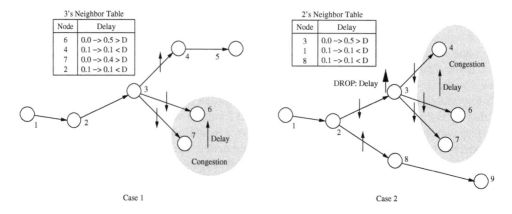

Figure 7.20   Two examples of back-pressure rerouting in SPEED.

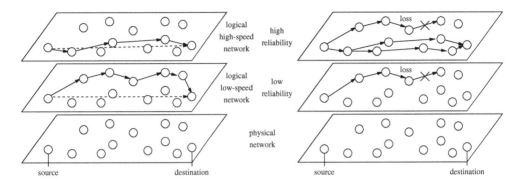

Figure 7.21   Service differentiation in the latency domain (left) and the reliability domain (right).

minimizing the protocol's overhead by making localized routing decisions without a priori route discovery or global network state updates. Similar to SPEED, the protocol relies on geographic locations of nodes to make forwarding decisions, where these locations are exchanged among neighboring nodes using periodic beacon messages.

With respect to timeliness, MMSPEED offers packets multiple delivery speed options that are guaranteed throughout the network. Conceptually, this protocol can be understood as a virtual overlay of multiple SPEED layers on top of a single physical layer (left graph in Figure 7.21). Each layer l is associated with a SetSpeed$_l$, which is a prespecified lower-bound speed. That is, when a node computes the relay speed for each of its neighbors (see Section 7.9.2), it then chooses a forwarding neighbor whose relay speed is at least the desired SetSpeed value. For example, assume that the minimum required speed level ReqSpeed(x) for packet x can be calculated as

$$\text{ReqSpeed}(x) = \frac{\text{dist}_{s,d}(x)}{\text{deadline}(x)} \tag{7.4}$$

where dist$_{s,d}$ is the distance from the source s to the destination d for packet x with a given (end-to-end) deadline(x). Then, the speed layer l for the packet is selected such that

$$\text{SetSpeed}_l = \min_{j=1}^{L}\{\text{SetSpeed}_j | \text{SetSpeed}_j \geq \text{ReqSpeed}(x)\} \tag{7.5}$$

where L is the number of available speed options. In this case, the node chooses a neighbor i whose progress speed estimation RelaySpeed $= |\text{dist}_{s,d} - \text{dist}_{i,d}|/\text{delay}_{s,i}$ is at least SetSpeed$_l$. It is possible that a packet's delays over a route differ from the delay estimations. Therefore, the layer selected at one node can differ from the layer selected at another node, for example, a slow packet can be *boosted* by using a higher layer at a subsequent node. Toward this end, it is necessary to determine a packet's remaining time to its deadline, which requires synchronized clocks in the network. Instead, MMSPEED measures the elapsed time at each node and piggybacks this information onto a packet such that subsequent nodes can determine the remaining time to the deadline.

Similarly, MMSPEED offers packets multiple levels of reliability. Toward this end, it exploits the fact that there exist multiple redundant paths from a source to a destination, even though these paths will differ in length and QoS (right graph in Figure 7.21). Each node i maintains the recent average of packet loss percentage $e_{i,j}$ to each neighbor j. Such losses include both intentional packet drops for congestion control and errors in the wireless channel. Based on these averages, a node computes an estimate of packet loss probability between itself and a packet's destination as

$$RP_{i,j}^{d} = (1 - e_{i,j})(1 - e_{i,j})^{\lceil dist_{j,d}/dist_{i,j} \rceil} \qquad (7.6)$$

where $\lceil dist_{j,d}/dist_{i,j} \rceil$ is the hop count estimation from node j to the destination d. This estimation is based on the assumption that subsequent nodes have a similar packet loss rate to node i and that the progress to the destination for each following hop will be similar to the current progress. Based on this computation, a node can determine the number of forwarding paths that satisfy the end-to-end reachability requirement of a packet. The total reaching probability (TRP) is originally set to zero and updated for each forwarding path that is being used, that is, the TRP is computed as

$$TRP = 1 - (1 - TRP)(1 - RP_{i,j}^{d}) \qquad (7.7)$$

Here, $(1 - TRP)$ is the probability that none of the current paths can successfully deliver the packet to the destination and $(1 - RP_{i,j}^{d})$ is the probability that the additional path will fail to deliver the packet. Therefore, the newly computed TRP is the probability that at least one path will successfully deliver the packet to the destination. A node adds paths to this TRP estimation until TRP becomes larger than the required end-to-end reachability P^{req}.

Both latency and reliability considerations can also be combined in MMSPEED. In this case, the protocol identifies the required speed level for a given packet and then it finds multiple forwarding nodes among those with sufficient progress speed such that the total reaching probability is at least as high as the required reaching probability.

7.10   Summary

While routing in general is a crucial component of any multi-hop network, routing is particularly challenging in wireless ad hoc and sensor networks due to their characteristics such as stringent resource constraints and unreliability of links and nodes. Specifically, routing protocols must operate efficiently to avoid premature exhaustion of the limited resources in a sensor network (most notably energy) and they must be able to adjust to changing and unpredictable network characteristics, including changes in the network topology and density. In this section, several classes of routing strategies (data-centric, hierarchical, location-based) and numerous examples of concrete routing protocols have been discussed. Table 7.1 summarizes some key characteristics of the protocols discussed in this chapter.

Although numerous routing solutions for sensor networks exist, the unique challenges and the many varieties of network deployment scenarios indicate that there still remain a variety of challenges, for example, with respect to resource efficiency and provision of QoS. For example, in the recent past, there has been an increased focus on wireless sensor networks that can support application-specific QoS requirements involving multiple performance metrics. Other areas of investigation of routing protocols for future sensor networks include the need for energy-efficient solutions that make localized decisions, protocols that effectively exploit redundancy for efficiency and reliability, protocols for newly emerging

Table 7.1 Network protocols summary

Protocol	Characteristics
SPIN	Flat topology, data-centric, query-based, negotiation-based
Directed diffusion	Flat topology, data-centric, query-based, negotiation-based
Rumor routing	Flat topology, data-centric, query-based
GBR	Flat topology, data-centric, query-based
DSDV	Flat topology with proactive route discovery
OLSR	Flat topology with proactive route discovery
AODV	Flat topology with reactive route discovery
DSR	Flat topology with reactive route discovery
LANMAR	Hierarchical with proactive route discovery
LEACH	Hierarchical, support of MAC layer
PEGASIS	Hierarchical
Safari	Hierarchical, hybrid route discovery (reactive near, proactive remote)
GPSR	Location-based, unicast
GAF	Location-based, unicast
SPBM	Location-based, multicast
GEAR	Location-based, geocast
GFPG	Location-based, geocast
SAR	Flat topology with QoS (real-time, reliability), multipath
SPEED	Location-based with QoS (real-time)
MMSPEED	Location-based with QoS (real-time, reliability)

topologies (e.g., architectures with multiple tiers), security-aware routing protocols, and integrated solutions to routing and in-network processing of sensor data.

Exercises

7.1 The previous chapter presented several MAC protocols, while this chapter introduced routing protocols. Can you think of examples how the choice of MAC protocol affects the design, performance, and efficiency of the routing protocol?

7.2 What is the difference between a proactive routing protocol and a reactive routing protocol? Name at least two examples for each category. Consider the following WSN scenarios and explain why you would choose either a proactive or a reactive routing solution:

(a) A WSN is used to monitor air pollution in a city where every sensor reports its sensor data once every minute to a single remote base station. Most sensors are mounted on lamp posts, but some are also mounted on city buses.

(b) A WSN is used to measure humidity in a field, where low-power sensors report measurements only when certain thresholds are exceeded.

(c) A WSN is used to detect the presence of vehicles, where each sensor locally records the times of vehicle detection. These records are delivered to the base station only when the sensor is explicitly queried.

7.3 What is data-centric routing? Why is data-centric routing feasible (or even necessary) compared to routing based on identities (addresses)?

7.4 Describe a WSN application for each of the following categories: time-driven, event-driven, and query-driven.

7.5 For the network topology shown in Figure 7.22, identify the optimal routes for source A to sink M according to the following criteria (describe how you compute the cost for the optimal route). The numbers X/Y along each link indicate the latency (X) and energy cost (Y) for transmitting a single packet over the link. The number Z under each node indicates the node's remaining energy capacity.

(a) Minimum number of hops

(b) Minimum energy consumed per packet

(c) Maximum average energy capacity (eliminate hops that would result in a higher average but unnecessarily add to the route length!)

(d) Maximum minimum energy capacity

(e) Shortest latency

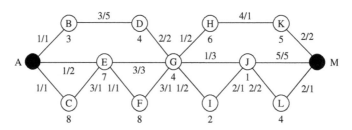

Figure 7.22 Topology for Exercise 7.5.

7.6 A WSN is modeled as a 5×5 grid as shown in Figure 7.23, with the base station placed at the center of the network (left topology) or at the bottom left corner (right topology). Assume that each node can communicate with only its immediate neighbors on the grid and that packet transmission or forwarding over a link costs exactly one unit of energy (packet reception and processing costs are neglected).

(a) For both topologies, find an energy optimal graph of routes, that is, the energy cost for each packet traveling through the network is a minimum.

(b) Consider the graphs shown in Figure 7.24. What is the average and total load in the network, when the per-node load is defined as the number of routes a node has to service (including its own)? Do not include the base station in your calculations.

(c) What is the lifetime of the network topologies in Figure 7.24 when during every second, each node generates and transmits its own packet and forwards all packets received during the previous second? Assume that each node has an initial energy budget of 100. Each transmission costs 1 unit of energy (there is no cost for reception, etc.). Consider the lifetime of a network to have expired once the first node depletes its energy budget. Compare the results and derive design principles for the network topology to optimize the lifetime of the network with respect to placement of the base station and the construction of routing trees.

(d) Assume that the first topology in Figure 7.24 is used and each sensor transmits exactly one packet to the base station. Then the topology is switched to the second one and each sensor transmits one packet to the base station in the bottom left corner. Then the topology is switched back to the first one and the process is repeated. Explain why the network lifetime changes and what other design principle can be derived from this insight. (To facilitate the comparison, focus on the case where each node has already reached its maximum load.)

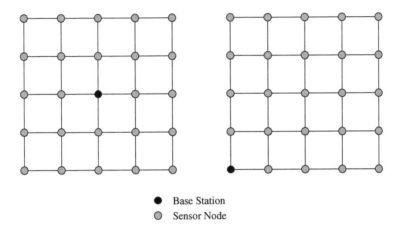

● Base Station
○ Sensor Node

Figure 7.23   Topologies for Exercise 7.6.

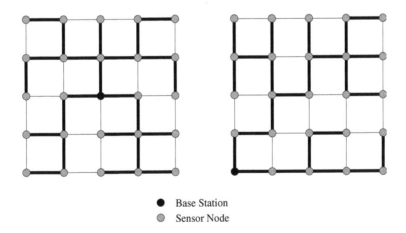

● Base Station
○ Sensor Node

Figure 7.24   Topologies and routes for Exercise 7.6.

7.7 Flooding is a simple strategy for distributing data to one specific node or all sensor nodes in a network. Answer the following questions:

(a) Explain the three challenges of flooding described in this chapter.

(b) Which of these can be addressed by gossiping and how can they be addressed?

(c) For the topologies shown in Figure 7.22 and Figure 7.23, what are good choices for the maximum hop count? Explain your answer.

(d) How do sequence numbers contribute to reducing unnecessary transmissions? Are sequence numbers alone sufficient and, if not, what other information is needed to use them correctly?

7.8 Using the topology in Figure 7.22, explain the problems of implosion, overlap, and resource blindness.

7.9 How does the SPIN family of protocols address the three challenges faced by flooding? What are the disadvantages of a negotiation-based protocol such as SPIN?

7.10 Explain the concept of directed diffusion. Can you imagine at least three strategies or goals for reinforcement?

7.11 Consider the network topology in Figure 7.22 and node G's routing table shown in Table 7.2.

(a) Describe how node G would send queries toward events E1, E2, and E3 using rumor routing (note that node G has no routing table entries for event E3).

(b) Assume that (i) I informs G that I can reach event E2 via 2 hops, (ii) F informs G that F can reach event E3 via 4 hops, (iii) E informs G that E can reach event E1 via 1 hop, (iv) D informs G that D can reach event E1 via 2 hops, (v) H informs G that H can reach event E2 via 2 hops, and (vi) D informs G that D can reach event E3 via 1 hop. What is the final table of node G? Can you identify the locations of all three events by the identity of the closest sensor?

Table 7.2   G's routing table (Exercise 7.11)

Event	Distance	Direction
E1	3	F
E2	4	I

7.12 What are the concepts behind distance vector routing and link state routing and how do they compare to each other with respect to overheads for maintaining routing tables?

7.13 Compare a proactive routing protocol such as DSDV with a reactive protocol such as DSR with respect to overheads and route optimality.

7.14 Does DSR incur larger or smaller overheads for route discovery compared to the AODV protocol? Justify your answer.

7.15 In AODV, is it possible that route discovery packets travel in the network forever? Why or why not?

7.16 Asymmetric (or unidirectional) links occur when node A can hear node B, but B cannot hear node A. Explain whether this is a problem for the AODV protocol and if so, how this can be addressed.

7.17 What is the concept behind hierarchical routing and what advantages does it have over other techniques?

7.18 Table 7.3 summarizes the routing information of all nodes in a WSN, that is, each row indicates the routing knowledge of that particular node. For example, the first row shows that node A knows that it can reach nodes B and C via 1 hop and nodes D and E via 2 hops. Given this information, draw the network topology and determine the landmark radius for each node.

Table 7.3 Routing information for Exercise 7.18

	A	B	C	D	E	F	G	H
A	0	1	1	2	2	–	–	–
B	1	0	1	1	1	2	–	–
C	1	1	0	2	1	–	2	–
D	–	1	2	0	1	1	2	2
E	2	1	1	1	0	–	1	–
F	–	2	–	1	2	0	1	1
G	–	2	2	2	1	1	0	1
H	–	–	3	2	–	1	1	0

7.19 What is the advantage of using Fisheye State Routing in the LANMAR protocol compared to the basic landmark routing technique?

7.20 Figure 7.25 shows a number of nodes as small dots. Each node has a radio range of 2 units. How would the gray node positioned at (0, 0) route a packet to the gray node at position (9, 9) using GPSR? Indicate the visited nodes.

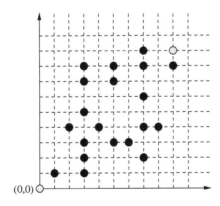

Figure 7.25 GPSR routing example (Exercise 7.20).

7.21 When does GPSR enter the perimeter routing mode and how does it use the right-hand rule in this mode?

7.22 Prove that it is false or show an example that the perimeter mode can cause a packet to traverse a network's entire outer boundary.

7.23 Consider the topology in Figure 7.26. Node A wishes to forward a packet toward destination L via one of its neighbors (its communication range is indicated with the circle). Which neighbor will A choose with each of the following forwarding strategies:

(a) greedy forwarding
(b) nearest with forwarding progress
(c) most forwarding progress within radius
(d) compass routing

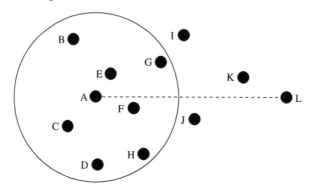

Figure 7.26   Forwarding strategies in GPSR (Exercise 7.23).

7.24 The cell size of the GAF virtual grid can be predetermined and each node knows to which cell it belongs. Discuss the consequences of choosing very large versus very small cell sizes.

7.25 How does the SPBM protocol ensure efficient multicast for large numbers of receivers?

7.26 What is the concept of RBMulticast and how does it address the shortcomings of the SPBM protocol?

7.27 The GEAR protocol uses two types of costs: learned and estimated. Explain how learned costs are used to route packets around holes (use a concrete example). What is the purpose of the estimated costs and what is the intuition behind calculating them as described in this chapter?

7.28 Figure 7.27 shows a sensor network topology, where each node's transmission range is two units. The node at position (0,0) wants to disseminate a packet to all nodes within the rectangle. Show how GFPG routes the packet toward the region and how it distributes it to all receivers within the rectangle. Clearly indicate which nodes (inside and outside the geocast region) will receive the packet.

7.29 Answer the following questions with respect to QoS-aware routing protocols:

(a) What advantages and disadvantages does multipath routing have?
(b) How does the SGNF component of SPEED work?

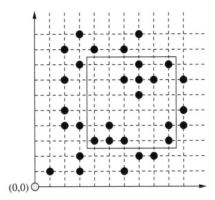

(0,0)

Figure 7.27   Geocast region with hole (Exercise 7.28).

(c) How does the back-pressure rerouting component of SPEED work?

(d) Why does MMSPEED change the speed of packets as they travel along a route?

(e) How can latency and reliability considerations be combined in MMSPEED?

References

Al-Karaki, J.N., and Kamal, A.E. (2004) Routing techniques in wireless sensor networks: A survey. *IEEE Wireless Communications* **11** (6), 6–28.

Braginsky, D., and Estrin, D. (2002) Rumor routing algorithm for sensor networks. *Proc. of the 1st ACM International Workshop on Wireless Sensor Networks and Applications*.

Clausen, T., Hansen, G., Christensen, L., and Behrmann, G. (2001) The optimized link state routing protocol, evaluation through experiments and simulation. *Proc. of the IEEE Symposium on Wireless Personal Mobile Communications*.

Couto, D.D., Aguayo, D., Bicket, J., and Morris, R. (2003) High throughput path metric for multi-hop wireless routing. *Proc. of the 9th Annual International Conference on Mobile Computing and Networking (MobiCom)*.

Draves, R., Padhye, J., and Zill, B. (2004) Routing in multi-radio, multi-hop wireless mesh networks. *Proc. of the 10th Annual International Conference on Mobile Computing and Networking (MobiCom)*.

Du, S., Khan, A., Pal-Chaudhuri, S., Post, A., Saha, A.K., Druschel, P., Johnson, D.B., and Riedi, R. (2008) Safari: A self-organizing, hierarchical architecture for scalable ad hoc networking. *Ad Hoc Networks* **6** (4), 485–507.

Felemban, E., Lee, C.G., and Ekici, E. (2006) MMSPEED: Multipath multi-SPEED protocol for QoS guarantee of reliability and timeliness in wireless sensor networks. *IEEE Transactions on Mobile Computing* **5** (6), 738–754.

Feng, C.H., and Heinzelman, W.B. (2009) RB Multicast: Receiver based multicast for wireless sensor networks. *Proc. of the IEEE Wireless Communications and Networking Conference (WCNC)*.

Gerla, M., Hong, X., and Pei, G. (2000) Landmark routing for large ad hoc wireless networks. *Proc. of the IEEE Global Communications Conference (GLOBECOM)*.

He, T., Stankovic, J.A., Lu, C., and Abdelzaher, T. (2003) SPEED: A real-time routing protocol for sensor networks. *Proc. of the International Conference on Distributed Computing Systems*.

Hedetniemi, S.H., Hedetniemi, S.T., and Liestman, A.L. (1988) A survey of gossiping and broadcasting in communication networks. *Networks* **18** (4), 319–349.

Heinzelman, W., Kulik, J., and Balakrishnan, H. (1999) Adaptive protocols for information dissemination in wireless sensor networks. *Proc. of the 5th ACM/IEEE International Conference on Mobile Computing and Networking (MobiCom)*.

Hou, T., and Li, V. (1986) Transmission range control in multi-hop packet radio networks. *IEEE Transactions on Communications* **34** (1), 38–44.

Intanagonwiwat, C., Govindan, R., and Estrin, D. (2000) Directed diffusion: A scalable and robust communication paradigm for sensor networks. *Proc. of the 6th Annual International Conference on Mobile Computing and Networking (MobiCom)*.

Johnson, D.B. (1994) Routing in ad hoc networks of mobile hosts. *Proc. of the IEEE Workshop on Mobile Computing Systems and Applications*.

Karp, B., and Kung, H.T. (2000) GPSR: Greedy perimeter stateless routing for wireless networks. *Proc. of the 6th Annual International Conference on Mobile Computing and Networking (MobiCom)*.

Kranakis, E., Singh, H., and Urrutia, J. (1999) Compass routing on geometric networks. *Proc. of the 11th Canadian Conference on Computational Geometry*.

Kulik, J., Heinzelman, W., and Balakrishnan, H. (2002) Negotiation-based protocols for disseminating information in wireless sensor networks. *Wireless Networks* **8** (2/3), 169–185.

Lindsey, S., and Raghavendra, C.S. (2002) PEGASIS: Power-efficient gathering in sensor information systems. *Proc. of the IEEE Aerospace Conference*.

Pei, G., Gerla, M., and Chen, T.W. (2000) Fisheye state routing in mobile ad hoc networks. *Proc. of the ICDCS Workshop on Wireless Networks and Mobile Computing*.

Perkins, C.E., and Bhagwat, P. (1994) Highly dynamic destination-sequenced distance-vector routing (DSDV) for mobile computers. *ACM SIGCOMM Computer Communication Review* **23** (4), 234–244.

Perkins, C.E., and Royer, E.M. (1999) Ad hoc on-demand distance vector routing. *Proc. of the 2nd IEEE Workshop on Mobile Computing Systems and Applications*.

Sanchez, J.A., Ruiz, P.M., and Stojmenovic, I. (2006) GMR: Geographic multicast routing for wireless sensor networks. *Proc. of the 3rd Annual IEEE Communications Society Conference on Sensor, Mesh and Ad Hoc Communications and Networks*.

Schurgers, C., and Srivastava, M.B. (2001) Energy efficient routing in wireless sensor networks. *Proc. of the IEEE Military Communications Conference (MILCOM)*.

Seada, K., and Helmy, A. (2004) Efficient geocasting with perfect delivery in wireless networks. *Proc. of the IEEE Wireless Communications and Networking Conference (WCNC)*.

Singh, S., Woo, M., and Raghavendra, C.S. (1998) Power-aware routing in mobile ad hoc networks. *Proc. of the 4th Annual International Conference on Mobile Computing and Networking (MobiCom)*.

Sohrabi, K., Gao, J., Ailawadhi, V., and Pottie, G. (2000) Protocols for self-organization of a wireless sensor network. *IEEE Personal Communications* **7** (5), 16–27.

Takagi, H., and Kleinrock, L. (1984) Optimal transmission ranges for randomly distributed packet radio terminals. *IEEE Transactions on Communications* **32** (3), 246–257.

Transier, M., Füssler, H., Widmer, J., Mauve, M., and Effelsberg, W. (2007) A hierarchical approach to position-based multicast for mobile ad hoc networks. *Wireless Networks* **13** (4), 447–460.

Tsuchiya, P.F. (1988) The landmark hierarchy: A new hierarchy for routing in very large networks. *Proc. of the ACM Symposium on Communications Architectures and Protocols*.

Xu, Y., Heidemann, J., and Estrin, D. (2001) Geography-informed energy conservation for ad hoc routing. *Proc. of the 7th Annual International Conference on Mobile Computing and Networking (MobiCom)*.

Yu, Y., Govindan, R., and Estrin, D. (2001) *Geographical and energy aware routing: A recursive data dissemination protocol for wireless sensor networks*. Technical Report. UCLA/CSDTR 010023, UCLA Computer Science Department.

Part Three

Node and Network Management

8

Power Management

The power consumption of a wireless sensor network (WSN) is of crucial concern because of the scarcity of energy. Whereas energy is a scarce resource in every wireless device, the problem in WSNs is amplified for the following reasons:

1. Compared to the complexity of the task they carry out – namely, sensing, processing, self-managing, and communication – the nodes are very small in size to accommodate high-capacity power supplies.
2. Ideally, a WSN consists of a large number of nodes. This makes manually changing, replacing or recharging batteries almost impossible.
3. While the research community is investigating the contribution of renewable energy and self-recharging mechanisms, the size of nodes is still a constraining factor.
4. The failure of a few nodes may cause the entire network to fragment prematurely.

The problem of power consumption can be approached from two angles. One is to develop energy-efficient communication protocols (self-organization, medium access, and routing protocols) that take the peculiarities of WSNs into account. The other is to identify activities in the networks that are both wasteful and unnecessary and mitigate their impact.

Wasteful and unnecessary activities can be described as local (limited to a node) or global (having a scope network-wide). In either case, these activities can be further considered as accidental side-effects or results of nonoptimal software and hardware implementations (configurations). For example, observations based on field deployment reveal that some nodes exhausted their batteries prematurely because of unexpected overhearing of traffic that caused the communication subsystem to become operational for a longer time than originally intended (Jiang *et al.* 2007). Similarly, some nodes exhausted their batteries prematurely because they aimlessly attempted to establish links with a network that had become no longer accessible to them.

Most inefficient activities are, however, results of nonoptimal configurations in hardware and software components. For example, a considerable amount of energy is wasted by an idle processing or a communication subsystem. A radio that aimlessly senses the media or overhears while neighboring nodes communicate with each other consumes a significant amount of power.

A dynamic power management (DPM) strategy ensures that power is consumed economically. The strategy can have a local or global scope, or both. A local DPM strategy aims to minimize the power consumption of individual nodes by providing each subsystem with the amount of power that is sufficient to carry out a task at hand. When there is no task to be processed, the DPM strategy forces some of the subsystems to operate at the most economical power mode or puts them into a sleeping mode. A global DPM strategy attempts to minimize the power consumption of the overall network by defining a network-wide sleeping state.

There are different ways to achieve this goal. One way is to let individual nodes define their own sleeping schedules and share these schedules with their neighbors to enable a coordinated sensing and an efficient internode communication. This is called synchronous sleeping. The problem with this approach is that neighbors need to synchronize time as well as schedules and the process is energy intensive. Another way is to let individual nodes keep their sleeping schedules to themselves; and a node that initiates a communication should send a preamble until it receives an acknowledgment from its receiving partner. This approach is known as asynchronous sleeping schedule and avoids the needs to synchronize schedules. But it can have a latency side-effect on data transmission. In both approaches, individual nodes wake up periodically to determine whether there is a node that wishes to communicate with them and to process tasks waiting in a queue.

The main focus of this chapter is on local dynamic power management strategies in WSNs.

8.1   Local Power Management Aspects

The first step toward developing a local power management strategy is the understanding of how power is consumed by the different subsystems of a wireless sensor node. This knowledge enables wasteful activities to be avoided and to frugally budget power. Furthermore, it enables one to estimate the overall power dissipation rate in a node and how this rate affects the lifetime of the entire network.

In the following subsections, a more detailed observation into the different subsystems of a node is made.

8.1.1   Processor Subsystem

Most existing processing subsystems employ microcontrollers, notably Intel's StrongARM and Atmel's AVR. These microcontrollers can be configured to operate at various power modes. For example, the ATmega128L microcontroller has six different power modes: idle, ADC noise reduction, power save, power down, standby, and extended standby. The *idle* mode stops the CPU while allowing the SRAM, Timer/Counters, SPI port, and interrupt system to continue functioning. The *power down* mode saves the registers' content while freezing the oscillator and disabling all other chip functions until the next interrupt or Hardware Reset. In the *power save mode*, the asynchronous timer continues to run, allowing the user to maintain a timer base while the remaining components of the device enter into a sleep mode. The *ADC noise reduction* mode stops the CPU and all I/O modules, except the asynchronous timer and the ADC. The aim is to minimize switching noise during ADC conversions. In *standby* mode, a crystal/resonator oscillator runs while the remaining hardware components enter into a sleep mode. This allows very fast start-up combined with low power consumption. In *extended standby* mode, both the main oscillator and the asynchronous timer continue to operate. Additional to the above configurations, the processing subsystem can operate with different supply voltages and clock frequencies.

While operating the processor subsystem at various power modes is energy-efficient, transiting from one power mode to another also has its own power and latency cost. This cost has to be considered before a decision for a particular operation of power mode is made.

8.1.2 Communication Subsystem

The power consumption of the communication subsystem can be influenced by several aspects, including the modulation type and index, the transmitter's power amplifier and antenna efficiency, the transmission range and rate, and the sensitivity of the receiver. Some of these aspects can be dynamically reconfigured. Moreover, the communication subsystem itself can activate or turn off the transmitter and the receiver, or both. Because of the presence of a large number of active components in the communication subsystem (amplifiers and oscillators), a significant amount of quiescent current flows even if the device is idle.

Determining the most efficient active state operational mode is not a simple decision. For example, the power consumption of a transmitter may not necessarily be reduced by simply reducing the transmission rate or the transmission power. The reason is that there is a tradeoff between the useful power required for data transmission and the power dissipated in the form of heat at the power amplifier. Usually, the dissipation power (heat energy) increases as the transmission power decreases. In fact most commercially available transmitters operate efficiently at one or two transmission power levels. Below a certain level, the efficiency of the power amplifier falls drastically. In some cheap transceivers, even when at the maximum transmission power mode, more than 60% of the supply DC power is dissipated in the form of useless heat.

For example, the Chipcon CC2420 transceiver has eight programmable output power levels ranging from -24 dBm to 0 dBm. This is described in Table 8.1. The columns of the table express the output power, the current consumption, and the power consumption at 1.8 V DC supply voltage. Figure 8.1 illustrates the normalized current consumption (taking the minimum current consumption as a reference) and the relationship between the transmission power levels and the current consumption. As can be seen in the figure, increasing the transmission power level by almost 55 dB scales the current consumption by double only.

Table 8.1 Chipcon CC2420: Output power settings and typical current consumption at 2.45 GHz

PA level	Output power		Current consumption	Power consumption*
	dBm	mW	mA	mW
31	0	1	17.4	31.32
27	−1	0.794328235	16.5	29.7
23	−3	0.501187234	15.2	27.36
19	−5	0.316227766	13.9	25.02
15	−7	0.199526231	12.5	22.5
11	−10	0.1	11.2	20.16
7	−15	0.031622777	9.9	17.82
3	−25	0.003162278	8.5	15.3

*$V_{dd} = 1.8$ V

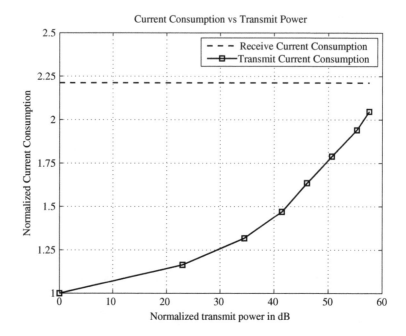

Figure 8.1 Relation between transmit power and current consumption in Chipcon CC2420 transceiver.

Figure 8.2 demonstrates the power amplifier's efficiency. The amplifier efficiency is defined as the ratio of the transmission power to the DC input power consumed by the amplifier.

An additional challenge to the power issue is the time needed by the communication subsystem to transit from an idle, or standby mode into an active mode. The transmission introduces latency and consumes power. For example, the Chipcon's transceiver frequency synthesizer's phase locked loop (PLL) requires 192 μs to lock up.

8.1.3 Bus Frequency and RAM Timing

The processor subsystem consumes power when it interacts with the other subsystems via the internal high-speed buses. The specific amount depends on the frequency and bandwidth of the communication. These two parameters can be optimally configured depending on the interaction type, but bus protocol timings are usually optimized for particular bus frequencies. Moreover, bus controller drivers require to be notified when bus frequencies change to ensure optimal performance.

8.1.4 Active Memory

The active memory is made up of electrical cells which are arranged in rows and columns, each row being a single memory bank. The cells have to be recharged periodically in order to store data. The refresh rate or refresh interval is a measure of the number of rows that

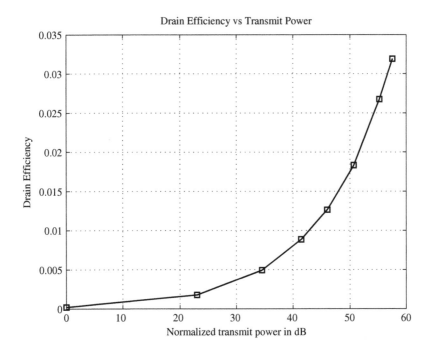

Figure 8.2   Amplifier efficiency in the Chipcon CC2420 transceiver.

must be refreshed. A low refresh interval corresponds to a low clock frequency that must elapse before a refreshing operation takes place. On the contrary, a higher refresh interval corresponds to a high clock frequency that must elapse before a refresh operation takes place. Consider two typical values: 2K and 4K. The lower refresh interval refreshes more cells at a low interval and completes the process faster, thus it consumes more power than the 4K refresh rate. The 4K refresh rate refreshes less cells at a slower pace, but it consumes less power.

A memory unit can be configured to operate in one of the following power modes: temperature-compensated self-refresh mode, partial array self-refresh mode, or power down mode. The standard refresh rate of a memory unit can be adjusted according to its ambient temperature. For this reason, some commercially available dynamic RAMs (DRAMs) already integrate temperature sensors. Apart from this, the self-refresh rate can be increased if the entire memory array is not needed to store data. The refresh operation can be limited to the portion of the memory array in which data will be stored. This approach is known as the partial array self-refresh mode. If no actual data storage is required, the supply voltage of most or the entire on-board memory array can be switched off.

The RAM timing is another parameter that affects the power consumption of the memory unit. It refers to the latency associated with accessing the memory unit. Before a processor subsystem accesses a particular cell in a memory, it should first determine the particular row or bank and then activate it with a row access strobe (RAS) signal. Once a row is activated, it can be accessed until the data is exhausted. The time required to activate a row in a memory is t_{RAS}, which is relatively small but could impact the system's stability if set incorrectly.

Table 8.2 Parameters of RAM timing

Parameter	Description
RAS	Row Address Strobe or Row Address Select
CAS	Column Address Strobe or Column Address Select
t_{RAS}	A time delay between the precharge and activation of a row
t_{RCD}	The time required between RAS and CAS access
t_{CL}	CAS latency
t_{RP}	The time required to switch from one row to the next row
t_{CLK}	The duration of a clock cycle
Command rate	The delay between Chip Select (CS)
Latency	The total time required before data can be written to or read from memory

A memory cell is activated through a column access strobe (CAS). The delays between the activation of a row as well as a cell and the writing of data into or reading of data from the cell is given as t_{RCD}. This time can be short or long, depending on how the memory cell is accessed. If it is accessed sequentially, it is insignificant. If, on the other hand, the memory is accessed in a random fashion, the current active row must first be deactivated before a new row is activated, in which case, t_{RCD} can cause significant latency.

The delay between the CAS signal and the availability of valid data on the data pins is called *CAS latency*. Low CAS latency means high performance but also high power consumption. The time required to terminate one row access and begin the next row access is t_{RP}. In conjunction with t_{RCD}, the time (or clock cycles) required to switch banks (rows) and select the next cell for reading, writing, or refreshing is expressed as $t_{RP} + t_{RCD}$. The duration of time required between the active and precharge commands is called t_{RAS}. It is a measure of how long the processor must wait before the next memory access can begin. Table 8.2 summarizes the quantities that express RAM timing.

When a RAM is accessed by clocked logic, the times are generally rounded up to the nearest clock cycle. For example, when accessed by a 100-MHz processor (with 10 ns clock duration), a 50-ns SDRAM can perform the first read in 5 clock cycles and additional reads within the same page every 2 clock cycles. This is generally described as "$5 - 2 - 2 - 2$" timing.

8.1.5 Power Subsystem

The power subsystem supplies power to all the other subsystems. It consists of the battery and the DC–DC converter. In some cases, it may consist of additional components such as a voltage regulator. The DC–DC converter is responsible for providing the right amount of supply voltage to each individual hardware component by transforming the main DC supply voltage into a suitable level. The transformation can be a step-down (buck), a step-up (boost), or an inversion (flyback) process, depending on the requirements of the individual subsystem. Unfortunately, a transformation process has its own power consumption and may be inefficient. In the following subsections, some of causes of power consumption and inefficiency will be discussed.

8.1.5.1 Battery

A wireless sensor node is powered by exhaustible batteries. Several factors affect the quality of these batteries, but the main factor is cost. In a large-scale deployment, the cost of hundreds and thousands of batteries is a serious deployment constraint.

Batteries are specified by a rated current capacity, C, expressed in *ampere-hour*. This quantity describes the rate at which a battery discharges without significantly affecting the prescribed supply voltage (or potential difference). Practically, as the discharge rate increases, the rated capacity decreases.

Most portable batteries are rated at $1C$. This means a $1000\,\mathrm{mAh}$ battery provides $1000\,\mathrm{mA}$ for 1 hour, if it is discharged at a rate of $1C$. Ideally, the same battery can discharge at a rate of $0.5C$, providing $500\,\mathrm{mA}$ for 2 hours; and at $2C$, $2000\,\mathrm{mA}$ for 30 minutes and so on. $1C$ is often referred to as a 1-hour discharge. Likewise, a $0.5C$ would be a 2-hour and a $0.1C$ a 10-hour discharge.

In reality, batteries perform at less than the prescribed rate. Often, the Peukert Equation is applied to quantifying the capacity offset (i.e., how long a battery lasts in reality):

$$t = \frac{C}{I^n} \tag{8.1}$$

where C is the theoretical capacity of the battery expressed in ampere-hours; I is the current drawn in Ampere (A); T is the time of discharge in seconds, and n is the Peukert number, a constant that directly relates to the internal resistance of the battery. The value of the Peukert number indicates how well a battery performs under continuous heavy current. A value close to 1 indicates that the battery performs well; the higher the number, the more capacity is lost when the battery is discharged at high current. The Peukert number of a battery is determined empirically. For example, for lead acid batteries, the number is typically between 1.3 and 1.4.

Drawing current at a rate greater than the discharge rate results in a current consumption rate higher than the rate of diffusion of the active elements in the electrolyte. If this process continues for a long time, the electrodes run out of active material even though the electrolyte has not yet exhausted its active materials. This situation can be overcome by intermittently drawing current from the battery.

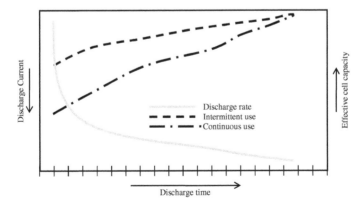

Figure 8.3 The Peukert curve displaying the relationship between the discharging rate and the effective voltage. The x-axis is a time axis.

Figure 8.3 shows how the effective battery capacity can be reduced at high and continuous discharge rates. By intermittently using the battery, it is possible during quiescent periods to increase the diffusion and transport rates of active ingredients and to match up the depletion created by excessive discharge. Because of this potential for recovery, the capacity reduction can be undermined and the operating efficiency can be enhanced. This is illustrated in the figure by the middle, dash-dotted line.

8.1.5.2 DC–DC Converter

The DC–DC converter transforms one voltage level into another. It is the equivalent of a transformer which performs AC–AC voltage transformation. The main problem with a DC–DC converter is its conversion efficiency. A typical DC–DC converter consists of a power supply, a switching circuit, a filter circuit, and a load resistor. Figure 8.4 illustrates the basic circuit structure of a DC–DC converter.

As can be seen in the figure, the circuit consists of a single-pole, double-throw (SPDT) switch that is connected to a DC supply voltage, V_g. Considering the inductor, L, as a short circuit and the capacitor, C, as an open circuit for the DC supply voltage, the switch's output voltage, $V_s(t)$ equals to V_g when the switch is in position 1 and 0 when it is in position 2. Varying the position of the switch at a frequency f_s yields a periodically varying square wave, $v_s(t)$, that has a period $T_s = 1/f_s$.

$v_s(t)$ can be expressed by a duty cycle D, which describes the fraction of time that the switch is in position 1, such that $0 \leq D \leq 1$. The output voltage of the switching circuit is displayed in Figure 8.5.

A DC–DC converter is realized by employing active switching components, such as diodes and power MOSFETs. Typically, the switching frequencies range from 1 kHz to 1 MHz, depending on the speed of the semiconductor devices.

Using the inverse Fourier transformation, the DC component of $v_s(t)$ (V_s) is described as:

$$V_s = \frac{1}{T_s} \int_0^{T_s} v_s(t) dt = D V_g \qquad (8.2)$$

which is the average value of $v_s(t)$.

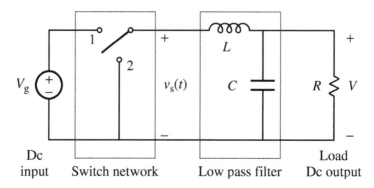

Figure 8.4 A DC–DC converter consisting of a supply voltage, a switching circuit, a filter circuit, and a load resistance.

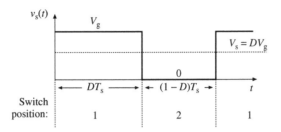

Figure 8.5 The output voltage of a switching circuit of a DC–DC converter.

In other words, the integral value represents the area under the waveform of Figure 8.5 for a single period, or the height of V_g multiplied by the time T_s. It can be seen that the switching circuit reduces the DC component of the supply voltage by a factor that equals to the duty cycle, D. Since $0 \leq D \leq 1$ holds, $V_s \leq V_g$ holds as well.

Ideally the switching circuit does not consume power. In practice, however, due to the existence of a resistive component in the switching circuit, there is power dissipation. As a result, the efficiency of a typical switching circuit is between 70 and 90%.

In addition to the desired DC voltage, $v_s(t)$ also contains undesired harmonics of the switching frequency, f_s. These harmonics must be removed so that the converter's output voltage $v(t)$ is essentially equal to the DC component $V = V_s$. For this purpose, a DC–DC converter employs a lowpass filter. In Figure 8.4, a first-order LC lowpass filter is connected to the switching circuit. The filter's cutoff frequency is given by:

$$f_c = \frac{1}{2\pi\sqrt{LC}} \tag{8.3}$$

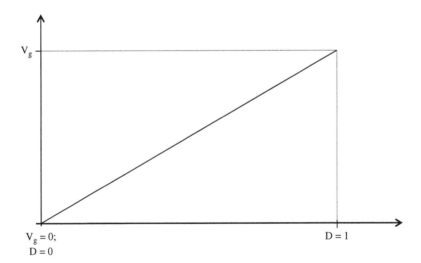

Figure 8.6 A linear relationship between a DC supply voltage and the duty cycle of a switching circuit.

The cutoff frequency, f_c, should be sufficiently less than the switching frequency, f_s, so that the lowpass filter allows only the DC component of $v_s(t)$ to pass while effectively attenuating all the harmonic components. Once again, in an ideal filter, there is no power dissipation because the passive components (inductors and capacitors) are energy storage components. Subsequently, the DC–DC converter produces a DC output voltage whose magnitude is controlled by the duty cycle, D, using circuit elements that (ideally) do not dissipate power.

The conversion ratio, $M(D)$, is defined as the ratio of the DC output voltage, V, to the DC input voltage, V_g, under a steady-state condition:

$$M(D) = \frac{V}{V_g} \tag{8.4}$$

For the buck converter shown in Figure 8.4, $M(D) = D$. Figure 8.6 illustrates the linear relationship between the input DC voltage, V_g and the switching circuit's duty cycle, D.

8.2 Dynamic Power Management

Wireless sensor nodes can be developed by taking the aspects discussed so far into account at design time. Once the design time parameters are fixed, a dynamic power management (DPM) strategy attempts to minimize the power consumption of the system by dynamically defining the most economical operation conditions. This condition takes the requirements of the application, the topology of the network, and the task arrival rate of the different subsystems into account. Whereas there are different approaches to a DPM strategy, they can be categorized in one of the following three approaches:

1. Dynamic operation modes.
2. Dynamic scaling.
3. Energy harvesting.

8.2.1 Dynamic Operation Modes

The subsystems of a wireless sensor node can be configured to operate in different power modes, depending on their present and anticipated activity. This has already been explained in the previous subsections. In general, a subcomponent can have n different power modes. If there are x hardware components that can have n distinct power consumption levels, a DPM strategy can define $x \times n$ different power mode configurations, P_n. Obviously, not all these configurations are plausible because of various constraints and system stability preconditions. Hence, the task of the DPM strategy is to select the optimal configuration that matches the activity of a wireless sensor node.

There are, however, two associated challenges in selecting a specific power configuration.

1. Transition between the different power configurations costs extra power.
2. A transition has an associated delay and the potential of missing the occurrence of an interesting event.

Table 8.3 demonstrates an example DPM strategy with six different power modes: $\{P_0, P_1, P_2, P_3, P_4, P_5\}$. Figure 8.7 shows corresponding potential transitions between five arbitrary power modes.

Table 8.3 Power saving configurations

Configuration	Processor	Memory	Sensing subsystem	Communication subsystem
P_0	Active	Active	On	Transmitting/receiving
P_1	Active	On	On	On (transmitting)
P_2	Idle	On	On	Receiving
P_3	Sleep	On	On	Receiving
P_4	Sleep	Off	On	Off
P_5	Sleep	Off	Off	Off

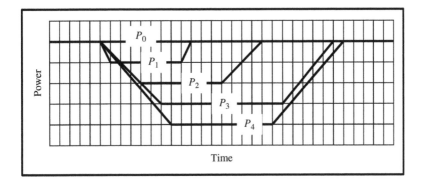

Figure 8.7 Transition between different power modes and the associated transition costs.

The decision for a particular power mode depends on the present as well as the anticipated task in the queues of the different hardware components. A realistic estimation of future tasks enables a node to determine the time it needs to put the required components in the right power mode, so that they can process the tasks with minimum latency. By the same token, failure to realistically estimate future tasks can cause a node to miss interesting events or to delay in response.

In a WSN, the events outside of the network (for example, a leak in a pipeline; a fracture in a structure; a pestilence in a farm; etc.) cannot be modeled as deterministic phenomena. Otherwise there is no need for setting up a monitoring system. Therefore, estimation of the arrival of events should be probabilistic. Knowledge of the sensing task can be useful to establish a realistic probabilistic model for estimating the arrival rate as well as the duration of events. An accurate event arrival model enables a DPM strategy to decide for the right configuration that has a long duration and minimal power consumption.

8.2.1.1 Transition Costs

Suppose each subsystem of a wireless sensor node operates in just two different power modes only, namely, it can be either *on* or *off*. Moreover, assume that the transition from *on* to *off* does not have an associated power cost, but the reverse transition (from *off* to *on*) has a cost in terms of both power and a time delay. These costs are justified if the power it saves in the *off* state is large enough. In other words, the amount of the *off* state power is considerably large and the duration of the *off* state is long. It is useful to quantify these costs and to set up a transition threshold.

Suppose the minimum time that a subsystem stays in an *off* state is t_{off}; the power consumed during this time is P_{off}; the transition time is $t_{off,on}$; the power consumed during the transition is $p_{off,on}$; and the power consumed in an *on* state is P_{on}. Hence:

$$P_{off} \cdot t_{off} + P_{off,on} \cdot t_{off,on} \geq P_{on} \cdot \left(t_{off} + t_{off,on} \right) \tag{8.5}$$

Therefore, t_{off} is justified if (Chiasserini and Rao 2003):

$$t_{off} \geq \max \left(0, \ \frac{\left(P_{on} - P_{off,on} \right) \cdot t_{off,on}}{P_{on} - P_{off}} \right) \tag{8.6}$$

Equations (8.5) and (8.6) can easily be generalized to describe a subsystem with n distinct operational power modes, in which case a transition from any state i into j is described as $t_{i,j}$. Hence, the transition is justified if Equation (8.7) is satisfied.

$$t_j \geq \max \left(0, \ \frac{\left(P_i - P_{j,k} \right) \cdot t_{i,j}}{P_i - P_j} \right) \tag{8.7}$$

where t_j is the duration of the subsystem in state j.

The equations above assume that the transition cost from a higher power mode (on) to a lower power mode (off) is negligible. If this is not the case, the energy that can be saved through a power transition (from state i to state j, $E_{saved,j}$) is expressed as:

$$E_{saved,j} = P_i \cdot \left(t_j + t_{i,j} + t_{j,i} \right) - \left(P_{i,j} \cdot t_{i,j} + p_{j,i} \cdot t_{j,i} + p_j \cdot t_j \right) \tag{8.8}$$

If the transition from state i to state j costs the same amount of power and time delay as the transition from state j to state i (a symmetric transition cost), Equation (8.8) can be expressed as:

$$E_{saved,j} = P_i \cdot \left(t_j + t_{i,j} + t_{j,i} \right) - \left(\frac{P_i + P_j}{2} \right) \left(t_{i,j} + t_{j,i} \right) - \left(P_i - P_j \right) \cdot t_j \tag{8.9}$$

Obviously, the transition is justified if $E_{saved,j} > 0$. This can be achieved in three different ways, namely, by:

1. increasing the gap between P_i and P_j;
2. increasing the duration of state j, (t_j); and
3. decreasing the transition times, particularly, $t_{j,i}$.

8.2.2 *Dynamic Scaling*

Dynamic voltage scaling (DVS) and dynamic frequency scaling (DFS) are complementary to the approach discussed in Section 8.2.1. These two approaches aim to adapt the performance of the processor core (as well as the memory unit and the communication buses) when it is in the active state. In most cases, the tasks scheduled to be carried out by the processor core do not require its peak performance. Rather, some tasks are completed ahead of their deadline and the processor enters into a low-leakage idle mode for the remaining time. Figure 8.8 shows a subsystem processing at peak performance. Even though the two tasks are completed ahead of their schedule, the processor still runs at peak frequency and supply voltage, which is wasteful.

Figure 8.9 displays the application of dynamic frequency and voltage scaling in which the performance of the processing subsystem is adapted (reduced) according to the criticality of the tasks it processes. As can be seen, each task is stretched to its planned schedule while the supply voltage and the frequency of operation are reduced.

The basic building blocks of the processor subsystem are transistors. Depending on their operation regions (namely, cut-off, linear, and saturation), transistors are classified into

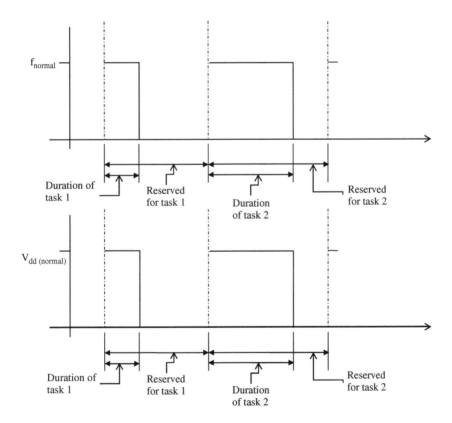

Figure 8.8   A processor subsystem operating at its peak performance.

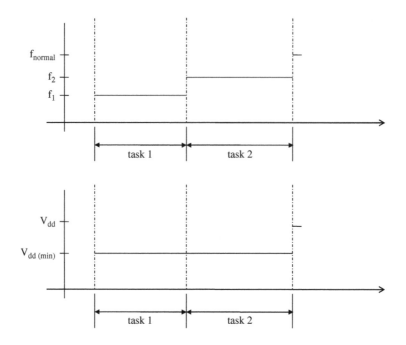

Figure 8.9 Application of dynamic voltage and frequency scaling.

analog and digital (switching) transistors. An analog transistor (amplifier) operates in the linear amplification region and there is a linear relationship between the input and output of the transistor. This is expressed as:

$$v_{\text{out}} = \frac{A}{1 - AB} v_{\text{in}} \tag{8.10}$$

where A is the open loop gain of the amplifier and B is a term that determines the portion of the output that should be fed back to the input in order to stabilize the amplifier.

A switching transistor, on the contrary, operates in either the cutoff or the saturation region, making the relationship between the input and the output voltage nonlinear. That is how the *zeros* and *ones* of a digital system are generated, represented or processed. The transition duration from the cutoff to the saturation region determines how good a transistor is as a switching element. In an ideal switching transistor, the transition takes place in no time. In practical transistors, however, the duration is greater than zero. The quality of the processor depends on, among other things, the switching time.

The switching in turn depends on many factors, one of them being the cumulative capacitance effect created in turn between the three joints of the transistors. Figure 8.10 displays a typical NAND gate made up of CMOS transistors.

Recall that a capacitor is created by two conductors that are separated by a dielectric material and there is a potential difference between the two conductors. The capacitance of a capacitor is proportional to the cross-sectional area of the conductors and inversely proportional to the separating distance.

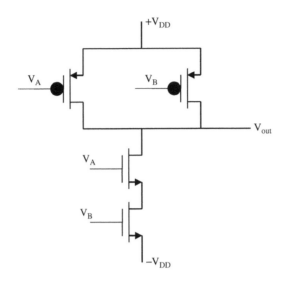

Figure 8.10   A schematic diagram of a NAND gate based on CMOS technology.

In a switching transistor, at a very high operating frequency, a capacitance is created at the contact points of the source, gate, and drain, affecting the transistor's switching response. The switching time can be approximated by the following equation:

$$t_{\text{delay}} = \frac{C_s \cdot V_{dd}}{I_{d_{\text{sat}}}} \tag{8.11}$$

where C_s is the source capacitance, V_{dd} is the biasing voltage of the drain, and $I_{d_{\text{sat}}}$ is the saturation drain current.

Switching costs energy and the magnitude of the energy depends on many factors, among which are the operating frequency and the biasing voltage. Sinha and Chandrakasan (2001) provide a first-order approximation that can be expressed as:

$$E(r) = C V_0 2 T_s f_{\text{ref}} r \left[\frac{V_t}{V_0} + \frac{r}{2} + \sqrt{ r \frac{V_t}{V_0} + \left(\frac{r}{2} \right)^2 } \right] \tag{8.12}$$

where, C is the average switching capacitance per cycle; T_s is the sampling period; f_{ref} is the operating frequency at V_{ref}; r is the normalized processing rate ($r = f/f_{\text{ref}}$); and $V_0 = (V_{\text{ref}} - V_t)^2 / V_{\text{ref}}$ with V_t being the threshold voltage.

From Equation (8.12), it can be deduced that reducing the operating frequency linearly reduces the energy cost, whereas reducing the biasing voltage reduces the energy cost quadratically. However, these two quantities cannot be reduced beyond a certain limit. For example, the minimum operating voltage for a CMOS logic to function properly was first derived by Swanson and Meindl (1972) and is expressed as:

$$V_{dd,\text{limit}} = 2 \cdot \frac{kT}{q} \cdot \left[1 + \frac{C_{fs}}{C_{ox} + C_d} \right] \cdot \ln \left(1 + \frac{C_d}{C_{ox}} \right) \tag{8.13}$$

where C_{fs} is the surface state capacitance per unit area; C_{ox} is the gate-oxide capacitance per unit area; and C_d is the channel depletion region capacitance per unit area. For a

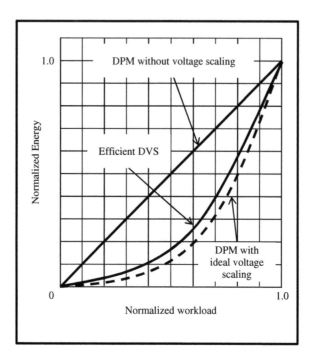

Figure 8.11   Application of dynamic voltage scaling based on workload estimation (Sinha and Chandrakasan 2001).

CMOS logic such as shown in Figure 8.10, Equation (8.13) yields, $V_{dd,\text{limit}} = 48\,\text{mV}$ at 300K. Finding the optimal voltage limit requires a tradeoff between the switching energy cost and the associated delay.

8.2.3   Task Scheduling

In a dynamic voltage and frequency scaling, the DPM strategy aims to autonomously determine the magnitude of the biasing voltage (V_{dd}) and the clock frequency of the processing subsystem. The decision for a particular voltage or frequency is based on several factors, including the application latency requirement and the task arrival rate. Ideally, these two parameters are adjusted so that a task is completed "just in time". This way, the processor does not remain idle and consume power unnecessarily. Practically, however, since the processor's workload cannot be known a priori, the estimation contains error and, as a result, idle cycles cannot be completely avoided. Comparison between an ideal and real dynamic voltage scaling strategies is shown in Figure 8.11.

8.3   Conceptual Architecture

A conceptual architecture for enabling a DPM strategy in a wireless sensor node should address three essential concerns:

1. In attempting to optimize power consumption, how much is the extra workload that should be produced by the DPM itself?

2. Should the DPM be a centralized or a distributed strategy?
3. If it is a centralized approach, which of the subcomponents should be responsible for the task?

A typical DPM strategy monitors the activities of each subsystem and makes decisions concerning the most suitable power configuration that optimizes the overall power consumption. This decision should reflect the application requirements, nevertheless. Since this process consumes a certain amount of power, it can be justified if the power that is saved as a result is significantly large. An accurate DPM strategy requires bench marking to estimate the task arrival and processing rate.

The decision whether a DPM strategy should be central or distributed depends on several factors. One advantage of a centralized approach is that it is easier to achieve a global view of the power consumption of a node and to implement a comprehensible adaptation strategy. On the other hand, a global strategy can add a computational overhead on the subsystem that does the management. A distributed approach scales well by authorizing individual subsystems to carry out local power management strategies. The problem with this approach is that local strategies may contradict with global goals. Given the relative simplicity of a wireless sensor node and the quantifiable tasks that should be processed, most existing power management strategies advocate a centralized solution.

In case of a centralized approach, the main question is which of the subsystems is responsible for handling the task – the processor subsystem or the power subsystem. Intuitively, the power subsystem should handle the management task, since it has complete information about the energy reserve of the node and the power budget of each subsystem. However, it requires vital information, such as the task arrival rate and priority of individual tasks, from the processing subsystems. Moreover, it needs to have some computational capability. Presently available power subsystems do not have these characteristics.

Most existing architectures for a wireless sensor node place the processor subsystem at the center and all the other subsystems communicate with each other through it. Moreover, the operating system (runtime environment) runs on the processing subsystem, managing, prioritizing, and scheduling tasks. Subsequently, the processing subsystem can have a more comprehensive knowledge about the activities of all the other subsystems, and these characteristics make the processing subsystem the appropriate place for executing a DPM.

8.3.1 Architectural Overview

Though the aim of a DPM strategy is to optimize the power consumption of a node, it should not affect the system's stability. Furthermore, the application requirements in terms of the quality of sensed data and latency should be satisfied. Fortunately, in most realistic situations, a WSN is deployed for a specific task. That task does not change, or changes only gradually. Therefore, the designer of a DPM has at his or her disposal the architecture of the wireless sensor node, the application requirements, and the network topology to devise a suitable strategy. The design space is illustrated in Figure 8.12.

The system's hardware architecture is the basis for defining multiple operational power modes and the possible transitions between them. A local power management strategy then defines rules to describe the behavior of the power mode transition according to a change in the activity of the node or based on a request from a global (network-wide) power management scheme, or from the application. This (see Figure 8.13) can be described as a circular

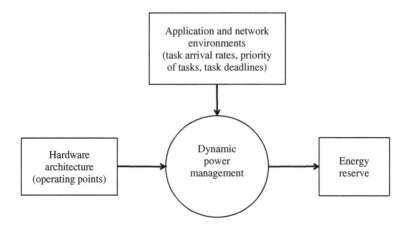

Figure 8.12   Factors affecting a dynamic power management strategy.

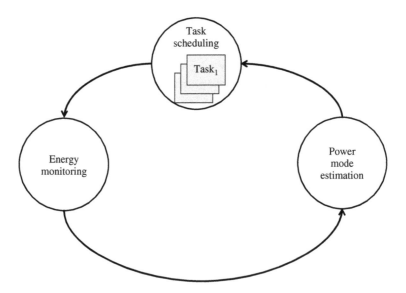

Figure 8.13   An abstract architecture for a dynamic power management strategy.

process consisting of three basic operations – energy monitoring, power mode estimation, and task scheduling.

Figure 8.13 illustrates how dynamic power management can be thought of as a machine that moves through different states in response to different types of events – tasks are scheduled in a task queue, and the execution time and energy consumption of the system are monitored. Depending on how fast the tasks are completed, a new power budget is estimated and transitions in power modes take place. In case of a deviation in the estimated power budget from the power mode that can be supported by the system, the DPM strategy decides the higher level of operating power mode.

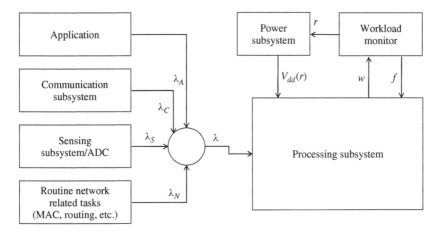

Figure 8.14   A conceptual architecture of a dynamic voltage scaling. (This architecture is the modified version of the one proposed by Sinha and Chandrakasan in (Sinha and Chandrakasan 2001)).

Figure 8.14 shows an implementation of the abstract architecture of Figure 8.13 to support dynamic voltage scaling. The processing subsystem receives tasks from the application, the communication subsystem, and the sensing subsystem. Additionally, it handles internal tasks pertaining to network management, such as managing a routing table and sleeping schedules. Each of these sources produces a task at a rate of λ_i. The overall task arrival rate, λ, is the summation of the individual tasks arrival rates, $\lambda = \sum \lambda_i$. The *workload monitor* observes λ for a duration of τ seconds and predicts the task arrival rate for the next β seconds. The estimated task arrival rate is represented by r in the figure. Based on the newly computed task arrival rate r, the processing subsystem estimates the supply voltage and the clock frequency it requires to process upcoming tasks.

Exercises

8.1 Give three reasons why dynamic power management is a crucial concern in wireless sensor networks.

8.2 What is the difference between local and global power management strategies? Give an example how a global power management can be realized at the link layer.

8.3 Give two examples for accidental causes of power consumption in wireless sensor networks.

8.4 How can a local power management strategy achieve an efficient power consumption in a wireless sensor node?

8.5 What is the main drawback of dynamic power management strategies that are based on a synchronous sleeping?

8.6 Explain the idea behind power management strategies that are based on an asynchronous sleeping.

8.7 Explain the six different operational modes of the ATmega128L microcontroller.

8.8 What is a refresh rate of an active memory?

8.9 Explain the following terms in the context of RAM timing:

   (a) RAS
   (b) CAS
   (c) t_{RCD}
   (d) t_{CL}

8.10 The RAM timing of a certain processor is configured as 2–3–2–6. Explain what it means.

8.11 Explain briefly how the following DC–DC converters function:

   (a) flyback
   (b) boost
   (c) buck

8.12 What is a rated current capacity?

8.13 Why do real batteries operate at a rate below the rated current capacity?

8.14 What is the side-effect of drawing current at a rate greater than the discharge rate?

8.15 Describe the components of a typical DC–DC converter.

8.16 Suppose the circuit shown in Figure 8.15 is used by a DC–DC converter. At what frequency is the voltage drop across the load resistor R_L maximum?

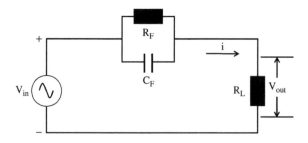

Figure 8.15   A conceptual architecture of a dynamic voltage scaling (Exercise 8.16).

8.17 Why does a transition from low power mode to high power mode cost some power in the following subsystems:

   (a) processor subsystem
   (b) communication subsystem

8.18 What conditions justify the power transition costs?

8.19 Why does the performance of a switching transistor deteriorate at high operation frequencies?

8.20 How does the cumulative capacitance affect the switching time of a CMOS transistor?

References

Chiasserini, C., and Rao, R. (2003) Improving energy saving in wireless systems by using dynamic power management. *IEEE Transactions on Wireless Communications* **2** (5), 1090–1100.

Jiang, X., Taneja, J., Ortiz, J., Tavakoli, A., Dutta, P., Jeong, J., Culler, D., Levis, P., and Shenker, S. (2007) An architecture for energy management in wireless sensor networks. *SIGBED Rev.* **4** (3), 31–36.

Sinha, A., and Chandrakasan, A. (2001) Dynamic power management in wireless sensor networks. *IEEE Des. Test* **18** (2), 62–74.

Swanson, R., and Meindl, J. (1972) Ion-implanted complementary MoS transistors in low-voltage circuits. *IEEE Journal of Solid State Circuits* **7** (2), 146–153.

9

Time Synchronization

In distributed systems, each node has its own clock and its own notion of time. However, a common time scale among sensor nodes is important to identify causal relationships between events in the physical world, to support the elimination of redundant sensor data, and to generally facilitate sensor network operation. Since each node in a sensor network operates independently and relies on its own clock, the clock readings of different sensor nodes will also differ. In addition to these random differences (phase shifts), the gap between clocks of different sensors will further increase due to the varying drift rates of oscillators. Therefore, time (or clock) synchronization is required to ensure that sensing times can be compared in a meaningful way. While time synchronization techniques for wired networks have received a significant amount of attention, these techniques are unsuitable for wireless sensors because of the unique challenges posed by wireless sensing environments. These challenges include the potentially large scale of wireless sensor networks, the necessity for self-configuration and robustness, the potential for sensor mobility, and the need for energy conservation (Sundararaman *et al.* 2005). This chapter introduces techniques for time synchronization that take these constraints and challenges into consideration.

9.1 Clocks and the Synchronization Problem

Computer clocks based on hardware oscillators are essential components of all computing devices. A typical clock consists of a quartz-stabilized oscillator and a counter that is decremented with every oscillation of the quartz crystal. Whenever the counter value reaches 0, it is reset to its original value and an interrupt is generated. Each interrupt, or *clock tick*, increments a software clock (another counter), which can be read and used by applications using a suitable application programming interface (API). Therefore, a software clock provides a *local time* for a sensor node, where $C(t)$ indicates the clock reading at some real time t. The time *resolution* is the distance between two increments (ticks) of the software clock.

Comparing the local times of two nodes, the *clock offset* indicates the difference between the times. Synchronization is required to adjust the time of one or both of these clocks such that their readings match. The *clock rate* indicates the frequency at which a clock progresses and the *clock skew* is the difference in the frequencies of two clocks. Perfect clocks have a clock rate $dC/dt = 1$ at all times, but various parameters affect

Fundamentals of Wireless Sensor Networks: Theory and Practice   Waltenegus Dargie and Christian Poellabauer
© 2010 John Wiley & Sons, Ltd

the actual clock rate, for example, the temperature and humidity of the environment, the supply voltage, and the age of the quartz. This deviation results in a *drift rate*, which expresses the rate by which two clocks can drift apart, that is, $dC/dt - 1$. The maximum drift rate of a clock is expressed as ρ with typical values for quartz-based clocks being 1 ppm to 100 ppm (1 ppm $= 10^{-6}$). This number is given by the manufacturer of the oscillator and guarantees that

$$1 - \rho \le \frac{dC}{dt} \le 1 + \rho \tag{9.1}$$

Figure 9.1 shows how the drift rate affects the clock reading with respect to real time, resulting in either a perfect, fast, or slow clock. This drift rate is responsible for inconsistencies in sensors' clock readings even after clocks have been synchronized, making it necessary to repeat the synchronization process periodically. Assuming identical clocks, any two clocks that are synchronized can drift from each other at a rate of at most $2\rho_{max}$. To limit the relative offset to δ seconds, the resynchronization interval τ_{sync} must meet the requirement:

$$\tau_{sync} \le \frac{\delta}{2\rho_{max}} \tag{9.2}$$

$C(t)$ must be piecewise continuous, that is, a strictly monotone function of time. Therefore, clock adjustments must be applied gradually, for example, using a linear compensation function that changes the slope of the local time. The consequences of simply having the clock jump forward or backward can be significant, for example, when a timer is set to trigger an interrupt at a certain time that may never occur on a clock that skips ticks due to the synchronization process.

We distinguish two types of synchronization: *external* and *internal*. External synchronization means that the clocks of all nodes are synchronized with an external source of time (or *reference clock*). The external reference clock is an accurate real-time standard such as Coordinated Universal Time (UTC). Internal synchronization means that the

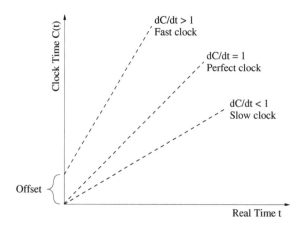

Figure 9.1   Relationship between local time $C(t)$ and real time t.

clocks of all nodes are synchronized with each other, without the support of an external reference clock. The goal of internal synchronization is to obtain a consistent view of time across all nodes in the network, even though this time may be different from any external reference times. External synchronization ensures both synchronization with an external source and consistency among all clocks within the network. When nodes are synchronized to an external reference clock, the *accuracy* of a clock describes the maximum offset of a clock with respect to the reference clock. When nodes in a network are internally synchronized, the *precision* indicates the maximum offset between any two clocks in the network (Kopetz 1997). Note that if two nodes are externally synchronized with an accuracy of Δ, they are also internally synchronized with a precision of 2Δ.

9.2   Time Synchronization in Wireless Sensor Networks

Time synchronization is a required service for many applications and services in distributed systems in general. Numerous protocols for time synchronization have been proposed for both wired and wireless systems, for example, the Network Time Protocol (NTP) (Mills 1991) is a widely deployed, scalable, robust, and self-configurable synchronization approach. Particularly in combination with the Global Positioning System (GPS), it has been shown to achieve accuracy in the order of a few microseconds. However, approaches such as NTP are not suitable for WSNs due to these networks' unique characteristics and constraints. This section describes a number of reasons time synchronization in WSNs is necessary and discusses challenges and constraints that must be met to achieve efficient and robust synchronization of clocks.

9.2.1   Reasons for Time Synchronization

Sensors in a WSN monitor objects in the physical world and report activities and events to interested observers. For example, proximity detecting sensors, such as magnetic, capacitive, or acoustic sensors, trigger an event when a moving object (e.g., a car) passes (see Figure 9.2). In dense sensor networks, multiple sensors will observe the same activity and trigger such events. Accurate temporal correlation of these events is crucial to answer questions such as *How many moving objects have been detected?*, *What is the direction of the moving object?*, and *What is the speed of the moving object?* As a consequence, it is important that an observer can establish the correct logical order of events; for example, when, in Figure 9.2, the real times have the ordering $t_1 < t_2 < t_3$, the sensor time stamps must reflect this order, that is, $C_1(t_1) < C_2(t_2) < C_3(t_3)$. Further, to accurately determine the velocity of the moving object, the time difference between sensor time stamps should correspond to the time difference of the real times, that is, $\Delta = C_2(t_2) - C_1(t_1) = t_2 - t_1$. This is an important requirement for *data fusion* in WSNs, which is concerned with the agglomeration of data coming from multiple sensors observing the same or related events. Further goals of data fusion include the suppression of duplicate sensor information, shorter response times to critical events, and reduction of resource requirements (e.g., energy consumption).

Time synchronization is also necessary for a variety of applications and algorithms in distributed systems in general, including communication protocols (e.g., at-most-once message delivery), security (e.g., to limit use of particular keys and to help to detect

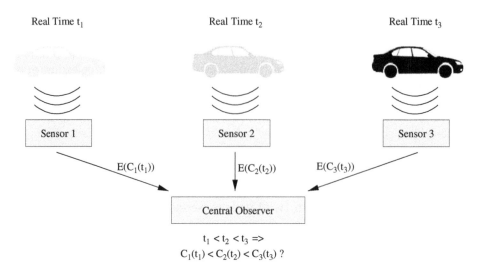

Real Time t_1 Real Time t_2 Real Time t_3

$t_1 < t_2 < t_3$ =>
$C_1(t_1) < C_2(t_2) < C_3(t_3)$?

Figure 9.2 Detection of speed and direction of moving objects using multiple sensors.

replayed messages in Kerberos-based authentication systems), data consistency (cache consistency and consistency of replicated data), and concurrency control (atomicity and mutual exclusion) (Liskov 1993).

Medium-access layer protocols such as time-division multiplexing (TDMA) allow multiple devices to share access to a common communication medium. Time is divided into slots that are allocated to wireless devices and each slot belongs to only one wireless device. The advantages of TDMA-based approaches are the predictability of medium access (every node is allowed to transmit data during one or more periodically recurring time slot) and the energy-efficiency enabled by this algorithm (a node can enter a power-saving sleep mode whenever it is not the sender or receiver of data during a slot). However, to implement TDMA, nodes must share a common view of time, that is, they need to be aware of the exact beginning and end of each slot.

With respect to energy, many WSNs rely on sleep/wake protocols that allow a network to selectively switch off sensor nodes or let them enter low-power sleep modes. Here, temporal coordination among sensors is essential for nodes to know when they can enter a sleep mode and when to reawake in order to ensure that neighboring nodes overlap in their wake periods to enable communication among them.

Finally, localization in WSNs is necessary to correctly position sensors or the objects they monitor. Many localization techniques (described in the next chapter) rely on ranging technologies to estimate distances between nodes and synchronization is required for time-of-flight measurements of radio or acoustic signals.

9.2.2 Challenges for Time Synchronization

Traditional time synchronization protocols have been designed for use in wired networks and do not consider the challenges inherit to low-cost low-power sensor nodes and the wireless medium. Similar to wired environments, time synchronization in WSNs is exposed to challenges such as clock glitches and varying clock drifts due to changes in temperature and humidity. However, time synchronization protocols for sensor networks must consider an array of additional challenges and constraints, some of which are discussed in this section.

9.2.2.1   Environmental Effects

Drift rates of clocks may differ with fluctuations in environmental temperature, pressure, and humidity. While typical wired computers are operated in rather stable environments (e.g., A/C-controlled cluster rooms or offices), wireless sensors are frequently placed outdoors and in harsh environments where these fluctuations in ambient properties are common. In controlled environments, oscillator frequency variations of up to 3 ppm (a deviation of 1 ppm amounts to an error of approximately 1 second every 12 days) due to room temperature changes have been reported (Mills 1998). For low-cost sensor nodes operating outdoors, these variations are likely to be much worse.

9.2.2.2   Energy Constraints

Wireless sensor nodes are typically driven by finite power sources, that is, either disposable or rechargeable (e.g., via solar panels) batteries. Battery replacement can add significantly to the cost of a WSN, particularly in large-scale networks and when the nodes are in difficult-to-service locations. Therefore, time synchronization protocols should not contribute significantly to the energy consumption of wireless nodes to ensure long battery life times. Since communication among sensor nodes is typically the basis for time synchronization, an energy-efficient synchronization protocol should aim for the minimum amount of the smallest possible messages necessary to obtain synchronized nodes.

9.2.2.3   Wireless Medium and Mobility

The wireless communication medium is known to be unpredictable and subject to fluctuations in performance due to changes in environmental properties caused by rain, fog, wind, and temperature (Otero *et al.* 2001). These fluctuations exacerbate the network throughput constraints, error rates, and wireless radio interferences experienced by wireless sensor nodes. Message exchanges between nodes can further be problematic when wireless links are asymmetric, that is, node A can receive node B's messages, while node A's messages are too weak to be correctly interpreted at node B. In general, the communication path from sensor node A to node B may differ significantly from the characteristics (delay) of the path from B to A, thereby resulting in asymmetric communication latencies. Further, communication interferences in wireless networks depend on the density of the network, the communication and interference ranges of wireless devices, and the level of activity of these devices. Numerous wireless sensors are mobile (e.g., mounted onto vehicles or carried by people), thereby causing significant and rapid changes in topology and connection quality. Finally, sensor nodes may fail or deplete their batteries, necessitating time synchronization that continues to remain functional even when network topology or density changes. In general, the consequence of these challenges is that time synchronization protocols must be designed for robustness and reconfigurability.

9.2.2.4 Additional Constraints

Besides energy limitations, low-power and low-cost sensor nodes are often constrained in their processor speeds and memory, further requiring that time synchronization protocols are lightweight. The small size and cost of sensor devices proscribe the use of large and expensive hardware to achieve synchronization (e.g., GPS receivers). Therefore, time synchronization protocols should be designed to operate in resource-constrained environments with little or no addition to the overall cost of a sensor device. Wireless sensor network deployments are often very large in scale and a synchronization protocol should scale well with increasing numbers of nodes or network density. Finally, different sensor applications will have differing requirements on clock accuracy or precision. For example, for applications such as object tracking, simple event and message ordering (without the help of external reference clocks) may suffice. However, the required precision may be in the range of a few microseconds. On the other hand, sensor networks that monitor foot traffic in public spaces during specific times of the day will require external synchronization, where a time accuracy in the range of seconds may be sufficient.

9.3 Basics of Time Synchronization

Synchronization is typically based on some sort of message exchange among sensor nodes. If the medium supports broadcast (as is the case in wireless systems), multiple devices can be synchronized simultaneously with a low number of messages. This section discusses the fundamental concepts behind most synchronization techniques.

9.3.1 Synchronization Messages

Most existing time synchronization protocols are based on *pairwise synchronization*, where two nodes synchronize their clocks using at least one synchronization message. *Network-wide synchronization* can be achieved by repeating this process among multiple node pairs until every node in a network has been able to adjust its clock.

9.3.1.1 One-Way Message Exchange

The simplest approach of pairwise synchronization occurs when only a single message is used to synchronize two nodes, that is, one node sends a time stamp to another node, illustrated in the left graph of Figure 9.3. Here, node i sends a synchronization message to node j at time t_1, embedding t_1 as *time stamp* into the message. Upon reception of this message, node j obtains a time stamp t_2 from its own local clock. The difference between the two time stamps is an indicator of the clock offset (between the clocks of nodes i and j) δ. More accurately, the difference between the two times is expressed as:

$$(t_2 - t_1) = D + \delta \tag{9.3}$$

where D is the unknown propagation time. Propagation times in the wireless medium are very small (a few microseconds) and are often ignored or assumed to be a certain constant value. Note that using this approach, node j is able to calculate an offset and adjust its clock to match the clock of node i.

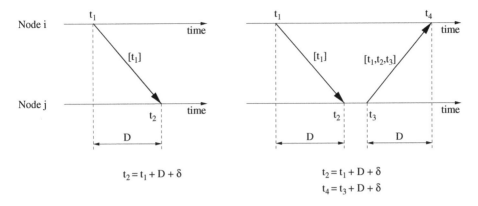

Figure 9.3 Concept of pairwise synchronization.

9.3.1.2 Two-Way Message Exchange

A somewhat more accurate approach is to use two synchronization messages as shown in the right graph of Figure 9.3. Here, node j responds with a message issued at time t_3, containing time stamps t_1, t_2, and t_3. Upon reception of this second message at time t_4, both nodes are able to determine the clock offset, again assuming a fixed value for the propagation delay. However, node i is now able to more accurately determine both the propagation delay and the offset as

$$D = \frac{(t_2 - t_1) + (t_4 - t_3)}{2} \tag{9.4}$$

$$\text{offset} = \frac{(t_2 - t_1) - (t_4 - t_3)}{2} \tag{9.5}$$

Note that this assumes that the propagation delay is identical in both directions and the clock drift does not change between measurements (which is feasible because of the brief time span). While only node i has sufficient information to determine the offset, node i can share the offset value with node j in a third message.

9.3.1.3 Receiver–Receiver Synchronization

A different approach is taken by protocols that apply the receiver–receiver synchronization principle, where synchronization is based on the time at which the same message arrives at each receiver. This is in contrast to the more traditional sender–receiver approach of most synchronization schemes. In broadcast environments, these receivers obtain the message at about the same time and then exchange their arrival times to compute an offset (i.e., the difference in reception times indicates the offset of their clocks). Figure 9.4 shows an example of this scheme. If there are two receivers, three messages are needed to synchronize both receivers. An example of such an approach is the RBS protocol discussed in Section 9.4.5. Note that the broadcast message does not carry a time stamp,

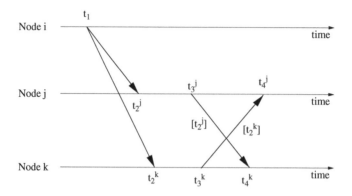

Figure 9.4   Receiver–receiver synchronization scheme.

instead the arrival times of the broadcast message at the different receivers is used to synchronize the receivers to each other.

9.3.2   Nondeterminism of Communication Latency

The nondeterminism of the communication latency significantly contributes to the precision that can be achieved. In general, this latency experienced by synchronization messages is the sum of several components (Kopetz and Ochsenreiter 1987), as illustrated in Figure 9.5:

1. *Send delay:* This is the time spent by the sender to generate the synchronization message and pass the message to the network interface. This includes delays caused by operating system behavior (system call interface, context switches), the network protocol stack, and the network device driver.
2. *Access delay:* This is the time spent by the sender to access the physical channel and is mostly determined by the medium access control (MAC) protocol in use. Contention-based protocols such as IEEE 802.11's CSMA/CA must wait for an idle channel before

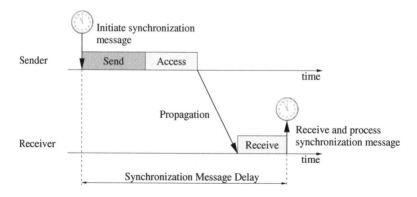

Figure 9.5   End-to-end delay experienced by a synchronization message.

access is allowed. When multiple devices access the channel at the same time, collisions occur that cause further delays (e.g., through the exponential backoff mechanism used in many MAC protocols). More predictable delays are experienced by protocols based on time-division (TDMA), where a device must wait for its periodic slot before transmission can occur.

3. *Propagation delay:* The actual time needed for the message to travel from the sender to the receiver is called propagation delay. When the nodes share the same physical medium, propagation delays are very small and are often negligible in critical path analysis.

4. *Receive delay:* This is the time spent by the receiver device to receive the message from the medium, to process the message, and to notify the host of its arrival. Host notification typically occurs via interrupts, at which the local time (i.e., the message arrival time) can be read. As a consequence, the receive time tends to be much smaller than the send time.

Many synchronization schemes for WSNs apply low-level techniques aimed at reducing the amount or variation of some of these components. For example, MAC-layer time stamping can reduce the send and receive delays on the sender and receiver, respectively.

9.4   Time Synchronization Protocols

Numerous time synchronization protocols for WSNs have been developed, where most of them are based on some variations of the message exchange concepts described in the previous section. This section provides an overview of some representative schemes and protocols.

9.4.1   Reference Broadcasts Using Global Sources of Time

The Global Positioning System (GPS) continuously broadcasts time measured from an epoch started at 0h 6 January, 1980 UTC. However, unlike UTC, GPS is not perturbed by leap seconds and is therefore ahead of UTC by an integer number of seconds (15 seconds as of 2009). Even inexpensive GPS receivers can receive GPS time with a precision of 200 ns (Dana 1997; Mannermaa *et al.* 1999). Time signals are also being transmitted by terrestrial radio stations, for example, the National Institute of Standards and Technology uses radio stations WWV/WWVH and WWVB (Lichtenecker 1997) to continuously broadcast time based on atomic clocks. However, such approaches exhibit a number of challenges that prohibit their use for many WSNs. For example, GPS is not ubiquitously available (underwater, indoors, under dense foliage, during Mars exploration), requires a relatively high-power receiver which may not be feasible for tiny low-cost sensor nodes, and may be too large and costly to be added to small sensor nodes. However, many sensor networks are hierarchical systems consisting of low-power sensor devices, but also more powerful devices that often serve as gateways or cluster heads. These devices may be able to support GPS or radio receivers, turning these nodes into master clocks that can be used to synchronize the rest of the network with any of the other sender–receiver approaches described in this section.

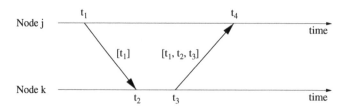

Figure 9.6   Pairwise synchronization with LTS.

9.4.2 Lightweight Tree-Based Synchronization

The primary goal of the Lightweight Tree-Based Synchronization (LTS) protocol (Van Greunen and Rabaey 2003) is to provide a specified precision (instead of a maximum precision) with as little overhead as possible. LTS can be used with different algorithms for both centralized and decentralized multi-hop synchronization. To understand the approach taken by LTS, let us first consider the message exchange for the synchronization of a pair of nodes. Figure 9.6 shows a graphical depiction of this scheme. First, node j transmits a synchronization message time-stamped with the transmission time t_1 to node k. Upon arrival of this message at node k at time t_2, node k responds with a message carrying a time stamp t_3 and the previously recorded times t_1 and t_2. This message is received by node j at time t_4. Note that times t_1 and t_4 are based on node j's clock, whereas times t_2 and t_3 are recorded using the clock of node k. Assuming a transmission delay D (which is further assumed to be the same in both directions) and an unknown clock offset between the clocks of nodes j and k, time t_2 of node k is equal to $t_1 + D +$ offset. Similarly, t_4 is then equal to $t_3 + D -$ offset. The offset can then be calculated as:

$$\text{offset} = \frac{t_2 - t_4 - t_1 + t_3}{2} \tag{9.6}$$

The centralized multi-hop version of LTS is based on a single reference node that is the root of a spanning tree comprising all nodes of the network. In order to maximize the synchronization accuracy, the depth of the tree should be minimized. This is due to the fact that the errors resulting from the pairwise synchronizations are additive and therefore increase along the branches of the tree as a function of the number of hops. In LTS, a tree construction algorithm such as breadth first search is executed each time the synchronization algorithm is executed. Once the tree has been established, the reference node initiates the synchronization by performing the pairwise synchronization with each of its children. Once synchronized, each child repeats this step with its own children until all nodes of the tree have been synchronized. Pairwise synchronization has a fixed overhead of 3 messages, therefore if a tree has n edges, the total message overhead is $3n - 3$.

The distributed multi-hop version of LTS does not require the construction of a spanning tree and the synchronization responsibility is moved from the reference node to the sensor nodes themselves. This version assumes the presence of one or more reference nodes, which are contacted by a sensor node whenever the sensor node requires synchronization. The decentralized approach allows nodes to determine their own desired resynchronization period. That is, nodes determine their resynchronization period based on their desired clock accuracy, the distance (in number of hops) from the nearest reference node, their clock drift ρ, and the time of their last synchronization. Finally, to eliminate potential inefficiencies, the distributed version of LTS strives to eliminate duplicate requests of neighboring nodes. Toward this end, a node can query its neighbors for pending synchronization requests and, if there are any, the node synchronizes with the one-hop neighbor instead of the reference node.

9.4.3 Timing-sync Protocol for Sensor Networks

The Timing-sync Protocol for Sensor Networks (TPSN) (Ganeriwal *et al.* 2003) is another traditional *sender–receiver synchronization* approach that uses a tree to organize a network. TPSN uses two phases for synchronization: the *level discovery phase* (executed during network deployment) and the *synchronization phase*.

9.4.3.1 Level Discovery Phase

The goal of this phase is to create a hierarchical topology of the network, where each node is assigned a level, with the root node (e.g., a GPS-equipped gateway to the external world) residing on level 0. The root node initiates this phase by broadcasting a `level_discovery` message that contains the level and the unique identity of the sender. Every immediate neighbor of the root node uses this message to identify its own level (i.e., level 1) and rebroadcasts the `level_discovery` message with its own identity and level. This process is repeated until every node in the network has identified its level. When a node receives multiple broadcasts from its neighbors, it simply discards them once it has established its level in the hierarchical structure. Situations may occur where nodes do not have an assigned level, for example, when MAC-layer collisions prevent a node from receiving a `level_discovery` message or when a node joins a network that has already concluded its level discovery phase. In this case, a node can issue a `level_request` message to its neighbors who reply with their assigned levels. Then, the node assigns itself a level that is one greater than the smallest level received from its neighbors. Node failures can be handled in the same way, that is, when a node at level i realizes that it does not have any neighbors at level $i - 1$ (through the communication steps in the synchronization phase described next), it also issues a `level_request` message to reinsert itself into the structure. Finally, if the root node dies, instead of issuing `level_request` messages, nodes in level 1 execute a leader election algorithm, which then restarts TPSN by beginning a new level discovery phase.

9.4.3.2 Synchronization Phase

During the synchronization phase, TPSN employs pairwise synchronization along the edges of the hierarchical structure established in the previous phase, that is, each i level node synchronizes its clock with nodes on level $i - 1$. The pairwise synchronization of TPSN shows similarity to the approach taken by LTS. A node j issues a *synchronization pulse* at time t_1, containing the node's level and a time stamp. This message is received by node k at time t_2 and node k responds with an acknowledgment at time t_3 (containing time stamps t_1, t_2, t_3, and node k's level). Finally, this packet is received by node j at time t_4. As with LTS, TPSN assumes that the propagation delay D and the clock offset do not change during the brief span of time. Since t_1 and t_4 are measured using node j's clock and t_2 and t_3 are measured using node k's clock, these times have the following relationships: $t_2 = t_1 + D + \text{offset}$ and $t_4 = t_3 + D - \text{offset}$. Based on these parameters, node j can calculate both the drift and propagation delay as:

$$D = \frac{(t_2 - t_1) + (t_4 - t_3)}{2} \tag{9.7}$$

$$\text{offset} = \frac{(t_2 - t_1) - (t_4 - t_3)}{2} \tag{9.8}$$

The synchronization phase is initiated by the root node issuing a `time_sync` packet. After waiting for some random time (to reduce contention during medium access), nodes in level 1 initiate the two-way message exchange with the root node. Once a node in level 1 receives an acknowledgment from the root, it computes its offset and adjusts its clock. Nodes on level 2 will overhear the synchronization pulses issued by their level 1 neighbors and after a certain backoff time they initiate their pairwise synchronization with nodes in level 1. The backoff time is necessary to give level 1 nodes time to receive and process the acknowledgment of their own synchronization pulses. This process is continued throughout the hierarchical structure until all nodes have synchronized to the root node.

Similar to LTS, the synchronization error of TPSN depends on the depth of the hierarchical structure and the end-to-end latencies experienced by messages during the pairwise synchronization. To minimize these latencies and to reduce the error, TPSN relies on time-stamping of packets at the MAC layer.

9.4.4 Flooding Time Synchronization Protocol

The goals of the Flooding Time Synchronization Protocol (FTSP) (Maróti *et al.* 2004) are to achieve network-wide synchronization with errors in the microsecond range, scalability up to hundreds of nodes, and robustness to changes in network topology including link and node failures. FTSP differs from other solutions in that it uses a single broadcast to establish synchronization points between sender and receivers while eliminating most sources of synchronization error. Toward this end, FTSP expands on the delay analysis described in Section 9.3 and decomposes the end-to-end delay into the components shown in Figure 9.7. In this analysis, the wireless radio of the sensor node informs the CPU using an interrupt at time t_1 that it is ready to receive the next piece of the message to be transmitted. After the *interrupt handling time* d_1, the CPU generates a time stamp at time t_2. The time needed by the radio to encode and transform the piece of the message into electromagnetic waves is described as *encoding time* d_2 (between t_1 and t_3). The propagation delay (between t_3 on node j's clock and t_4 on node k's clock) is followed by the *decoding time* d_4 (between t_4 and t_5). This is the time the radio requires to decode the message from electromagnetic waves back into binary data. The *byte alignment time* d_5 is a delay caused by the different byte alignments (*bit offsets*) of nodes j and k, that is, the receiving radio has to determine the offset from a known synchronization byte and then

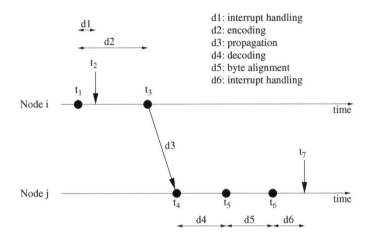

Figure 9.7   End-to-end delay of synchronization message.

shift the incoming message accordingly. Finally, the radio on node k issues an interrupt at time t_6, which allows the CPU to obtain a final time stamp at time t_7.

The impact of these delays on the overall end-to-end delay varies significantly, for example, the propagation delay (d_3) is typically small ($< 1 \mu s$) and deterministic. Similarly, the encoding and decoding times (d_2 and d_4) are also deterministic and in the low hundreds of microseconds. The byte alignment delay (d_5) depends on the bit offset and also reaches several hundreds of microseconds. Finally, the interrupt handling time (d_1 and d_6) is nondeterministic and is typically a few microseconds.

9.4.4.1   Time-Stamping in FTSP

In FTSP, a sender synchronizes one or more receivers with a single radio broadcast, where the broadcast message contains the sender's time stamp (which is the estimated global time at the transmission of a given byte of the message). Upon arrival, a receiver extracts the time stamp from the message and time stamps the arrival using its own local clock. The global–local time pair provides a *synchronization point*. The sender's time stamp must be embedded into the currently transmitted message, therefore the time stamping must occur before the bytes containing the time stamp are transmitted over the medium. In FTSP, the synchronization message begins with a number of preamble bytes followed by several SYNC bytes, a data field, and a cyclic redundancy check (CRC) for error detection (Figure 9.8). The preamble bytes are used to synchronize the receiver radio to the carrier frequency and the SYNC bytes are used to calculate the bit offset, which is needed to correctly reassemble the message. FTSP uses multiple time stamps at both the sender and the receiver to reduce the jitter of interrupt handling and encoding/decoding times. These time stamps are recorded at each byte boundary after the SYNC bytes as they are transmitted or received. The time stamps are normalized by subtracting an appropriate integer multiple of the nominal byte transmission time (e.g., approximately $417 \mu s$ on Mica2 platforms). The jitter caused by the interrupt handling time can be removed by taking the minimum of these normalized time stamps. Further, the jitter caused by the

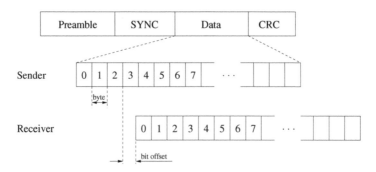

Figure 9.8   Synchronization message format and bit offset between sender and receiver.

encoding and decoding steps can be reduced by averaging these corrected normalized time stamps. Only the final (error-corrected) time stamp is added into the data part of the message. At the receiver side, the time stamp must be further corrected by the byte alignment time (which can be determined from the transmission speed and the bit offset).

9.4.4.2   Multi-Hop Synchronization

Similar to TPSN, FTSP relies on an elected synchronization root to synchronize the network, where root election is based on unique node IDs (i.e., the node with the lowest ID is elected as the root node). The root node maintains the global time and all other nodes in the network synchronize their clocks to that of the root. Synchronization is triggered through a broadcast message by the root node containing its time stamp. All nodes within the communication range of the root can establish synchronization points directly from the broadcast message. Other nodes collect synchronization points from broadcasts of synchronized nodes that are closer to the root.

   Similar to TPSN, FTSP relies on a root election algorithm to ensure that there is exactly one synchronization root in the network. Every broadcast message contains the unique ID of the root (rootID) and a sequence number (besides the already discussed time stamp). Whenever a node does not receive a synchronization message for a certain amount of time, it declares itself to be the new root. Whenever a node receives a synchronization message with a rootID lower than its own ID, it gives up its root status. A new node joining a network with a lower ID than the rootID will not immediately declare itself as root, but instead wait for a certain period of time to collect synchronization messages and adjust its own clock to the current global time. These techniques ensure that TPSN can handle network topology changes, including mobile nodes.

9.4.5   Reference-Broadcast Synchronization

The Reference-Broadcast Synchronization (RBS) protocol (Elson *et al.* 2002) relies on broadcast messages among a set of receivers to synchronize them with each other. In the wireless medium, broadcast messages will arrive at multiple receivers at approximately the same time. The variability in message delay will be dominated by the propagation delays and the time needed by the receivers to receive and process the incoming broadcast message. The strength of RBS lies in the removal of nondeterministic synchronization errors caused by the sender. Since all synchronization methods are based on some form of message exchange, the nondeterministic delays experienced by these messages limit the granularity of time synchronization that can be obtained. Figure 9.9 compares the critical paths of traditional synchronization protocols with RBS (Elson *et al.* 2002). Exploiting the broadcast nature of the wireless medium, the send delay and access delay of broadcast messages are identical for both receivers, that is, their actual message arrival times will differ only due to variations of the propagation and receive delays. As a consequence, the *RBS critical path* is much shorter than the critical path of traditional synchronization techniques.

For example, in a scenario with two receivers, each receiver will record when a beacon was received (using their local clocks). Next, the two receivers exchange their recorded information, allowing them to calculate an offset (i.e., the difference of the local beacon arrival times). With more than two receivers, the maximum phase error between all receiver pairs is expressed as *group dispersion*. Increasing the number of receivers increases the likelihood that at least one receiver will be poorly synchronized, leading to larger group dispersion. On the other hand, increasing the number of reference broadcasts can decrease the group dispersion. The reason for this is that a receiving node may experience variations in the receive time of messages and using multiple reference broadcasts can increase the synchronization precision. That is, a receiver j can compute its offset to any other receiver i as the average of phase offsets for all m packets received by receivers i and j:

$$\text{offset}[i, j] = \frac{1}{m} \sum_{k=1}^{m} (T_{j,k} - T_{i,k}) \tag{9.9}$$

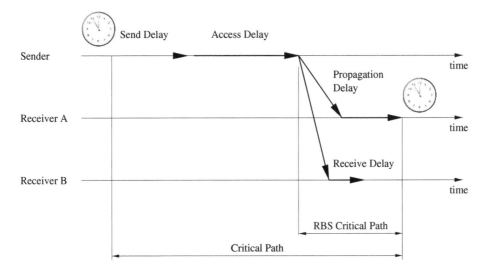

Figure 9.9   Critical path analysis for synchronization message exchanges.

RBS can be extended to multi-hop scenarios by establishing multiple reference beacons, each with its own broadcast domain. These domains can overlap and nodes within overlapping regions serve as *bridges* to allow synchronization across domains. For example, if nodes A and B are in range of reference node C and nodes C and D are in range of reference E, C is the bridge node between the two broadcast domains.

The extensive amount of message exchanges needed to synchronize sensor nodes appears to make RBS a costly synchronization technique. However, RBS is a candidate protocol for a synchronization scheme called *post-facto synchronization* (Elson and Estrin 2001). Here, nodes do not synchronize with each other until an event of interest happens. If synchronization occurs quickly after such an event occurs, sensor nodes can reconcile their clocks only when required, thereby preventing them from wasting energy on unnecessary synchronization messages.

9.4.6 Time-Diffusion Synchronization Protocol

The Time-Diffusion Synchronization (TDP) protocol (Su and Akyildiz 2005) allows a sensor network to reach an *equilibrium time*, that is, nodes agree on a network-wide time and maintain their clocks within a small bounded deviation from this equilibrium. Nodes in the network dynamically structure themselves in a tree-like configuration using two types of elected roles: *master nodes* and *diffused leader nodes*. TDP's *Time Diffusion Procedure* (TP) is responsible for diffusing timing information messages from master nodes to their neighboring nodes, some of which become diffused leader nodes responsible for further propagating the master nodes' messages. TDP distinguishes between two phases of operation: during the *active* phase, master nodes are elected every τ seconds (based on an Election/Reelection Procedure or ERP) such that the workload in the network is balanced and the network is able to agree on an equilibrium time. Every active phase is followed by an *inactive* phase where no time synchronization takes place. Every interval of τ seconds is further divided into intervals of δ seconds, each beginning with the election of diffused leader nodes. The ERP eliminates leaf nodes and nodes whose clocks deviate from their neighboring clocks by more than a certain threshold value. This is achieved through message exchanges of neighboring nodes, allowing them to compare their clock readings. Further, the ERP ensures that master node and diffused leader node election considers the energy status of the sensor nodes.

Figure 9.10 illustrates the concept of synchronization with TDP. First, an elected master node broadcasts a timing information message to its neighbors. All diffused leader nodes

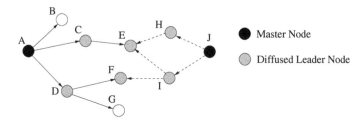

Figure 9.10 Concept of synchronization with TDP (with $n = 2$ for both masters).

receiving this message (in the figure, nodes C and D are diffused leader nodes for node A) respond with an ACK message, allowing the master node to determine a round-trip delay Δ_j for each neighbor j, an estimated one-way delay for all neighbors ($\Delta/2$ where Δ is the average of all round-trip delays), and the standard deviation of the round-trip delays. The standard deviation is sent in another time-stamped message from the master node to each neighboring diffused leader node. A diffused leader node adjusts its clock using the time stamp, the one-way delay estimation, and the standard deviation, and then repeats this diffusion process with its neighbors. The process is continued for n times, where n is the distance from the master node in terms of hops (e.g., in Figure 9.10, $n = 2$). Nodes that receive timing information messages from multiple master nodes use the standard deviations as weighted ratio of their time contribution to the adjusted time.

9.4.7 Mini-Sync and Tiny-Sync

Two closely related protocols, called Mini-sync and Tiny-sync, provide pairwise synchronization (that can be used as basic building blocks to synchronize an entire sensor network) with low bandwidth, storage, and processing requirements (Yoon *et al.* 2007). The relationship of the clocks of two nodes in a sensor network can be expressed as:

$$C_1(t) = a_{12}C_2(t) + b_{12} \tag{9.10}$$

where a_{12} expresses the relative drift and b_{12} the relative offset between the clocks of nodes 1 and 2. In order to determine this relationship, nodes can use the two-way messaging scheme described in Section 9.3.1, for example, node 1 sends a time-stamped probe message at time t_0 to node 2 and node 2 responds immediately with a time-stamped reply message at time t_1. Node 1 records the arrival time of the second message (t_2) to obtain a 3-tuple of time stamps (t_0, t_1, t_2), which is called a *data point*. Since t_0 happened before t_1 and t_1 happened before t_2, the following inequalities should hold:

$$t_0 < a_{12}t_1 + b_{12} \tag{9.11}$$

$$t_2 > a_{12}t_1 + b_{12} \tag{9.12}$$

This procedure is repeated multiple times, resulting in a series of data points and new constraints on the admissible values of a_{12} and b_{12} (thereby increasing the precision of the algorithms).

The two versions of the protocol are based on the observation that not all data points are useful. Every data point results in two constraints for the relative drift and offset. The Tiny-sync algorithm maintains only four of these constraints, that is, whenever a new data point has been obtained, the current four and the two new constraints are compared and only the four constraints that result in the best estimates of offset and drift are kept. A downside of this approach is that constraints may be eliminated that may provide better estimates if combined with other data points that have yet to occur. Therefore, the Mini-sync protocol only discards a data point if it is certain that this data point will be useless. This results in larger computational and storage costs compared to Tiny-sync, but the advantage is an increased precision.

Exercises

9.1 Why is time synchronization needed in a WSN? Name at least three concrete examples.

9.2 Explain the difference between external and internal time synchronization and name at least one concrete example for each type of synchronization.

9.3 Consider two nodes, where the current time at node A is 1100 and the current time at node B is 1000. Node A's clock progresses by 1.01 time units once every 1 s and node B's clock progresses by 0.99 time units once every 1 s. Explain the terms clock offset, clock rate, and clock skew using this concrete example. Are these clocks fast or slow, and why?

9.4 Assume that two nodes have a maximum drift rate from the real time of 100 ppm each. Your goal is to synchronize their clocks such that their relative offset does not exceed 1 s. What is the necessary resynchronization interval?

9.5 You need to design a wireless sensor node and you have three choices for clocks with maximum drift rates of $\rho_1 = 1$ ppm, $\rho_2 = 10$ ppm, and $\rho_3 = 100$ ppm. Clock 1 costs significantly more than clock 2, which in turn costs significantly more than clock 3. Explain why one would choose clock 1 instead of clock 2 or clock 3 and vice versa.

9.6 A network of five nodes is synchronized to an external reference time with maximum errors of 1, 3, 4, 1, and 2 time units, respectively. What is the precision that can be obtained in this network?

9.7 Node A sends a synchronization request to node B at 3150 (on node A's clock). At 3250, node A receives the reply from node B with a time stamp of 3120.

(a) What is node A's clock offset with respect to the time at node B (you can ignore any processing delays at either node)?

(b) Is node A's clock going too slow or too fast?

(c) How should node A adjust the clock?

9.8 Node A issues a synchronization request simultaneously to nodes B, C, and D (Figure 9.11). Assume that nodes B, C, and D are all perfectly synchronized to each other. Explain why the offsets between node A and the three other nodes may still differ.

9.9 Describe the reasons for nondeterminism of communication latencies and why this nondeterminism affects time synchronization.

9.10 Explain why the depth of the synchronization tree in centralized LTS should be small.

9.11 Discuss the differences and similarities in the design of the TPSN and the LTS synchronization protocols.

9.12 Explain the six different types of time stamps that characterize the communication in FTSP. How does FTSP remove the jitter of the interrupt handling and the encoding/decoding times?

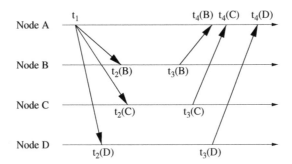

Figure 9.11   Pair-wise synchronization with multiple neighboring nodes (Exercise 9.8).

9.13 Explain the concept behind the RBS protocol. How can RBS be extended to work in multi-hop scenarios?

9.14 Describe the term "post-facto synchronization".

9.15 Compare the TPSN and RBS time synchronization protocols.

9.16 Compare the broadcast approach used by RBS with the pair-wise synchronization approach by TPSN and other protocols for the following scenarios:

(a) synchronization messages experience send and access delays with high variance and all other delays are negligible;

(b) synchronization messages are sent using acoustic signals and the distances between nodes are unknown;

(c) synchronization messages experience send and access delays without variance and all other delays are negligible;

(d) synchronization messages experience significant receive delays that may differ from node to node.

9.17 Two nodes A and B use RBS to receive periodic acoustic synchronization signals from a reference node. Node A's clock shows 10 s when it receives the last synchronization beacon, while node B's clock shows 15 s. Node A detects an event at time 15 s, while node B detects the same event at time 19.5 s. Assume that node A is 100 m away from the synchronization source and node B is 400 m away from the synchronization source. Which node detected the event sooner and by how much? Assume a signal speed of 300 m/s.

References

Dana, P.H. (1997) Global Positioning System (GPS) time dissemination for real-time applications. *Real-Time Systems* **12** (1), 9–40.

Elson, J., and Estrin, D. (2001) Time synchronization for wireless sensor networks. *Proc. of the 15th International Parallel and Distributed Processing Symposium (IPDPS)*.

Elson, J., Girod, L., and Estrin, D. (2002) Fine-grained network time synchronization using reference broadcasts. *Proc. of the 5th Symposium on Operating Systems, Design, and Implementation*.

Ganeriwal, S., Kumar, R., and Srivastava, M. B. (2003) Timing-sync protocol for sensor networks. *Proc. of the 1st International Conference on Embedded Networked Sensor Systems*.

Kopetz, H. (1997) *Real-Time Systems: Design Principles for Distributed Embedded Applications*. The International Series in Engineering and Computer Science: Springer.

Kopetz, J., and Ochsenreiter, W. (1987) Clock synchronization in distributed real-time systems. *IEEE Transactions on Computers* **36** (8), 933–939.

Lichtenecker, R. (1997) Terrestrial time signal dissemination. *Real-Time Systems* **12** (1), 41–61.

Liskov, B. (1993) Practical uses of synchronized clocks in distributed systems. *Distributed Computing* **6** (4), 211–219.

Mannermaa, J., Kalliomäki, K., Mansten, T., and Turunen, S. (1999) Timing performance of various GPS receivers. *Proc. of the 1999 Joint Meeting of the European Frequency and Time Forum and the IEEE International Frequency Control Symposium*.

Maróti, M., Kusy, B., Simon, G., and Lédeczi, A. (2004) The flooding time synchronization protocol. *Proc. of the 2nd International Conference on Embedded Networked Sensor Systems*.

Mills, D.L. (1991) Internet time synchronization: The network time protocol. *IEEE Transactions on Communications* **39** (10), 1482–1493.

Mills, D.L. (1998) Adaptive hybrid clock discipline algorithm for the network time protocol. *IEEE/ACM Transactions on Networking (TON)* **6** (5), 505–514.

Otero, J., Yalamanchili, P., and Braun, H.W. (2001) *High performance wireless networking and weather*. White Paper, University of California at San Diego. Available online (6 pages).

Su, W., and Akyildiz, I.F. (2005) Time-diffusion synchronization protocol for wireless sensor networks. *IEEE/ACM Transactions on Networking (TON)* **13** (2), 384–397.

Sundararaman, B., Buy, U., and Kshemkalyani, A.D. (2005) Clock synchronization for wireless sensor networks: A survey. *Ad Hoc Networks* **3** (3), 281–323.

Van Greunen, J., and Rabaey, J. (2003) Lightweight time synchronization for sensor networks. *Proc. of the International Workshop on Wireless Sensor Networks and Applications*.

Yoon, S., Veerarittiphan, C., and Sichitiu, M.L. (2007) Tiny-sync: Tight time synchronization for wireless sensor networks. *ACM Transactions on Sensor Networks (TOSN)*.

10

Localization

Sensors monitor phenomena in the physical world and the spatial relationships between them and the objects and events of the physical world are an essential component of the sensor information. Without knowing the position of a sensor node, its information will only tell part of the story. For example, sensors deployed in a forest to raise alarms whenever wildfires occur gain significantly in value if they are able to report the spatial relationship between them and the monitored event. Further, accurate location information is needed for various tasks such as routing based on geographic information, object tracking, and location-aware services. Localization is the task of determining the physical coordinates of a sensor node (or a group of sensor nodes) or the spatial relationships among objects. It comprises a set of techniques and mechanisms that allow a sensor to estimate its own location based on information gathered from the sensor's environment. While the Global Positioning System (GPS) is undoubtedly the most well-known location-sensing system, it is not accessible in all environments (e.g., indoors or under dense foliage) and may incur resource costs unacceptable for resource-constrained wireless sensor networks (WSNs). Therefore, this chapter discusses various techniques and case studies for localization and location services targeted at WSNs.

10.1 Overview

Wireless sensor networks are often deployed in an ad hoc fashion, that is, their location is not known a priori. Localization is necessary to provide a physical context to sensor readings, for example, in many applications such as environmental monitoring, sensor readings without knowledge of the location where the readings were obtained are meaningless. Location information is further necessary for services such as intrusion detection, inventory and supply chain management, and surveillance. Finally, localization is fundamental for sensor network services that rely on the knowledge of sensor positions, including geographic routing (Stojmenovic 2002) and coverage area management (Siqueira *et al.* 2007).

The location of a sensor node can be expressed as a *global* or *relative* metric. A global metric is used to position nodes within a general global reference frame, for example, as provided by the GPS (longitudes and latitudes) and the Universal Transverse Mercator (UTM) coordinate systems (zones and latitude bands). In contrast, relative metrics are based on arbitrary coordinate systems and reference frames, for example, a sensor's location expressed as distances to other sensors without any relationship to global coordinates. Two important qualities of localization information are the *accuracy* and *precision* of a position. For

Fundamentals of Wireless Sensor Networks: Theory and Practice   Waltenegus Dargie and Christian Poellabauer
© 2010 John Wiley & Sons, Ltd

example, for a GPS sensor indicating a position that is true within 10 m for 90% of all measurements, the accuracy of the GPS reading is 10 m (how close is the reading to the ground truth?) and the precision is 90% (how consistent are the readings?). Apart from these *physical* positions discussed so far, many applications (e.g., indoor tracking systems) may only require *symbolic* locations (Hightower and Borriello 2001) such as "office 354", "mile marker 17 on Highway 23", or "bathroom".

While it may be infeasible for all sensor nodes in a WSN to have knowledge of their global coordinates, many sensor networks rely on a subset of nodes that know their global positions. These anchor nodes are then used by all other nodes to perform localization. Techniques that rely on such anchors are called anchor-based localization (as opposed to anchor-free localization). A large number of localization techniques (including many anchor-based approaches) are based on range measurements, that is, estimations of distances between several sensor nodes. These techniques, called range-based localization techniques, require sensors to monitor measurable characteristics such as received signal strengths of wireless communications or time difference of arrival of ultrasound pulses. The following sections discuss the basics of different localization techniques based on these concepts.

10.2 Ranging Techniques

The foundation of numerous localization techniques is the estimation of the physical distance between two sensor nodes. Estimates are obtained through measurements of certain characteristics of the signals exchanged between the sensors, including signal propagation times, signal strengths, or angle of arrival.

10.2.1 Time of Arrival

The concept behind the time of arrival (ToA) method (also called time of flight method) is that the distance between the sender and receiver of a signal can be determined using the measured signal propagation time and the known signal velocity. For example, sound waves travel 343 m/s (in 20 °C), that is, a sound signal takes approximately 30 ms to travel a distance of 10 m. In contrast, a radio signal travels at the speed of light (about 300 km/s), that is, the signal requires only about 30 ns to travel 10 m. The consequence is that radio-based distance measurements require clocks with high resolution, adding to the cost and complexity of a sensor network. The *one-way* time of arrival method measures the *one-way* propagation time, that is, the difference between the sending time and the signal arrival time (Figure 10.1(a)), and requires highly accurate synchronization of the clocks of the sender and

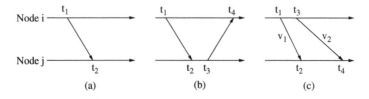

Figure 10.1 Comparison of different ranging schemes (one-way ToA, two-way ToA, and TDoA).

receiver. Therefore, the *two-way* time of arrival method is preferred, where the round-trip time of a signal is measured at the sender device (Figure 10.1(b)). In summary, for one-way measurements, the distance between two nodes i and j can be determined as:

$$\text{dist}_{ij} = (t_2 - t_1) \times v \qquad (10.1)$$

where t_1 and t_2 are the sending and receive times of the signal (measured at the sender and receiver, respectively) and v is the signal velocity. Similarly, for the two-way approach, the distance is calculated as:

$$\text{dist}_{ij} = \frac{(t_4 - t_1) - (t_3 - t_2)}{2} \times v \qquad (10.2)$$

where t_3 and t_4 are the sending and receive times of the response signal. Note that with one-way localization, the receiver node calculates its location, whereas in the two-way approach, the sender node calculates the receiver's location. Therefore a third message will be necessary in the two-way approach to inform the receiver of its location.

10.2.2 Time Difference of Arrival

The time difference of arrival (TDoA) approach uses two signals that travel with different velocities (Figure 10.1(c)). The receiver is then able to determine its location similar to the ToA approach. For example, the first signal could be a radio signal (issued at t_1 and received at t_2), followed by an acoustic signal (either immediately or after a fixed time interval $t_{\text{wait}} = t_3 - t_1$). Therefore, the receiver can determine the distance as:

$$\text{dist} = (v_1 - v_2) \times (t_4 - t_2 - t_{\text{wait}}) \qquad (10.3)$$

TDoA-based approaches do not require the clocks of the sender and receiver to be synchronized and can obtain very accurate measurements. The disadvantage of the TDoA approach is the need for additional hardware, for example, a microphone and speaker for the above example.

Another variant of this approach uses TDoA measurements of a single signal to estimate the location of the sender using multiple receivers with known locations. The propagation delay d_i for the signal to receiver i depends on the distance between sender and receiver i. Correlation analysis can then provide a time delay $\delta = d_i - d_j$ which corresponds to the difference in path length to receivers i and j (Gustafsson and Gunnarsson 2003). The main disadvantage of this approach is that the clocks of the receivers must be tightly synchronized.

10.2.3 Angle of Arrival

Another technique used for localization is to determine the direction of signal propagation, typically using an array of antennas or microphones. The angle of arrival (AoA) is then the angle between the propagation direction and some reference direction known as *orientation* (Peng and Sichitiu 2006). For example, for acoustic measurements, several spatially separated microphones are used to receive a single signal and the differences in arrival time, amplitude, or phase are used to determine an estimate of the arrival angle, which in turn can be used to determine the position of a node. While the appropriate hardware can obtain

accuracies within a few degrees, AoA measurement hardware can add significantly to the
size and cost of sensor nodes.

10.2.4 Received Signal Strength

The concept behind the received signal strength (RSS) method is that a signal decays with the
distance traveled. A commonly found feature in wireless devices is a received signal strength
indicator (RSSI), which can be used to measure the amplitude of the incoming radio signal.
Many wireless network card drivers readily export RSSI values, but their meaning may
differ from vendor to vendor and there is no specified relationship between RSSI values and
the signal's power levels. Typically, RSSI values are in the range of $0 \ldots RSSI\_Max$, where
common values for RSSI_Max are 100, 128, and 256. In free space, the RSS degrades with
the square of the distance from the sender. More specifically, the Friis transmission equation
expresses the ratio of the received power P_r to the transmission power P_t as:

$$\frac{P_r}{P_t} = G_t G_r \frac{\lambda^2}{(4\pi)^2 R^2} \tag{10.4}$$

where G_t is the antenna gain of the transmitting antenna and G_r is the antenna gain
of the receiving antenna. In practice, the actual attenuation depends on multipath
propagation effects, reflections, noise, etc., therefore a more realistic model replaces R^2 in
Equation (10.4) with R^n with n typically in the range of 3 and 5.

10.3 Range-Based Localization

10.3.1 Triangulation

Triangulation uses the geometric properties of triangles to estimate sensor locations. Specifically, triangulation relies on the gathering of angle (or *bearing*) measurements as described
in the previous section. A minimum of two bearing lines (and the locations of the anchor
nodes or the distance between them) are needed to determine the location of a sensor node
in two-dimensional space. Figure 10.2(a) illustrates the concept of triangulation using three

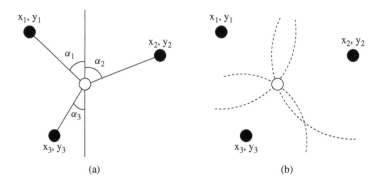

Figure 10.2 Triangulation (a) and trilateration (b).

anchor nodes with known locations (x_i, y_i) and measured angles α_i (expressed relative to a fixed baseline in the coordinate system, for example, the vertical line in the figure). If more than two bearings are measured, the presence of noise in the measurements may prevent them from intersecting in a single point. Therefore statistical algorithms or fixing methods have been developed to obtain a single location (Stansfield 1947).

Assume that the unknown receiver location is $\mathbf{x_r} = [x_r, y_r]^T$, the bearing measurements from N anchor points are expressed as $\beta = [\beta_1, ..., \beta_N]^T$, and the known anchor locations are $\mathbf{x_i} = [x_i, y_i]^T$. The measured bearings do not perfectly reflect the actual bearings $\theta(\mathbf{x}) = [\theta_1(\mathbf{x}), ..., \theta_N(\mathbf{x})]^T$ due to some noise, that is, the relationship between measured and actual bearings is:

$$\beta = \theta(\mathbf{x_r}) + \delta\theta \tag{10.5}$$

where $\delta\theta = [\delta\theta_1, ..., \delta\theta_N]^T$ is the Gaussian noise with zero-mean and $N \times N$ covariance matrix $S = \text{diag}(\sigma_1^2, ..., \sigma_N^2)$ (Gavish and Weiss 1992). In two-dimensional space, the relationship between the bearings of N anchors and their locations can be expressed as (Mao et al. 2007; Tekdas and Isler 2007):

$$\tan\theta_i(\mathbf{x}) = \frac{y_i - y_r}{x_i - x_r} \tag{10.6}$$

Various statistical methods have been applied to estimating a sensor's location. For example, the maximum likelihood (ML) estimator of the receiver location is:

$$\hat{\mathbf{x}}_\mathbf{r} = \arg\min \frac{1}{2}[\theta(\hat{\mathbf{x}}_\mathbf{r}) - \beta]^T S^{-1}[\theta(\hat{\mathbf{x}}_\mathbf{r}) - \beta] \tag{10.7}$$

$$= \arg\min \frac{1}{2}\sum_{i=1}^{N} \frac{(\theta_i(\hat{\mathbf{x}}_\mathbf{r}) - \beta_i)^2}{\sigma_i^2} \tag{10.8}$$

This nonlinear least squares minimization can be performed using Newton–Gauss iterations:

$$\hat{\mathbf{x}}_{\mathbf{r},i+1} = \hat{\mathbf{x}}_{\mathbf{r},i} + (\theta_\mathbf{x}(\hat{\mathbf{x}}_{\mathbf{r},i})^T S^{-1}\theta_\mathbf{x}(\hat{\mathbf{x}}_{\mathbf{r},i}))^{-1}\theta_\mathbf{x}(\hat{\mathbf{x}}_{\mathbf{r},i})^T S^{-1}[\beta - \theta_\mathbf{x}(\hat{\mathbf{x}}_{\mathbf{r},i})] \tag{10.9}$$

where $\theta_\mathbf{x}(\hat{\mathbf{x}}_{\mathbf{r},i})$ is the partial derivative of θ with respect to $\mathbf{x}$ evaluated at $\hat{\mathbf{x}}_{\mathbf{r},i}$. Equation (10.9) requires an initial estimate (e.g., obtained from prior information) that is close enough to the true minimum of the cost function.

10.3.2 Trilateration

Trilateration refers to the process of calculating a node's position based on measured distances between itself and a number of anchor points with known locations. Given the location of an anchor and a sensor's distance to the anchor (e.g., estimated through RSS measurements), it is known that the sensor must be positioned somewhere along the circumference of a circle centered at the anchor's position with a radius equal to the sensor–anchor distance. In two-dimensional space, distance measurements from at least three noncollinear anchors are required to obtain a unique location (i.e., the intersection of three circles). Figure 10.2(b) illustrates an example for the two-dimensional case. In three dimensions, distance measurements to at least four noncoplanar anchors are required.

Assume that the locations of n anchor nodes are given as $\mathbf{x_i} = (x_i, y_i)$ $(i = 1...n)$ and that the distances between an unknown sensor location $\mathbf{x} = (x, y)$ and these anchor nodes are also known $(r_i, i = 1...n)$. This information leads to a matrix expressing the relationships among anchor/sensor positions and distances:

$$
\begin{bmatrix}
(x_1 - x)^2 + (y_1 - y)^2 \\
(x_2 - x)^2 + (y_2 - y)^2 \\
\vdots \\
(x_n - x)^2 + (y_n - y)^2
\end{bmatrix}
=
\begin{bmatrix}
r_1^2 \\
r_2^2 \\
\vdots \\
r_n^2
\end{bmatrix}
\tag{10.10}
$$

While the example shown here is for two dimensions, the same process can be used for localization in more than two dimensions by increasing matrix dimensions. After some rearrangements and subtracting the last matrix equation from all previous ones (to remove the square of the unknown sensor location (x, y)), we obtain:

$$
A\mathbf{x} = b \tag{10.11}
$$

with the coefficient matrix

$$
A =
\begin{bmatrix}
2(x_n - x_1) & 2(y_n - y_1) \\
2(x_n - x_2) & 2(y_n - y_2) \\
\vdots & \vdots \\
2(x_n - x_{n-1}) & 2(y_n - y_{n-1})
\end{bmatrix}
\tag{10.12}
$$

and the right side vector

$$
b =
\begin{bmatrix}
r_1^2 - r_n^2 - x_1^2 - y_1^2 + x_n^2 + y_n^2 \\
r_2^2 - r_n^2 - x_2^2 - y_2^2 + x_n^2 + y_n^2 \\
\vdots \\
r_{n-1}^2 - r_n^2 - x_{n-1}^2 - y_{n-1}^2 + x_n^2 + y_n^2
\end{bmatrix}
\tag{10.13}
$$

This least squares system can now be used to obtain an estimation of the position (x, y) using:

$$
\mathbf{x} = (A^T A)^{-1} A^T b \tag{10.14}
$$

Anchor positions and distance measurements are rarely perfect, therefore, if the positions and distances are based on Gaussian distributions, each equation i can have a weight:

$$
w_i = 1/\sqrt{\sigma_{\text{distance}_i}^2 + \sigma_{\text{position}_i}^2} \tag{10.15}
$$

where $\sigma_{\text{distance}_i}^2$ is the variance of the distance measurement between $\mathbf{x}$ and anchor i and $\sigma_{\text{position}_i}^2 = \sigma_{x_i}^2 + \sigma_{y_i}^2$. The least squares system is then $A\mathbf{x} = b$ with

$$A = \begin{bmatrix} 2(x_n - x_1) \times w_1 & 2(y_n - y_1) \times w_1 \\ 2(x_n - x_2) \times w_2 & 2(y_n - y_2) \times w_2 \\ \vdots & \vdots \\ 2(x_n - x_{n-1}) \times w_{n-1} & 2(y_n - y_{n-1}) \times w_{n-1} \end{bmatrix} \tag{10.16}$$

and

$$b = \begin{bmatrix} (r_1^2 - r_n^2 - x_1^2 - y_1^2 + x_n^2 + y_n^2) \times w_1 \\ (r_2^2 - r_n^2 - x_2^2 - y_2^2 + x_n^2 + y_n^2) \times w_2 \\ \vdots \\ (r_{n-1}^2 - r_n^2 - x_{n-1}^2 - y_{n-1}^2 + x_n^2 + y_n^2) \times w_{n-1} \end{bmatrix} \tag{10.17}$$

The covariance matrix of $\mathbf{x}$ is then given by $\mathrm{Cov}_x = (A^T A)^{-1}$.

10.3.3 Iterative and Collaborative Multilateration

While the lateration technique relies on the presence of at least three anchor nodes to position a fourth unknown node, this technique can be extended to determine locations of nodes without three neighboring anchor nodes. Once a node has identified its position using the beacon messages from the anchor nodes, it becomes an anchor and broadcasts beacon messages containing its estimated position to other nearby nodes. This *iterative multilateration* process (Savvides *et al.* 2001) repeats until all nodes in a network have been localized. Figure 10.3(a) visualizes this process: in the first iteration, the gray node estimates its location with the help of the three black anchor nodes and in the second iteration, the white nodes estimate their respective locations with the help of two original anchor nodes and the gray node. The drawback of iterative multilateration is that the localization error accumulates with each iteration.

In ad hoc deployments of sensor and anchor nodes, it is possible that a node will not have three neighboring anchor nodes, therefore preventing it from determining its own location. In this case, a node can use a process called *collaborative multilateration* to estimate its position using location information obtained over multiple hops. Figure 10.3(b) shows a simple example with six nodes: four anchor nodes A_i (black) and two nodes with unknown locations S_i (white). The goal of collaborative multilateration is to construct a graph of

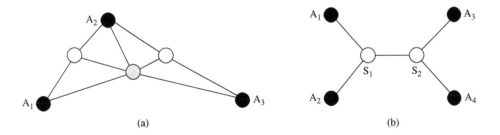

Figure 10.3 (a) Iterative multilateration and (b) collaborative multilateration.

participating nodes, that is, nodes that are anchors or have at least three participating neighbors (e.g., all nodes in Figure 10.3(b) are participants). A node can then try to estimate its position by solving the corresponding system of overconstrained quadratic equations relating the distances among the node and its neighbors.

10.3.4 GPS-Based Localization

The Global Positioning System (GPS) (Hofmann-Wellenhof *et al.* 2008) is the most widely publicized location-sensing system, providing an excellent lateration framework for determining geographic positions (Hightower and Borriello 2001). GPS (formally known as NAVSTAR – Navigation Satellite Timing and Ranging) is the only fully operational global navigation satellite system (GNSS) and it consists of at least 24 satellites orbiting the earth at altitudes of approximately 11,000 miles. It began as a test program in 1973 and became fully operational in 1995. In the meantime, GPS has established itself as a widely used aid to civilian navigation, surveying, tracking and surveillance, and scientific applications. GPS provides two levels of service (Dana 1997):

1. The **Standard Positioning Service (SPS)** is a positioning service available to all GPS users on a continuous worldwide basis without restrictions or direct charge. High-quality GPS receivers based on SPS are able to attain accuracies of 3 m and better horizontally.
2. The **Precise Positioning Service (PPS)** is used by US and Allied military users and is a more robust GPS service that includes encryption and jam resistance. For example, it uses two signals to reduce radio transmission errors, while SPS only uses one signal.

GPS satellites are uniformly distributed in a total of six orbits (i.e., there are four satellites per orbit) and they circle the earth twice a day at approximately 7000 miles per hour. The number of satellites and their spatial distribution ensure that at least eight satellites can be seen simultaneously from almost anywhere on the planet. Each satellite constantly broadcasts coded radio waves (known as *pseudorandom code*) that contain information on the identity of the particular satellite, the location of the satellite, the satellite's status (i.e., whether it is working properly), and the date and time a signal has been sent. In addition to the satellites, GPS further relies on infrastructure on the ground to monitor satellite health, signal integrity, and orbital configuration. At least six *monitor stations* located around the world constantly receive the data sent by the satellites and forward the information to a *master control station* (MCS). The MCS (located near Colorado Springs, Colorado) uses the data from the monitor stations to compute corrections to the satellites' orbital and clock information, which are then sent back to the appropriate satellites via *ground antennas*.

A GPS receiver (e.g., embedded into a mobile device) receives the information transmitted by the satellites that are currently in view by the receiver. The basic principle of GPS positioning is illustrated in Figure 10.4. Satellites and receivers use very accurate and synchronized clocks so that they generate the same code at exactly the same time. The GPS receiver compares its generated code with the code received from the satellite, thereby determining the actual generation time (e.g., t_0 in Figure 10.4) of the code at the satellite and the time difference Δ between the code generation time and the current time. Therefore, Δ then expresses the travel time of the code from the satellite to the receiver. Note that the received satellite data is attenuated due to the satellite–earth path even if no obstructions

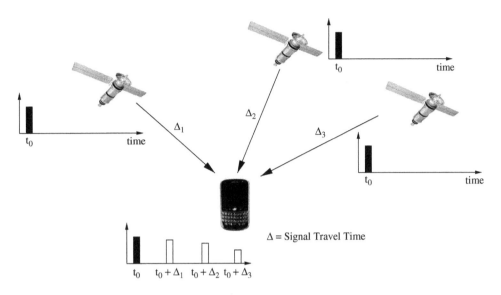

Figure 10.4 GPS positioning principle.

occur. Radio waves travel at the speed of light (about 186 000 miles per second), so if Δ is known, the distance from the satellite to the receiver (distance = speed × time) can be determined. Once the distance has been determined, the receiver knows that it is located somewhere on a sphere centered on the satellite with a radius equal to the computed distance. Repeating this process with two more satellites, the position of the receiver can be narrowed down to the two points where the three spheres intersect. Typically, one of the two points can be eliminated very easily, for example, because it would position the receiver far out in space or the receiver would travel at a virtually impossible velocity.

While three satellites appear to be sufficient for localization, a fourth satellite is needed to obtain an accurate position. Positioning via GPS relies on correct timing to make accurate measurements, that is, the clocks of the satellites and the receivers must be synchronized precisely. Satellites are equipped with four atomic clocks (synchronized to each other within a few nanoseconds), providing highly accurate time readings. However, the clocks used for GPS receivers are not nearly as accurate as the atomic clocks onboard the satellites, introducing measurement errors that can have a significant impact on the quality of localization. Because radio waves travel at very high speeds (and therefore require very little time to travel), small errors in the timing can result in large deviations in position measurements. For example, a clock error of 1 ms would result in a position error of about 300 km. Therefore, a fourth measurement is required, where the fourth sphere should ideally intersect the other three spheres at the exact location of the receiver. Because of timing errors, the fourth sphere may not intersect with all other spheres, even though we know that they are supposed to align. If the spheres are too large, we can reduce their sizes by adjusting the clock (by moving it forward) until the spheres are small enough to intersect in one point. Similarly, if the spheres are too small, we adjust the clock by moving it backwards. That is, because the timing error is the same for all measurements, a receiver can calculate the required clock adjustment to obtain a single intersection point among all four spheres. In addition to providing a means for clock synchronization, a fourth measurement also allows a receiver to obtain a three-dimensional position, that is, latitude, longitude, and elevation.

While most GPS receivers available today are able to provide position measurements with accuracies of 10 m or less, advanced techniques to further increase the accuracy are available. For example, Differential GPS (DGPS) (Monteiro *et al.* 2005) relies on land-based receivers with exactly known locations to receive GPS signals, compute correction factors, and broadcast them to GPS receivers that are then able to correct their own GPS measurements. While it is possible to build wireless sensor networks where each sensor has its own GPS receiver, constraints such as high power consumption, cost, and the need for line-of-sight make a fully GPS-based solution impractical for most sensor networks. However, GPS receivers deployed on a few nodes in a WSN may be sufficient to provide location services based on reference points as described in the following section.

10.4   Range-Free Localization

The localization approaches discussed in the previous sections are based on distance estimations using ranging techniques (RSS, ToA, TDoA, and AoA) and belong therefore to the class of *range-based* localization algorithms. In contrast, *range-free* techniques estimate node locations based on connectivity information instead of distance or angle measurements. Range-free localization techniques do not require additional hardware and are therefore a cost-effective alternative to range-based techniques. This section describes various different approaches to localization without reliance on ranging techniques.

10.4.1   Ad Hoc Positioning System (APS)

APS (Niculescu and Nath 2001) is an example of a distributed connectivity-based localization algorithm that estimates node locations with the support of at least three anchor nodes, where localization errors can be reduced by increasing the number of anchors. Each anchor node propagates its location to all other nodes in the network using the concept of distance vector (DV) exchange (Lu *et al.* 2003), where nodes in a network periodically exchange their routing tables with their one-hop neighbors. In the most basic scheme of APS, called DV-hop, each node maintains a table $\{X_i, Y_i, h_i\}$, where $\{X_i, Y_i\}$ is the location of node i and h_i is the distance in hops between this node and node i. When an anchor obtains distances to other anchors, it then determines an average size for one hop (called the correction factor), which too is then propagated throughout the network. The correction factor c_i of anchor i is determined as:

$$c_i = \frac{\sum \sqrt{(X_i - X_j)^2 + (Y_i - Y_j)^2}}{\sum h_i} \qquad (10.18)$$

for all landmarks j ($i \neq j$). Given the locations of the anchors and the correction factor, a node is then able to perform trilateration to estimate its own location. Figure 10.5 presents an example with three anchor nodes A_1, A_2, and A_3. Anchor A_1, knowing its Euclidean distances (50 m and 110 m) and hop distances (two hops and six hops) to the other two anchor nodes, computes a correction of $(50 + 110)/(2 + 6) = 20$, which represents the estimated distance of a hop in meters. In a similar fashion, A_2 computes a correction factor of 18.6 and A_3 computes a correction factor of 17.3. Corrections are propagated via controlled flooding

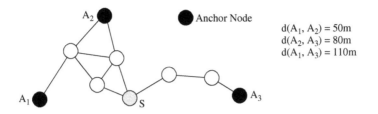

Figure 10.5   Example of DV-hop localization.

(i.e., once a node receives a correction, it ignores subsequent ones) to ensure that each node will only use one correction factor, typically from the closest anchor. For example, sensor node S in Figure 10.5 uses the correction factor obtained from A_2, that is, 18.6, to estimate its distances to the three anchors by multiplying the correction factor with the hop counts (leading to distances 3×18.6 to A_1, 2×18.6 to A_2, and 3×18.6 to A_3). Given these distances, triangulation (as described in 10.3.1) can be used to determine the position of S.

In a variation of this approach, called the DV-distance method, distances between neighboring nodes are determined using radio signal strength measurements and distributed to other nodes in meters instead of hops. While this approach provides finer granularity (not all hops are estimated to be the same size), it is also more sensitive to measurement errors. Finally, in the Euclidean method, true Euclidean distances to anchors are used. A node must have at least two neighbors that have distance measurements to an anchor, where the distance between the two neighbors is known. Based on this information, simple trigonometric relationships can be used to determine the distance of a node to an anchor.

10.4.2   Approximate Point in Triangulation

The Approximate Point In Triangulation (APIT) approach is an *area-based* range-free localization scheme (He *et al.* 2003). Similar to APS, APIT relies on the presence of several anchor nodes that know their own location (e.g., via GPS). Any combination of three anchors forms a triangular region and a node's presence inside or outside such a region allows a node to narrow down its possible locations. The key step in APIT localization is the Point In Triangulation (PIT) test that allows a node to determine the set of triangles within which the node resides. After a node M has received location messages from a set of anchors, it evaluates all possible triangles formed by the anchors. A node is outside a given triangle ABC formed by anchors A, B, and C, if there exists a direction such that a point adjacent to M is either further or closer to all points A, B, and C simultaneously. Otherwise, M is inside the triangle and triangle ABC can be added to the set of triangles in which M resides. This concept is illustrated in Figure 10.6. Unfortunately, this *perfect PIT* test is infeasible in practice since it would require that nodes can be moved in any direction. However, an APIT test can be used in networks with sufficient node density. The idea is to emulate the node movement in the perfect PIT test using neighbor information that is exchanged via beacon messages. For example, signal strengths between nodes and an anchor can be used to estimate which node is closer to the anchor. Then, if no neighbor of node M is further from or closer to the three anchors A, B, and C simultaneously, M assumes that it is inside the triangle ABC; otherwise M assumes that it is outside the triangle. Figure 10.7 illustrates this concept. In the

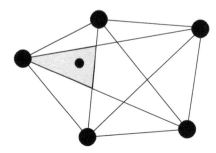

Figure 10.6   Location estimation based on the intersection of anchor triangles.

left graph, node M has four neighbors, none of which is simultaneously closer to or further away from any of the three anchor nodes. M therefore correctly concludes that it is inside the ABC triangle. The situation is different in the right graph. For example, neighbor 4 is closer to all three anchor nodes than node M, while node 2 is further away from the anchor nodes than node M. Therefore, node M concludes that it must be outside the ABC triangle. In this scheme, a node can make incorrect decisions because only a finite number of directions (the number of neighbors) can be evaluated. For example, in the left graph, if node 4's RSS measurements indicate that it is further from node B than node M (e.g., because there is an obstacle between anchor B and node 4), node M would conclude that it must be outside the triangle. Once the APIT test completes, a position estimate can be computed as the center of gravity of the intersection of all triangles in which M resides.

10.4.3   Localization Based on Multidimensional Scaling

Multidimensional scaling (MDS) has its roots in psychometrics and psychophysics and is a set of data analysis techniques that display the structure of distance-like data as a geometrical picture. Applied to localization (Shang *et al.* 2004), MDS can be used in centralized localization techniques, where a powerful central device (e.g., base station) collects

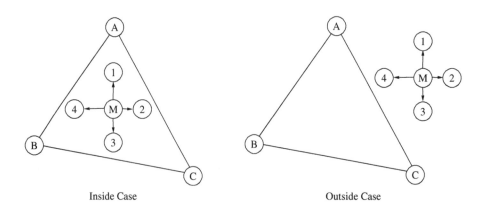

Figure 10.7   Examples of APIT test scenarios.

information from the network, determines the nodes' locations, and propagates this information back into the network. The network is represented as an undirected graph of n nodes
($m < n$ of which are anchors and know their locations) and edges representing the connectivity information. Given the distances between all pairs of nodes, the goal of MDS is to
preserve the distance information such that the network can be recreated in the multidimensional space. The result of MDS will be an arbitrarily rotated and flipped version of the
original network layout.

While there are many variations of MDS, the simplest version (called *classical* MDS) has
a closed form solution allowing for efficient implementations. Assume a matrix of squared
distances between nodes written as:

$$D^2 = c1' + 1c' - 2SS' \qquad (10.19)$$

where 1 is an $n \times 1$ vector of ones, S is the similarity matrix for n points, where each row
represents the coordinates of point i along m coordinates, SS' is called the scalar product
matrix, and c is a vector consisting of the diagonal elements of the scalar product matrix.
Multiplying both sides of Equation (10.19) by the centering matrix $T = I - 11'/n$, where
I is the identity matrix and 1 is again a vector of ones, leads to:

$$TD^2T = T(c1' + 1c' - 2SS')T = Tc1'T + T1c'T - T(2B)T \qquad (10.20)$$

where $B = SS'$. Centering a vector of ones yields a vector of zeros, therefore:

$$TD^2T = -T(2B)T \qquad (10.21)$$

Further multiplying both sides with $-1/2$ results in:

$$B = -\frac{1}{2}TD^2T \qquad (10.22)$$

B is a symmetric matrix and can therefore be decomposed into:

$$B = Q\Lambda Q' = \left(Q'\Lambda^{1/2}\right)\left(Q\Lambda^{1/2}\right)' = SS' \qquad (10.23)$$

Once B has been obtained, the coordinates S can be computed by eigendecomposition:

$$S = Q\Lambda^{1/2} \qquad (10.24)$$

Based on this concept, a localization method for sensor networks called MDS–MAP
(Shang *et al.* 2004) can be applied. First, a distance matrix D is constructed using an
all pairs shortest path algorithm (e.g., Dijkstra's), with d_{ij} being the distance (i.e., the
minimum number of hops) between nodes i and j. Next, classical MDS as described above
is applied to this matrix and an approximate value of the relative coordinate of each node is
obtained. Finally, these relative coordinates are then transformed to absolute coordinates by
aligning the estimated relative coordinates of anchors with their absolute coordinates.
These location estimates can further be refined using least-squares minimization.

An extension to this approach divides the entire sensor network into overlapping regions,
where localization is performed in individual regions using the approach described above.
These local maps are then patched together to form a global map by using common nodes
shared between adjacent regions. This results in better performance in irregularly-shaped

networks by avoiding the use of distance information between far away nodes. While the approach described here is a centralized solution relying on global information, a distributed implementation is also possible (Shang and Ruml 2004).

10.5 Event-Driven Localization

10.5.1 The Lighthouse Approach

A third category of localization schemes is based on *events* that can be utilized to determine distances, angles, and positions. Such events can be the arrival of radio waves, beams of light, or acoustic signals at a sensor node. In the lighthouse location system (Römer 2003), sensor nodes can estimate their location with high accuracy without the need for additional infrastructure components besides a base station equipped with a light emitter. Figure 10.8 illustrates the concept using an idealistic light source, which has the property that the emitted beam of light is parallel, that is, the width b remains constant. The light source rotates and when the parallel beam passes by a sensor, it will see the flash of light for a certain period of time t_{beam}. The main idea behind this concept is that t_{beam} varies with the distance between the sensor and the light source (since the beam is parallel). The distance d between the sensor and the light source can be expressed as:

$$d = \frac{b}{2\sin(\alpha/2)} \tag{10.25}$$

where α expresses the angle under which the sensor sees the beam of light as follows:

$$\alpha = 2\pi \frac{t_{beam}}{t_{turn}} \tag{10.26}$$

Here, t_{turn} is the time the light source takes to perform a complete rotation. While b is given and constant, a sensor can calculate $t_{beam} = t_2 - t_1$ and $t_{turn} = t_3 - t_1$, where t_1 is the time the sensor sees the light for the first time, t_2 is the time the sensor no longer sees the light, and t_3 is the time when the sensor sees the light again.

A key assumption so far has been that the width b of the beam stays constant for all distances from the light source. However, perfectly parallel light beams are difficult to realize in practice and even small beam spreads can result in significant localization errors, for example, a beam with $b = 10$ cm and a beam spread of $1°$ would result in a beam width of 18.7 cm at a distance of 5 m. An additional requirement is that the beam width should be as large as possible to keep inaccuracies small. To achieve this, two laser beams can be

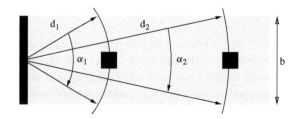

Figure 10.8   The lighthouse localization approach (top view).

used to create the outline of a "virtual" parallel beam (the sensor nodes are only interested in detecting the edges of the virtual beam which are represented by the two laser beams).

10.5.2 Multi-Sequence Positioning

The Multi-Sequence Positioning (MSP) approach (Zhong and He 2007) works by extracting relative location information from multiple simple one-dimensional orderings of sensor nodes. For example, Figure 10.9 shows a small sensor network with five nodes with unknown locations and two anchor nodes. Events are generated by event generators at different locations one at a time (e.g., ultrasound propagations or laser scans with diverse angles). The nodes in the sensor field observe these events at different times, depending on their distances to the event generators. For each event, we can establish a node sequence, that is, a node ordering (including both the sensor and the anchor nodes) based on the sequential detection of the event. Then, a multisequence processing algorithm narrows the potential locations for each node to a small area and, finally, a distribution-based estimation method estimates the exact locations.

The basic concept of the MSP algorithm is to split a sensor network area into small pieces by processing node sequences. For example, in Figure 10.9, performing a straight-line scan from top to bottom results in a node sequence 2, B, 1, 3, A, 4, 5. The basic MSP algorithm uses two straight lines to scan an area from different directions, treating each scan as an event. In Figure 10.9, a left-to-right scan results in a node sequence 1, A, 2, 3, 5, B, 4. Since the anchor locations are known, the two anchors split the area into nine parts. This process can be extended to cut the area into smaller pieces by increasing the number of anchors and scans (from different angles). The basic MSP algorithm processes each node sequence to determine the boundaries of a node (by searching for the predecessor and successor anchor nodes for the node) and shrinks the location area of this node according to the newly obtained boundary information. Finally, a centroid estimation algorithm sets the center of gravity of the resulting polygon as the estimated location of the target node.

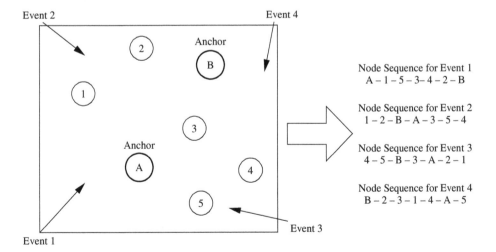

Figure 10.9   Basic concept of MSP.

Exercises

10.1 Why is localization needed in wireless sensor networks? Name at least two concrete scenarios or applications where localization is required.

10.2 A node's position in two-dimensional space is $(x, y) = (10, 20)$ with a maximum error of 2 in the x direction for 95% of all measurements and a maximum error of 3 in the y direction for 90% of all measurements. What is the accuracy and the precision of this location information?

10.3 Explain the difference between physical and symbolic positions and name at least two examples for each type.

10.4 Define the terms anchor-based localization and range-based localization.

10.5 Time of Arrival (ToA) is one example of a ranging technique. Answer the following questions (assume a propagation time of 300 m/s):

  (a) What is the advantage of two-way ToA over one-way ToA?

  (b) In a synchronized network with unknown synchronization error, an anchor node periodically broadcasts an acoustic signal to sensor nodes in its range. At time 1000 ms on the anchor node's clock, the anchor node issues a beacon, which is received by node A at time 2000 ms (on node A's clock). What is the distance that A can now compute?

  (c) Instead of computing the distance itself, node A also responds with an acoustic signal issued at time 2500 ms, which is received by the anchor node at time 3300 ms. What is the distance computed by the anchor node? What can you say about the synchronization of anchor node and node A?

10.6 What is the main disadvantage for both TDoA and AoA ranging techniques?

10.7 RSS-based localization techniques are often combined with a process called RF profiling, that is, the mapping of the effects of objects in the environment on signal propagation. Why is this necessary and can you think of examples of such objects?

10.8 Two nodes A and B are known to be positioned at locations $(0, 0)$ (node A) and $(1, 1)$ (node B) in two-dimensional space. A third node C wishes to determine its position using trilateration. Based on ranging techniques, node C knows its distances to node A $(d(A,C) = \sqrt{0.75})$ and node B $(d(B,C) = \sqrt{0.75})$. What are the two possible positions of C?

10.9 Three nodes A, B, and C are known to be positioned at locations $(0, 0)$, $(10, 0)$, and $(4, 15)$, respectively. Node D is estimated to be a distance of 7 from A, a distance of 7 from B, and a distance of 10.15 from C. Determine the location of D using trilateration.

10.10 Consider the two-dimensional topology in Figure 10.10. The sensor node in the center can select three of the six anchor nodes as basis for trilateration. Which nodes should the sensor node select? Justify your answer, that is, what guideline for anchor selection should be considered? What would this guideline be in three-dimensional space?

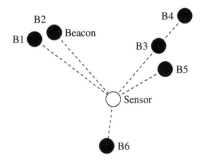

Figure 10.10   Exercise 10.10.

10.11 Two nodes A and B do not know their own positions, but they can hear beacons in their proximities. Node A can hear beacons located at $(4, 2)$ and $(2, 5)$. Node B can hear beacons located at $(2, 5)$ and $(3, 7)$. All nodes have a radio range of 2 units.

(a) Are either $(3, 3.5)$ or $(3, 4.5)$ possible locations for node A?

(b) Are either $(2, 6)$ or $(4, 5)$ possible locations for node B?

10.12 What are the differences between iterative and collaborative multilateration?

10.13 Explain the concept of GPS localization and answer the following questions:

(a) Why are three satellites enough to obtain a position on the globe?

(b) Why is it preferred to have at least four satellites available for localization?

(c) What is the purpose of the monitor stations and the master control station?

(d) Why is it typically not feasible to have all wireless sensor nodes equipped with a GPS receiver?

10.14 Explain the difference between range-based and range-free localization.

10.15 Figure 10.11 shows a network topology with three anchor nodes. The distances between anchors A_1 and A_2, anchors A_1 and A_3, and anchors A_2 and A_3 are 40 m, 110 m, and 35 m, respectively. Use the Ad Hoc Positioning System to estimate the location of the gray sensor node (show each step of your process).

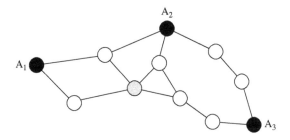

Figure 10.11   Exercise 10.15.

10.16 For the APIT test, can you show a concrete scenario where a node M would come to the wrong conclusion that it must be inside a triangle? Use a scenario where node M has at least three neighbors. Can you also show an example where node M would come to the wrong conclusion that it must be outside a triangle?

10.17 A sensor node in a WSN using the lighthouse approach for localization detects the first beam of light at time 0 s and the second beam of light at time 0.25 s. The next time the first beam of light is detected is 7 s. The distance of the two light sources (beam width) is 10 cm. What is the distance of the sensor to the light emitter?

References

Dana, P.H. (1997) Global Positioning System (GPS): Time-dissemination for real-time applications. *Real-Time, Systems* **12** (1), 9–40.

Gavish, M., and Weiss, A.J. (1992) Performance analysis of bearing-only target location algorithms. *IEEE Transactions on Aerospace and Electronic Systems* **28** (3), 817–828.

Gustafsson, F., and Gunnarsson, F. (2003) Positioning using time-difference of arrival measurements. *Proc. of the IEEE International Conference on Acoustics, Speech, and Signal Processing*.

He, T., Huang, C., Blum, B.M., Stankovic, J.A., and Abdelzaher, T. (2003) Range-free localization schemes for large scale sensor networks. *Proc. of the 9th Annual International Conference on Mobile Computing and Networking (MobiCom)*.

Hightower, J., and Borriello, G. (2001) Location systems for ubiquitous computing. *Computer* **34** (8), 57–66.

Hofmann-Wellenhof, B., Lichtenegger, H., and Collins, J. (2008) *Global Positioning System: Theory and Practice* (5th edn). Springer.

Lu, Y., Wang, W., Zhong, Y., and Bhargava, B. (2003) Study of distance vector routing protocols for mobile ad hoc networks. *Proc. of the 1st IEEE International Conference on Pervasive Computing and Communications (PerCom)*.

Mao, G., Fidan, B., and Anderson, B.D.O. (2007) Wireless sensor network localization techniques. *Computer Networks: The International Journal of Computer and Telecommunications Networking* **51** (10), 1389–1286.

Monteiro, L.S., Moore, T., and Hill, C. (2005) What is the accuracy of DGPS? *The Journal of Navigation* **58** (2), 207–225.

Niculescu, D., and Nath, B. (2001) Ad hoc positioning system (APS). *Proc. of the IEEE Global Telecommunications Conference (GLOBECOM)*.

Peng, R., and Sichitiu, M.L. (2006) Angle of arrival localization for wireless sensor networks. *Proc. of the 3rd Annual IEEE Communications Society Conference on Sensor and Ad Hoc Communications and Networks*.

Römer, K. (2003) The lighthouse location system for smart dust. *Proc. of the 1st International Conference on Mobile Systems, Applications and Services* (pp. 15–30).

Savvides, A., Han, C.C., and Strivastava, M.B. (2001) Dynamic fine-grained localization in ad hoc networks of sensors. *Proc. of the 7th Annual International Conference on Mobile Computing and Networking*.

Shang, Y., and Ruml, W. (2004) Improved MDS-based localization. *Proc. of the 23rd Annual Joint Conference of the IEEE Computer and Communications Societies (INFOCOM)*.

Shang, Y., Ruml, W., Zhang, Y., and Fromherz, M. (2004) Localization from connectivity in sensor networks. *IEEE Transactions on Parallel and Distributed Systems* **15** (11), 961–974.

Siqueira, I.G., Ruiz, L.B., Loureiro, A.A.F., and Nogueira, J.M. (2007) Coverage area management for wireless sensor networks. *International Journal of Network Management* **17** (1), 17–31.

Stansfield, R.G. (1947) Statistical theory of DF fixing. *Journal of IEE* **14**, Pt. III A (15), 762–770.

Stojmenovic, I. (2002) Position based routing in ad hoc networks. *IEEE Communications Magazine* **40** (7), 128–134.

Tekdas, O., and Isler, V. (2007) Sensor placement algorithms for triangulation based localization. *Proc. of the IEEE International Conference on Robotics and Automation* (pp. 4448–4453).

Zhong, Z., and He, T. (2007) MSP: Multi-sequence positioning of wireless sensor nodes. *Proc. of the 5th International Conference on Embedded Networked Sensor Systems* (pp. 15–28).

11

Security

Security and privacy are enormous challenges in all types of wired and wireless networks. These challenges are of even greater importance in wireless sensor networks, where the unique characteristics of these networks and the application purposes they serve make them attractive targets for intrusions and other attacks. In applications such as battlefield surveillance and assessment, target tracking, monitoring civil infrastructure such as bridges and tunnels, and assessment of disaster zones to guide emergency response activities, any breach of security, compromise of information, or disruption of correct application behavior can have very serious consequences. Sensor networks are frequently used in remote areas, left to operate unattended and therefore providing an easy target for physical attacks, unauthorized access, and tampering. Sensor nodes are typically very resource-constrained and operate in harsh environments, which further facilitates compromises and makes it often difficult to distinguish security breaches from node failures, varying link qualities, and other commonly found challenges in sensor networks. Finally, these resource constraints require security mechanisms that are customized for WSN applications, such that the limited resources are used efficiently. This chapter provides an overview of the security concerns of WSNs and illustrates possible solutions to providing security and privacy protection. Note that the terms *attacker*, *intruder*, and *adversary* are used interchangeably to describe an entity (person or device) that performs an attack on a network or system.

11.1 Fundamentals of Network Security

Computer and network security is the collection of all policies, mechanisms, and services that afford a computer system or network the required protection from unauthorized access or unintended uses. Most security mechanisms are built to address three well-known services in the CIA security model: *Confidentiality, Integrity*, and *Availability*. The following describes these services in more detail:

1. *Confidentiality:* Security mechanisms must ensure that only the intended receiver can correctly interpret a message and that unauthorized access and usage is prevented. For example, confidentiality ensures that sensitive information such as a person's social security number or credit card information are not obtained by an unauthorized individual.
2. *Integrity:* Security mechanisms must ensure that a message cannot be modified as it propagates from the sender to the receiver, that is, unauthorized individuals should not be able to destroy or alter the contents of sensitive information.

Fundamentals of Wireless Sensor Networks: Theory and Practice   Waltenegus Dargie and Christian Poellabauer
© 2010 John Wiley & Sons, Ltd

3. *Availability:* Security mechanisms must ensure that a system or network and its applications are able to perform their tasks at any time without interruption. Availability is often measured in terms of percentages of up or down time.

Figure 11.1 illustrates examples of attacks on a transmission between a sender and its intended receiver. *Eavesdropping* refers to the reception of a message by an unauthorized individual, which can be prevented using confidentiality measures. A *man-in-the-middle* attack refers to a situation where an unauthorized individual or system positions itself between the sender and receiver such that the sender's messages are intercepted, modified, and retransmitted to the receiver (where the receiver believes the received message came directly from the original sender). This illustrates the need for integrity mechanisms. Finally, a *denial-of-service* attack refers to an adversary's attempt to disrupt the transmission or service provided by the sender. For example, the adversary can overload the sender with requests and tasks such that the sender is not able to transmit its message (in a timely fashion) to the receiver. This type of attack necessitates security mechanisms that ensure availability.

In addition to the three components of the CIA triad, *authentication* refers to the process of establishing or confirming the identify of a user or a device, ensuring that a message came from who it claims to have come from. Also, *nonrepudiation* refers to the process of proving that a person or device has performed a transaction or transmission. *Digital signatures* are often used to support both authentication and nonrepudiation, but are also used to provide confidence that a message has not been altered (i.e., integrity).

In all types of communication networks, there are several fundamental security mechanisms that can be used to provide confidentiality, integrity, and availability. *Cryptography* is the process of hiding and protecting information using encoding and decoding mechanisms. In *symmetric key cryptography*, a single key between two communicating parties is used for the encryption and decryption of a message. For example, a simplistic encoding strategy could be to replace each plaintext letter with another letter that is a certain number of positions down the alphabet. For example, using a shift of 2 would replace the letter A

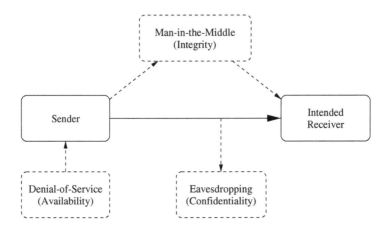

Figure 11.1   Examples of attacks and the CIA model.

with the letter C. In this *shift cipher*, the fixed shift value is then the symmetric key. A major challenge in the use of symmetric cryptographic techniques is the secure distribution of the shared key between the two communicating parties. Popular examples of symmetric key cryptographic mechanisms include DES, AES, and IDEA (Menezes *et al.* 1996).

In contrast to this approach, *public key cryptography*, such as the well-known RSA algorithm (Rivest *et al.* 1983) or the Diffie–Hellman key agreement protocol (Menezes *et al.* 1996), rely on a pair of keys. A node generates both a *secret key* and a *public key*, where the secret key will never be communicated with any other node. The public key, on the other hand, can be shared freely with anyone in the network. Any message encrypted with the secret key can only be deciphered using the corresponding public key (e.g., this can be used to authenticate the identity of the sender). Any message encrypted with the public key can only be deciphered using the corresponding secret key (e.g., this can be used to provide confidentiality).

11.2 Challenges of Security in Wireless Sensor Networks

Security has been a challenge in computing systems and networks for several decades, during which the types of attacks and the security measures and mechanisms to counter them have advanced and developed significantly, particularly because of the rapid growth of the Internet. Compared to the traditional attacks and security mechanisms developed for the Internet, WSNs exhibit a variety of unique challenges that must be considered when addressing the security concerns that may arise in sensor network applications:

1. *Resource constraints:* Traditional security mechanisms that have high overheads are not suitable for resource-constrained WSNs. Many security mechanisms are computationally expensive or require communication with other nodes or "remote" devices (e.g., for authorization purposes), thereby leading to energy overheads. Small sensor devices are also constrained in their available memory and storage capacities. Common sensor devices have very limited amounts of memory, for example, TelosB devices only have 10 kbytes RAM and 48 kbytes flash memory available. Traditional security algorithms that require a significant amount of memory and storage space are therefore infeasible for such sensors.
2. *Lack of central control:* It is often infeasible to have a central point of control in sensor networks, for example, because of their large scale, resource constraints, and network dynamics (topology changes, network partitioning). Therefore, security solutions should be decentralized and nodes must collaborate to achieve security.
3. *Remote location:* The first line of defense against security attacks is to provide only controlled physical access to a sensor node. Many WSNs are left unattended, because they are operated in remote and hard-to-reach locations, deployed in environments open to public access, or so large that it would be infeasible to continuously monitor and protect sensor nodes from attacks. These challenges make it difficult to prevent unauthorized physical access and to detect tampering with the sensor devices, particularly since the low cost of many sensor nodes may prohibit advanced (and expensive) protective measures.
4. *Error-prone communication:* Packets in WSNs may be lost or corrupted due to a variety of reasons, including channel errors, routing failures, and collisions. This may interfere with some security mechanisms or their ability to obtain critical event reports. Furthermore, this may make it difficult to distinguish "benign" erroneous communications or node and link failures from malicious attacks.

Certain characteristics of sensor networks, on the other hand, facilitate the provision of security. For example, the self-managing and self-repairing nature of a WSN may allow it to continue to operate even if a sensor or entire regions of the sensor network have been compromised. Redundancy in a sensor network allows it to gather information about events in the environment even when some sensors are unavailable due to an attack. Furthermore, this redundancy can be used to detect, isolate, and mask potentially compromised nodes.

Data collected by sensors may contain sensitive information and should not be leaked to unauthorized devices. Further, encryption keys and information about sensors themselves (e.g., identity, location, etc.) must be protected to prevent eavesdropping and attacks based on traffic analysis. These challenges require measures that provide data confidentiality for sensor networks. Integrity is required to prevent adversaries from modifying sensor data, for example, with the purpose of injecting false readings and therefore affecting the response to the sensor readings. Authentication is necessary to ensure that any data disseminated in a sensor network originates from the correct source, particularly when a single node controls the entire network (e.g., a base station establishing routes or distributing multicast tree information). Further, many security attacks in sensor networks have the goal to disrupt the correct functioning of the network altogether, necessitating measures that ensure network availability. An additional requirement in sensor networks is the need for *data freshness*, which ensures that sensor data are recent and no old recordings of such data are being replayed. This is particularly important for key distribution schemes, for example, an attacker could record shared keys that are being exchanged in a network and replay these key distribution messages at a later time. Finally, many node and network management responsibilities found in WSNs provide adversaries with opportunities for attacks. For example, sensor node localization is important for correctly interpreting sensor data, for geographic routing protocols, and for redundancy elimination. However, many localization techniques require the exchange of information among sensors (e.g., beacons carrying positions, time stamps, and identity information) that may necessitate encryption. Similarly, time synchronization in sensor networks is based on message exchange among sensor nodes, where an adversary could inject false time stamps to increase synchronization errors among sensors.

11.3 Security Attacks in Sensor Networks

Sensor networks are vulnerable to a variety of attacks that attempt to compromise the network's operation and the data the sensor nodes generate. Specifically when sensor networks serve application purposes such as battlefield assessments and monitoring of civil infrastructure, they require protection from unauthorized access and tampering. This section describes a variety of security attacks that could occur in a WSN.

11.3.1 Denial-of-Service

A *Denial-of-Service* (DoS) attack can be characterized as an attempt of an adversary to stop a network from functioning or to disrupt the services a network provides. In wireless sensor networks, DoS attacks can occur at various layers of the protocol stack, where some may affect multiple layers simultaneously or attempt to exploit interactions between them (Wood and Stankovic 2002).

11.3.1.1 Physical Layer DoS

The wireless medium used in a WSN facilitates a variety of attacks. A *jamming* attack occurs when an adversary interferes with the radio frequencies of a WSN. If well positioned, a few attacking nodes can disable an entire network, even if the number of attacking nodes is much smaller than the number of nodes in the network. Even a single attacking node could disable an entire network if it is positioned close to a "critical" node (e.g., a gateway, therefore preventing any sensor data from leaving the sensor network) or its transmission power is large such that all nodes in a network may be prevented from correctly receiving any meaningful data. A common technique against jamming is to use *spread-spectrum* communication, as found in well-known standards such as IEEE 802.11 and Bluetooth. For example, in *frequency-hopping spread spectrum* (FHSS), communicating devices frequently hop between frequencies according to a certain hopping sequence. A jammer either must know this sequence to be able to jam the correct frequency for continuous disruption, or must jam a large frequency band. In addition, sensor networks should be able to detect and respond to jamming attacks in the network, for example, by switching nodes into low-power sleep modes (in order to preserve energy), while awakening them periodically to check if the jamming attack is still active. Nodes may also want to alert a gateway or base station to report the attack. Toward this end, nodes detecting a jamming attack could issue brief alerts to their neighbors and if at least one of these neighbors is outside the region of the attack (i.e., it is able to receive the alert message without interference), the message can be propagated to other nodes including the base station.

A *tampering* attack in a sensor network occurs when an adversary obtains physical access to a sensor node, allowing the attacker to destroy or modify the device, gain access to sensitive information (e.g., cryptographic keys), or use the device as an entry point for further attacks into the network. Possible strategies to protect a device from tampering and the consequences thereof include using tamper-proof materials and enclosures and to disable a device or delete its information when an attack is detected. For example, a technique often used in systems handling sensitive information (e.g., credit card payment terminals) is to erase all such data whenever a light sensor activates (e.g., due to the terminal's enclosure being opened).

11.3.1.2 Link Layer DoS

A *collision* attack at the link layer (Wood and Stankovic 2002) attempts to interfere with packet transmissions, thereby causing costly exponential backoff procedures and retransmissions in some MAC protocols. While error-correcting codes can be used to recover from corrupted bits in a packet, they may not be able to recover from all types of interferences (e.g., if too many bits have been corrupted) and they incur additional resource and energy overheads. An attacker could also attempt to cause collisions near the end of a frame, causing a node to repeatedly retransmit the entire packet. The goal of an attacker could be to cause the premature depletion of the node's energy resources (*exhaustion* attack). Similarly, a malicious node could exploit certain handshake techniques often found in MAC protocols. For example, an attacker could continuously issue an RTS message (IEEE 802.11 protocol) to prompt a CTS response from another node, eventually exhausting the energy resources of both nodes.

11.3.2   Attacks on Routing

One example of an attack on routing protocols of sensor networks is the *blackhole attack* (Karlof and Wagner 2003). With this type of attack, an adversary attempts to be a forwarder of data for one or more routes across the network. A malicious node can then simply drop all traffic that should pass through this node, therefore, such traffic never reaches the desti-nation. A similar attack is called the *selective forwarding attack* (Karlof and Wagner 2003), where only packets that match certain criteria are dropped instead of dropping all packets indiscriminately. Selective forwarding attacks are more difficult to detect or react upon than blackhole attacks since they are harder to distinguish from packet losses due to mobility or channel errors.

A *rushing attack* (Hu *et al.* 2003) on a sensor network exploits the nature of the route dis-covery procedure of on-demand routing protocols, for example, as found in protocols such as AODV and DSR. In this type of attack, a malicious node immediately forwards incoming route request messages to its neighbors, therefore "rushing" these messages without consid-eration of any protocol rules (e.g., that specify certain timeout or queuing procedures before forwarding). As a consequence, the node has an increased probability of being part of the chosen route between source and destination.

A *sinkhole attack* (Karlof and Wagner 2003) is another variant of the blackhole attack. However, to attract as much traffic as possible, the malicious node attempts to position itself on the path of as many network flows as possible. Traffic is therefore drawn toward this sinkhole, providing an attacker with an opportunity to disrupt or tamper with as much traffic as possible.

A *Sybil attack* occurs when an attacker claims to have several identities in the network. Similarly, in location-based routing protocols, an attacker claims to be at several locations simultaneously. If many nodes believe that this malicious node is their neighbor, there is a good chance that they will choose this node as forwarding node for their network traffic.

Another attack on the routing procedure of a sensor network is the *wormhole attack*. This attack is performed by nodes that have more resources available than typical sensor nodes in the network. For example, two collaborating attackers may attempt to deceive the rest of the network by possessing an out-of-band (and often bandwidth-rich) communication channel between themselves. To the rest of the network, this appears to be a fast, high-bandwidth link, which is desirable for many routing techniques. With this approach, the attacker nodes can fake an efficient and short path to the gateway of a network, therefore attracting a significant amount of traffic and enabling a variety of other attacks, such as the blackhole or sinkhole attacks.

11.3.3   Attacks on Transport Layer

The transport layer of the network protocol stack is responsible for managing end-to-end connections, for example, two well-known transport layer protocols are Transmission Control Protocol (TCP) for reliable stream-based communication and User Datagram Protocol (UDP) for unreliable packet-based communication. The *flooding* attack (Wood and Stankovic 2002) exploits the fact that many transport protocols (such as TCP) maintain state information and are therefore vulnerable to memory exhaustion. For example, an attacker may repeatedly make new connection requests, each adding more state information at the affected node and potentially leading to the node refusing further connections due to resource exhaustion. This in turn prevents connection requests from legitimate nodes from succeeding.

In the *desynchronization* attack, an adversary attempts to disrupt the communication between two legitimate nodes by repeatedly forging messages to these nodes. For example, reliable transport-layer protocols may use sequence numbers to keep track of successfully received packets, identify packet loss, and detect duplicates. Fake packets issued by an adversary can use these sequence numbers to make a node believe that its packets have not arrived at the destination, thereby eliciting resource-costly retransmissions.

11.3.4 Attacks on Data Aggregation

Data aggregation and data fusion are often used to combine multiple sensor data and to elim-inate redundant information. Aggregation can often have beneficial effects on the resource requirements of sensor flows, for example, by reducing the frequency of transmissions or the packet sizes. Even simple aggregation functions can easily be influenced by an attacker such that a network's behavior can be altered (Wagner 2004). For example, the *average* function $f(x_1, \ldots, x_n) = (x_1 + \cdots + x_n)/n$ is insecure even in the presence of a single malicious node. By replacing one real measurement x_1 with a fake reading x_1^*, the average is changed from $y = f(x_1, ..., x_n)$ to $y^* = f(x_1^*, x_2, ..., x_n) = y + (x_1^* - x_1)/n$. An attacker can freely choose the value of x_1^* and, therefore, can control the outcome of the aggregation.

Similarly, the *sum*, *minimum*, and *maximum* functions are also insecure. The sum $f(x_1, ..., x_n) = x_1 + \cdots + x_n$ can be modified at will by maliciously replacing a real mea-surement x_1 with a fake reading x_1^*. The minimum function $f(x_1, ..., x_n) = \min(x_1, ..., x_n)$ is also insecure, even though replacing a real measurement with a fake value does not always affect the function's outcome. That is, replacing x_1 with x_1^* only raises the minimum if x_1 is the unique smallest sensor reading among all x_i. However, an attacker can modify the computed minimum by choosing x_1^* to be very small compared to all correct readings. By symmetry, the maximum function is also insecure, since an attacker can raise the maximum value by hijacking a single sensor reading.

In contrast, the effect of hijacking a single sensor reading may be comparably small for the *count* operation if the number of correct readings is sufficiently large. The count function is similar to the sum function, except that each sensor reading only contributes 0 or 1 to the result of the operation. That is, an attacker with control over k compromised nodes can change the outcome of the function by at most k, which may be negligible if k is small compared to the total number of sensor inputs.

11.3.5 Privacy Attacks

While the security threats described so far are mainly targeted at disrupting a network from correct operation, the vast amount of information collected in a WSN itself is also at risk of potential abuse. That is, an adversary may attempt to obtain sensitive information by accessing information stored on a sensor node or by *eavesdropping* on the network (Gruteser *et al.* 2003). The broadcast nature of wireless networks makes it easy to monitor and capture the transmissions between nodes, particularly when no cryptographic mechanisms are used to protect the sensor data. Eavesdropping can also be combined with *traffic analysis* (Deng *et al.* 2005a), which can be used by an adversary to identify sensor nodes of interest in a network. For example, increases in communications between certain nodes can indicate an increased level of activity (and therefore the presence of data that could be compromised) in those parts of the network. Similarly, traffic analysis can be used to identify nodes that may be more important to network operation than others, such as base stations and gateways.

11.4 Protocols and Mechanisms for Security

In order to defend against the many possible attacks in a WSN, a variety of security protocols and other defense mechanisms can be used. This section presents and discusses a variety of such protocols and mechanisms with a particular focus on their applicability in sensor networks.

11.4.1 Symmetric and Public Key Cryptography

While public key cryptography can be used to provide confidentiality, integrity, and authentication, public key algorithms are computationally very expensive, which may limit their use in resource-constrained sensor networks (Gaubatz et al. 2004). Symmetric key cryptography approaches can be significantly more resource-efficient, which makes them the more common choice in WSNs, even though implementations of RSA (Rivest et al. 1983) and ECC (elliptic curve cryptography) (Menezes et al. 1996) for resource-constrained sensors do exist. A major disadvantage of symmetric key approaches is the problem of key distribution, that is, the shared symmetric key must first be known to both communicating nodes before they can exchange data securely.

11.4.2 Key Management

Symmetric cryptographic schemes are the common choice for sensor networks when resource constraints prohibit the use of the more complex public key schemes. However, a major shortcoming of symmetric cryptography is the need for key management, that is, the reliable and secure establishment of shared cryptographic keys among neighboring nodes in a WSN. For example, the Peer Intermediaries for Key Establishment (PIKE) approach (Chan and Perrig 2005) is a technique that uses sensor nodes as trusted intermediaries for the distribution of keys. In this approach, every sensor shares a different pairwise key with each of $O(\sqrt{n})$ other nodes, where n is the number of nodes in the network. Furthermore, the keys are deployed such that for any pair of nodes A and B, there exists at least one node, C, that shares a pairwise key with both A and B. Each sensor in PIKE has an ID of the form (x, y), where $x, y \in \{0, 1, 2, ..., \sqrt{n} - 1\}$. That is, the sensor network is represented as a matrix with $\sqrt{n}$ rows and columns, where a node's position in the matrix is the node's ID. Then, each node (x, y) shares a pairwise key with each node in the two following sets:

$$(i, y) \, \forall i \in \{0, 1, 2, ..., \sqrt{n} - 1\} \tag{11.1}$$

$$(x, j) \, \forall j \in \{0, 1, 2, ..., \sqrt{n} - 1\} \tag{11.2}$$

For example, node (x, y) will share a key $K_{(x,y),(1,y)}$ with node $(1, y)$ and another key $K_{(x,y),(2,y)}$ with node $(2, y)$. Altogether, a node will maintain $2(\sqrt{n} - 1)$ keys. Figure 11.2

shows a sample virtual ID space for 100 nodes, where each number represents the ID of a node (note that this representation does not reflect the actual physical positions of the sensor nodes). The dark shaded boxes identify all nodes that share a key with node 91, while the light shaded boxes indicate all nodes that share a key with node 14. Due to this approach, any two nodes in the network will be able to find two node IDs which will share pairwise keys with both of them. Specifically, if node A has ID (x_A, y_A) and node B has ID (x_B, y_B), then the nodes with IDs (x_A, y_B) and (x_B, y_A) will share pairwise keys with both A and B. If node A (e.g., node 14 in Figure 11.2) wants to perform key establishment with another node (e.g., node 91), A can identify the identities of potential intermediaries by looking for the intersections of the shaded boxes. For example, node 94 is in the same row as 91 and in the same column as 14, therefore it shares keys with both of them and can serve as intermediary. Node 14 then encrypts the new key to be shared with node 91 using the existing key shared with node 94 and then sends the encrypted key to node 94. Node 94 decrypts the message, encrypts it again with the key shared with node 91 and sends the new message to node 91. Node 91 decrypts the message, obtains the new key, and confirms the receipt of the new key by replying to node 14.

11.4.3   Defenses Against DoS Attacks

Denial-of-service (DoS) attacks in sensor networks are common and require effective measures to avoid them or prevent them from spreading throughout the network. For example, when a jamming attack is detected or suspected, a sensor network can attempt to isolate the affected region by routing traffic around the disabled parts of the network. Another technique to limit the damage from jamming attacks is to use spread-spectrum techniques as described in Section 11.3.1. At the link layer, collision and exhaustion attacks can be addressed using error-correcting codes (which add processing and communication overheads) and rate-limiting schemes that allow a device to ignore requests that could lead to premature energy depletion. Spoofing and alteration can be addressed at the network layer by using message authentication code or MAC (not to be confused with *medium access control*), which can be viewed as the cryptographically secure checksum of a message. These checksums allow a receiver to verify whether a message has been spoofed or altered (Sen 2009).

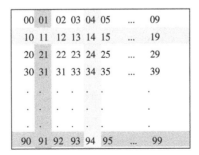

Figure 11.2   Virtual ID space in PIKE.

The path-based denial-of-service attack (PDoS) is an attack in which the attacker overwhelms the nodes in a remote sensor network by flooding a multi-hop end-to-end communication path with either replayed packets or randomly injected packets (Deng *et al.* 2005b). One-way hash chains are sequences of numbers where it is trivial to compute $y = F(x)$, but computationally infeasible to compute $x = F^{-1}(y)$. Each node in the network utilizes the hash chain to validate a received packet, that is, a node systematically cycles through the chain to determine whether the packet is from a trusted source. If a packet cannot be validated, it is dropped.

11.4.4 Defenses Against Aggregation Attacks

As previously discussed, many simple aggregation functions such as sum, minimum, and maximum are inherently insecure. However, several techniques for improving the resilience of aggregation functions can be used, for example, two such techniques are delayed aggregation and delayed authentication (Hu and Evans 2003).

In these techniques, it is assumed that the base station generates a one-way key chain using a public one-way function F, where $K_i = F(K_{i+1})$. Each device stores key K_0 before deployment where $K_0 = F^n(K)$ (i.e., F applied to a secret key n times). Then, the first base station transmissions will be encrypted using key $K_1 = F^{n-1}(K)$. Once all messages transmitted using K_1 have been received, the base station reveals K_1. As a consequence, all nodes can compute $F(K_1) = F(F^{n-1}(K))$ and verify that it matches $K_0 = F^n(K)$. Further, sensor nodes can then decrypt the messages that were previously transmitted encrypted with K_0. In a similar manner, successive keys can be revealed until $K_n = K$ is reached (if more keys are needed, the base station can then start a new sequence).

Assume that four sensor nodes A–D are sending messages to the base station in a network structured as a tree as shown in Figure 11.3. Each node's message contains the

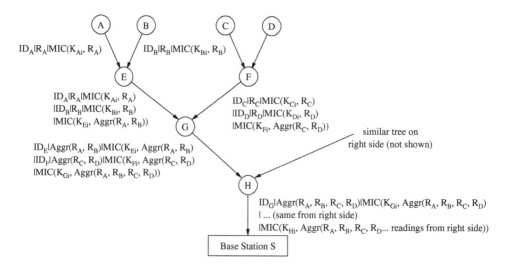

Figure 11.3   Secure aggregation example.

sender's ID, the sensor data, and a MAC calculated over the data using a temporary key. The parent node of the sensor node is not yet able to verify the MAC since the child's key has not been revealed to the parent. The parent node (e.g., node E in Figure 11.3) stores this message and retransmits it to its own parent after a certain timeout value. E's message to its parent G contains the messages received from its children (e.g., nodes A and B) and a MAC computed over the aggregate of A's and B's data using E's key. This process continues, that is, every intermediate node combines the data coming from its children and adds its own MAC over the aggregate of all data using its own key. Once the base station receives messages from its children, it can compute the final aggregate value.

The base station has a shared temporary key with each sensor node, therefore it can verify whether a received message was transmitted by H by calculating the MAC of the aggregation using K_{Hi} and comparing it to the MAC in the message. While this validates that H sent the final message, it does not validate that the message correctly reflects the readings from the other nodes. To validate data, the base station reveals temporary node keys to the network by sending each key (along with a MAC) to all sensor nodes using its own current key K_i. After sending out all the node keys, the base station sends out its current key K_i such that nodes are able to check the transmitted MAC values and to advance to the next key in the chain for future messages.

In summary, the described process delays both aggregation and authentication, for example, aggregation does not take place at the first hop that would be able to perform this aggregation, but at the second hop. While this may increase resource expenditures, it may also enable integrity guarantees where consecutive nodes have not been compromised.

11.4.5 Defenses Against Routing Attacks

Most attacks from the "outside" of a network can be prevented using simple link-layer encryption and authentication using a globally shared key (Karlof and Wagner 2003). Because the adversary is prevented from joining the network, attacks such as selective forwarding or sinkholes are not possible. However, when networks are attacked from the "inside", for example, using a compromised node, this approach is ineffective and more sophisticated solutions are needed.

Sybil attacks can be addressed by verifying the identities of sensor nodes. For example, each sensor node could share a unique symmetric key with a trusted base station, which can be used to verify each other's identity. A base station can also limit the number of neighbors a node is allowed to have, that is, even when a node is compromised, it can only communicate with its verified neighbors.

Sinkholes are difficult to defend against in protocols where routes are established on the basis of information that is hard to verify, for example, reliability or energy measurements. Routes based on minimum hop counts are easier to verify, but the hop count can be misrepresented through a wormhole (Karlof and Wagner 2003). One category of protocols that is resistant to these attacks is geographic routing, because networks using location-based routing techniques establish a topology on demand based on localized interactions and information, without the initiation from a base station. Since traffic is "naturally" routed toward the physical location of the base station, it is difficult to redirect traffic elsewhere to create a sinkhole.

In a rushing attack, a node's goal is to exploit the route discovery process in on-demand routing protocols to position itself on as many routes as possible. However, to prevent such attacks, a combination of several protective measures can be used. For example, some attackers may forward route requests beyond the normal radio transmission range (e.g., using high transmission power), thereby suppressing subsequent request messages from this route discovery. A *secure neighbor detection* approach (Hu *et al.* 2003) can be used to allow both the sender and the receiver of a route request to verify that the other party is in fact within the normal transmission range. For example, a three-round mutual authentication protocol with tight delay timing can be deployed. In the first round, a node sends a *neighbor solicitation* packet (either via broadcast or via unicast to a specific node). In the second round, a node receiving the solicitation packet responds with a *neighbor reply* message and in the third round, the initiator of this handshake communication sends a *neighbor verification* message, which includes broadcast authentication of a timestamp and the link from the source to the destination.

11.4.6 Security Protocols for Sensor Networks

The Security Protocols for Sensor Networks (SPINS) project makes two main contributions to defending against attacks: the Secure Network Encryption Protocol (SNEP) and a "micro" version of the Timed, Efficient, Streaming, Loss-tolerant Authentication (μTESLA) protocol (Perrig *et al.* 2002). The main goal of the SNEP protocol is to provide confidentiality, two-party data authentication, and data freshness, while μTESLA provides authentication for data broadcast. Each node is assumed to have a secret key shared with the base station.

11.4.6.1 Secure Network Encryption Protocol

SNEP takes the resource limitations of typical sensor nodes into consideration by relying on simple algorithms for encryption, authentication, and random number generation. The key properties of SNEP are its symmetric security, replay protection, and low communication overhead. Symmetric security refers to the fact that the same message is encrypted differently each time. To achieve two-party authentication and integrity, SNEP uses a MAC, where the larger the MAC the more difficult it is for an adversary to guess the appropriate code for a message. On the other hand, large codes also mean larger packet sizes.

Two communicating nodes A and B share a secret *master key*, which is used to derive four independent keys using a pseudorandom function. Two of these keys are used for encryption of messages in each direction (K_{AB} and K_{BA}) and two keys are used as message integrity codes, again one for each direction (K'_{AB} and K'_{BA}). A complete encrypted message has then the following format:

$$A \rightarrow B : \{D\}_{\langle K_{AB}, C_A \rangle}, \text{MAC}(K'_{AB} C_A || \{D\}_{\langle K_{AB}, C_A \rangle}) \tag{11.3}$$

where D is the data encrypted with the encryption key K and the counter is C. The MAC is computed in the form $M = \text{MAC}(K', C || E)$. SNEP provides data authentication (using the MAC), replay protection (using the counter value in the MAC), freshness (the counter values enforce a message ordering), semantic security (since the counter is encrypted with each message, the same message will be encrypted differently each time), and low communication overhead (assuming that the counter state is kept at each end point and is not sent in the message). Data freshness under SNEP is considered to be *weak* only since SNEP enforces a sending order within a node B, but no absolute assurance to node A that a message was created by B in response to an event in A. In order to achieve strong freshness, a nonce (i.e., a random number so long that an exhaustive search for all possible nonces is infeasible) can be included in the protocol. Node A randomly generates nonce N_A and sends it along with a request message to node B. Node B then returns the nonce with the response message in an authenticated protocol that operates as follows:

$$A \rightarrow B : N_A, R_A \tag{11.4}$$

$$B \rightarrow A : \{R_B\}_{\langle K_{BA}, C_B \rangle}, \text{MAC}(K'_{BA}, N_A || C_B || \{R_B\}_{\langle K_{BA}, C_B \rangle}) \tag{11.5}$$

If the MAC verifies correctly, A knows that node B generated its response after A's request.

The μTESLA protocol focuses on the need for authenticated broadcast in wireless sensor networks. It relies on the symmetric mechanisms provided by SNEP to authenticate the first packet in a broadcast message. It is an extension of TESLA (Perrig *et al.* 2000), which was not designed for use in environments with limited computing resources. TESLA uses digital signatures to authenticate the initial packet and has an overhead of 24 bytes per packet, which can be significant for sensor networks, where messages are typically very small. Authenticated broadcast requires an asymmetric mechanism (otherwise any compromised receiver could forge messages from the sender), but asymmetric cryptographic mechanisms are often high in resource requirements. Instead, μTESLA emulates asymmetry through a delayed disclosure of symmetric keys. μTESLA assumes that the base station and the sensor nodes are loosely time synchronized and each node knows an upper bound on the maximum synchronization error. When the base station sends a message, it authenticates it by computing a MAC on the packet with a key that is secret at this point. When a node receives the packet and the key is unknown, the node knows that the MAC key is known only to the base station. The node stores the packet until the base station, at the time of key disclosure, broadcasts the verification key to all receivers. The node can now use the key to authenticate the stored packet.

11.4.7 TinySec

The TinySec architecture is a lightweight and generic link-layer security package that developers can easily integrate into sensor network applications (Karlof *et al.* 2004). It supports two different security options: (1) authenticated encryption (TinySec-AE), where data payload is encrypted and a MAC is used to authenticate a packet, and (2) authentication only (TinySec-Auth), where an entire packet is authenticated with a MAC (but the payload is left unencrypted). TinySec relies on cipher block chaining (CBC) and a specially formatted 8-byte initialization vector (IV) for encryption. For authentication, TinySec relies on efficient and fast cipher block chaining construction (CBC-MAC) for computing and verifying MACs. An advantage of CBC-MAC is that since it relies on a block cipher, it minimizes the number of cryptographic primitives that must be implemented, which is beneficial for sensor nodes with limited storage capacities. The length of the MAC is chosen to be only 4 bytes, that is, an adversary can repeatedly attempt blind forgeries, which would lead to success after at most 2^{32} attempts. While this number appears small, it must be noted that an adversary must assess the validity of a code by sending it to an authorized receiver. That further means that up to 2^{32} messages must be transmitted, which provides a sufficient level of security for sensor networks (Boyle and Newe 2008).

11.4.8 Localized Encryption and Authentication Protocol

The Localized Encryption and Authentication Protocol (LEAP) (Zhu *et al.* 2003) is a key management protocol for sensor networks, designed to support in-network processing. A key motivation for this protocol is the observation that different types of messages (e.g., control packets versus data packets) in a sensor network have different security requirements. A single keying mechanism may not suitable for meeting these different requirements, for example, while authentication may be needed for all types of packets, confidentiality may only be required for certain types of messages (e.g., aggregated sensor readings).

LEAP provides four keying mechanisms: *individual keys*, *group keys*, *cluster keys*, and *pairwise shared keys*. In the individual key mechanism, every node has its own unique key shared with the base station. This key is used for confidential communication or for computing message authentication codes if a node wants the base station to verify its sensed readings. A group key is a globally shared key that is used by the base station for the transmission of encrypted messages to the entire sensor network. Common examples of such messages include queries or interests. A cluster key is a key shared between a sensor node and its neighbors and is used for securing local broadcast messages (e.g., routing control messages). Finally, a pairwise shared key is a key shared by a sensor node and one of its immediate neighbors. LEAP uses these keys for secure communications among a pair of nodes, for example, allowing a node to securely distribute its cluster key to its neighbors or to securely transmit its sensor readings to an aggregation node.

LEAP also provides a technique for local broadcast authentication. Toward this end, every node generates a one-way key chain of certain length and transmits the first key in the chain to each neighbor, encrypted with the pairwise shared key. Whenever a node sends a message, it takes the next key from the chain (each key is called an AUTH key) and attaches it to the message. These keys are disclosed in the reverse order of their generation and a receiver can verify the message based on the first received key or a recently disclosed AUTH key.

11.5   IEEE 802.15.4 and ZigBee Security

The IEEE 802.15.4 standard and the ZigBee specification are popular protocol choices for WSNs. Therefore, this chapter concludes with a discussion of the security measures available in these protocols.

The IEEE 802.15.4 standard provides four basic security models: *access control, message integrity, message confidentiality*, and *replay protection* (Sastry and Wagner 2004). Security in IEEE 802.15.4 is handled by the MAC layer and an application can choose specific security requirements by setting appropriate parameters in the radio stack (by default, security is not enabled). The standard distinguishes between eight security suites (outlined in Table 11.1), each with different levels of protection for the transmitted data. The first suite offers no security, the second suite offers encryption only (AES – CTR), followed by a group of suites with authentication only (AES – CBC – MAC), and a group of suites with both authentication and encryption (AES – CCM). Suites that offer authentication differ in the sizes of the MAC, which varies from 32 to 128 bits. For every suite that offers encryption,

Table 11.1 Security suites supported in IEEE 802.15.4
(Sastry and Wagner 2004)

Name	Description
Null	No security
AES – CTR	Encryption only, CTR mode
AES – CBC – MAC – 128	128-bit MAC
AES – CBC – MAC – 64	64-bit MAC
AES – CBC – MAC – 32	32-bit MAC
AES – CCM – 128	Encryption and 128-bit MAC
AES – CCM – 64	Encryption and 64-bit MAC
AES – CCM – 32	Encryption and 32-bit MAC

IEEE 802.15.4 also offers optional replay protection consisting of monotonically increasing sequence numbers for messages to allow a recipient to detect replay attacks.

The first suite Null does not provide any security. All other security suites use the Advanced Encryption Standard (AES) block cipher, which is also known as Rijndael. The National Institute of Standards and Technology defines five modes of operation, including the counter (CTR) and cipher block chaining (CBC) modes (Sastry and Wagner 2004). When authentication is needed, one of the three AES – CBC – MAC variants can be used, which compute a message integrity code using a block cipher in CBC mode. The three AES – CCM suites combine encryption and authentication by using the counter mode and the CBC mode (CCM is short for Counter with CBC – MAC).

In addition to the security features of IEEE 802.15.4, the ZigBee specification also introduces the concept of a *trust center*, a responsibility typically assumed by the ZigBee coordinator. The trust center is responsible for authentication of devices wishing to join a network (*trust manager*), maintaining and distributing keys (*network manager*), and enabling end-to-end security between devices (*configuration manager*).

ZigBee also differentiates between a *residential* and a *commercial* mode (Boyle and Newe 2008). In the residential mode, the trust center allows nodes to join the network, but it does not establish keys with the network devices. In the commercial mode, it generates and maintains keys and freshness counters with every device in the network. The disadvantage of the commercial mode is its memory cost, which grows with the size of the network.

The ZigBee specification uses the CCM* mode for its security services, which is also a combination of CTR mode and CBC – MAC mode. Compared to the CCM mode, CCM* offers encryption-only and integrity-only capabilities. Similar to the specifications in the IEEE 802.15.4 standard, ZigBee has several levels of security, including no security, encryption only, authenticated only, and both encryption and authentication. Levels that provide authentication use a MAC that can vary from 4 to 16 bytes.

11.6 Summary

Like every other computer network, wireless sensor networks are exposed to a variety of threats and attacks and like most other networks, sensor networks require support for confidentiality, integrity, and authentication to protect sensor nodes and sensor data.

However, several unique characteristics of WSNs, such as remote deployment (which facilitates an adversary's physical access to sensor nodes) and resource constraints, make it easier to compromise sensors and sensor data. Further, many sensor networks are attractive targets for attackers due to the nature of many WSN applications and the sensitive data they generate (e.g., military applications, emergency response, health care). This chapter provided a brief overview of several types of attacks commonly found in sensor networks and techniques and protocols to defend a network or to detect an intrusion or compromised node. As WSNs continue to become more commonplace, it is to expect that security challenges will increase, the types and number of threats will evolve, and new solutions to protect sensor networks and sensor data will be required.

Exercises

11.1 Describe the CIA security model. Which service(s) described in this model do you think are essential for the following scenarios? Justify your answers.

   (a) A WSN that allows emergency response teams to avoid risky and dangerous areas and activities.
   (b) A WSN that collects biometric information collected at an airport.
   (c) A WSN that measures air pollution in a city for a research study.
   (d) A WSN that alerts a city of an impending earthquake.

11.2 What is a man-in-the-middle attack? Can you imagine a concrete WSN scenario where such an attack could be catastrophic?

11.3 Explain the concepts of symmetric and asymmetric keys. This chapter mentioned a *shift cipher* as a simple example of a cryptographic technique. Is this cipher a symmetric or an asymmetric key cryptography technique? What are the problems with such a simple cipher?

11.4 Why do you think authentication can be a particularly significant problem in a WSN?

11.5 Explain some of the characteristics of a WSN that make routing security difficult to implement.

11.6 While "typical" computers are in homes, offices, labs, etc., wireless sensor nodes are often placed in places that are publicly open and accessible. What kind of attacks could an adversary initiate by accessing a single sensor node in a large-scale WSN?

11.7 What is "data freshness" and why is it important in sensor networks?

11.8 What is a denial-of-service attack? Explain the following attacks:

   (a) Jamming attack
   (b) Exhaustion attack
   (c) Tampering attack

11.9 Consider routing attacks such as selective forwarding, sinkhole, blackhole, Sybil, rushing, and wormhole attacks. Describe briefly each type of attack and discuss how these attacks could take place in the following types of networks:

(a) A network using a table-based routing protocol such as OLSR.

(b) A network using an on-demand routing protocol such as DSR.

(c) A network using a location-based routing protocol such as GEAR.

11.10 In this chapter, data aggregation functions such as *average*, *sum*, and *minimum* were called "insecure". What does this mean and which technique can be used to increase the resilience of aggregation functions?

11.11 Consider the virtual ID space for the PIKE scheme in Figure 11.2. In this example, how many options does node 3 have to establish a key with node 15? Describe each option.

11.12 What is a "nonce"? How does SPINS use them and what services are provided by the SNEP protocol?

11.13 What are the security models provided by IEEE 802.15.4? What is the purpose of the trust center in ZigBee?

References

Boyle, D., and Newe, T. (2008) Securing wireless sensor networks: Security architectures. *Journal of Networks* **3** (1), 65–77.

Chan, H., and Perrig, A. (2005) PIKE: Peer intermediaries for key establishment in sensor networks. *Proc. of the 24th Annual Joint Conference of the IEEE Computer and Communications Societies (INFOCOM), Miami, FL.*

Deng, J., Han, R., and Mishra, S. (2005a) Countermeasures against traffic analysis attacks in wireless sensor networks. *Proc. of the IEEE Conference on Security and Privacy for Emerging Areas in Communication Networks (SecureComm), Athens, Greece.*

Deng, J., Han, R., and Mishra, S. (2005b) Defending against path-based DoS attacks in wireless sensor networks. *Proc. of the 3rd ACM Workshop on Security of Ad Hoc and Sensor Networks (SANS), Alexandria, VA.*

Gaubatz, G., Kaps, J.P., and Sunar, B. (2004) Public key cryptography in sensor networks revisited. *Proc. of the 1st European Workshop on Security in Ad Hoc and Sensor Networks, Heidelberg, Germany.*

Gruteser, M., Schelle, G., Jain, A., Han, R., and Grunwald, D. (2003) Privacy aware location sensor networks. *Proc. of the 9th USENIX Workshop on Hot Topics in Operating Systems (HotOS IX), Lihue, HI.*

Hu, L., and Evans, D. (2003) Secure aggregation for wireless networks. *Proc. of the Workshop on Security and Assurance in Ad Hoc Networks, Orlando, FL.*

Hu, Y.C., Perrig, A., and Johnson, D. B. (2003) Rushing attacks and defense in wireless ad hoc network routing protocols. *Proc. of the 2nd ACM Workshop on Wireless Security, San Diego, CA.*

Karlof, C., and Wagner, D. (2003) Secure routing in wireless sensor networks: Attacks and countermeasures. *Ad Hoc Networks* **1** (23), 293–315.

Karlof, C., Sastry, N., and Wagner, D. (2004) TinySec: A link layer security architecture for wireless sensor networks. *Proc. of the 2nd International Conference on Embedded Networked Sensor Systems, Baltimore, MD.*

Menezes, A.J., Vanstone, S.A., and Oorschot, P.C.V. (1996) *Handbook of Applied Cryptography*. CRC Press, Boca Raton, FL.

Perrig, A., Canetti, R., Tygar, J., and Song, D. (2000) Efficient authentication and signing of multicast streams over lossy channels. *Proc. of the IEEE Symposium on Security and Privacy, Berkeley, CA.*

Perrig, A., Szewczyk, R., Tygar, J.D., Wen, V., and Culler, D.E. (2002) SPINS: Security protocols for sensor networks. *Wireless Networks* **8**, 521–534.

Rivest, R.L., Shamir, A., and Adleman, L. (1983) A method for obtaining digital signatures and public key cryptosystems. *Communications of the ACM* **26** (1), 96–99.

Sastry, N., and Wagner, D. (2004) Security considerations for IEEE 802.15.4 networks. *Proc. of the 3rd ACM Workshop on Wireless Security, Philadelphia, PA*.

Sen, J. (2009) A survey on wireless sensor network security. *International Journal on Communications Networks and Information Security (IJCNIS)* **1** (2), 59–82.

Wagner, D. (2004) Resilient aggregation in sensor networks. *Proc. of the 2nd ACM Workshop on Security of Ad Hoc and Sensor Networks, Washington, DC*.

Wood, A.D., and Stankovic, J.A. (2002) Denial of service in sensor networks. *Computer* **35** (10), 54–62.

Zhu, S., Setia, S., and Jajodia, S. (2003) LEAP: Efficient security mechanism for large-scale distributed sensor networks. *Proc. of the 10th ACM Conference on Computer and Communications Security, Washington, DC*.

12

Sensor Network Programming

Application development for sensor networks differs in many ways from programming "traditional" distributed computing systems. Examples of such differences include the continuous interaction of sensor nodes with their physical environment, the stringent resource constraints of sensor nodes, the ad hoc deployment of many sensor networks, and the frequent changes in network topology due to failures or mobility. This chapter discusses aspects of programming large sensor networks that consider these challenges. From the network developer's perspective, the goal is to design and program a reliable and efficient wireless sensor network that can cope with the dynamics and uncertainties present in sensing systems. From the user's perspective, the network is often viewed as a database and the users interact with sensor nodes via queries, which must be responded to in a reliable and efficient fashion. Many simulation tools and techniques are closely tied to the operating system used on sensor nodes. The reader is referred to Chapter 4 for a discussion of operating systems for wireless sensor nodes.

Sensor network programming approaches can be classified as either *node-centric* or *application-centric*. Node-centric languages and programming tools focus on the development of sensor software on a per-node level. In contrast, programming using an application-centric approach considers parts or all of the network as one single entity (Sugihara and Gupta 2008). This chapter presents representative examples for both categories.

12.1 Challenges in Sensor Network Programming

A sensor network differs from traditional computing environments in various aspects, thereby necessitating programming frameworks and tools that consider a sensor network's unique characteristics. Specifically, the following characteristics significantly affect the design of sensor network programming tools:

1. *Reliability:* Wireless sensor networks are inherently more unreliable than other distributed systems. Therefore, sensor networks are built to adapt to changing dynamics and node and link errors such that the network continues to serve its intended purpose even when parts of the network have failed. While many faults in a network will never be noticed by an application (e.g., a routing protocol autonomously reroutes traffic around a failed node), resilience to failures and topology changes should be supported by a programming environment.

Fundamentals of Wireless Sensor Networks: Theory and Practice   Waltenegus Dargie and Christian Poellabauer
© 2010 John Wiley & Sons, Ltd

2. *Resource constraints:* Wireless sensor networks are typically very resource-constrained, which affects the programming approach, maximum code size, and other aspects of application development. Most notably, energy efficiency is particularly critical in WSNs and penetrates every aspect of sensor network design, from duty cycles to routing protocols to in-network data processing. Therefore, programming tools and models should allow a developer to effectively exploit energy-saving techniques and approaches, while details should be hidden from the programmer.

3. *Scalability:* Sensor networks can scale up to hundreds and thousands of sensor nodes, therefore programming models should support developers in designing applications and software for large-scale (and possibly heterogeneous) networks. Manual configuration, maintenance, and repair of individual sensor nodes will be infeasible due to the large number of devices, therefore necessitating support for self-management and self-configuration. The scale of a network can also be addressed by using programming models that consider the entire network as one whole entity instead of focusing on each individual device.

4. *Data-centric networks:* In many wireless sensor networks, not only are the individual sensor nodes of interest, but also the data they generate and disseminate. Sensor network applications are therefore concerned about obtaining useful information in a timely fashion, where it is irrelevant which sensor node(s) generated this information. Many applications are only concerned with the collection of data at a central point, for example, a server that stores, analyzes, or visualizes the sensor data. Other applications require immediate processing and analysis of data within the network, for example, to eliminate redundant information, to aggregate data from multiple sensors, and to quickly identify if sensor data should be propagated further or acted upon. Each category will require different programming models, where the latter category will also require support for collaboration, that is, programming a network results in generating distributed algorithms that must work across many or all nodes in a resource-efficient manner.

12.2   Node-Centric Programming

Under the node-centric model, programming abstractions, languages, and tools focus on the development of sensor software on a *per-node* level. The overall network-wide sensing application is then described as a collection of pairwise interactions of individual sensor nodes. This section describes examples of programming models that focus on software development for individual nodes.

12.2.1   nesC Language

The combination of the TinyOS operating system and the nesC (Gay *et al.* 2003) programming language has become the de facto standard for node-centric programming in WSNs. The programming language nesC is an extension to the popular C programming language and provides a set of language constructs to implement code for distributed embedded systems such as motes. TinyOS is a component-based OS written in nesC and is described in Section 4.3.1. Unlike traditional programming languages, nesC must address the unique challenges of WSNs. For example, activities in a sensor network (e.g., sensor acquisition, message transmission and arrival) are initiated by *events* such as the detection of a change in the physical environment. These events may occur while a node is processing data, that is, sensor nodes must be able to concurrently perform their processing tasks while responding to events. In addition, as discussed many times throughout this book, sensor nodes are typically very resource-constrained and prone to hardware failures; therefore, programming languages for sensor nodes should take these characteristics into consideration.

Applications based on nesC consist of a collection of *components*, where each component *provides* and *uses* interfaces. A "provides" interface in nesC is a set of method calls that are exposed to higher layers, while a "uses" interface is a set of method calls that hide details of lower-layer components. An interface describes the use of some kind of service (e.g., sending a message). The following code shows a concrete example from the TinyOS timer service. This example provides the StdControl and Timer interfaces and uses a Clock interface (Gay *et al.* 2003).

```
module TimerModule {
    provides {
        interface StdControl;
        interface Timer;
    }
    uses interface Clock as Clk;
}

interface StdControl {
    command result_t init ();
}

interface Timer {
    command result_t start (char type, uint32_t interval);
    command result_t stop ();
    event result_t fired ();
}

interface Clock {
    command result_t setRate (char interval, char scale);
    event result_t fire ();
}

interface Send {
    command result_t send (TOS_Msg *msg, uint16_t length);
    event result_t sendDone (TOS_Msg *msg, result_t success);
}

interface ADC {
    command result_t getData ();
    event result_t dataReady (uint16_t data);
}
```

This example also shows the definitions for the Timer, StdControl, Clock, Send (communication), and sensor (ADC) interfaces. The Timer interface defines two types of *commands* (which are essentially functions): start and stop. The Timer interface further defines an *event*, which is also a function. While commands are implemented by the providers of an interface, events are implemented by the users. Similarly, all other interfaces in this example define both commands and events.

Besides the interface specification, components in nesC also have an implementation. *Modules* are components implemented by application code, while *configurations* are components that are implemented by connecting interfaces of existing components. Every nesC application has a *top-level configuration* that describes how components are wired together. Functions (i.e., commands and events) in nesC are described as $f.i$, where f is a function in an interface i. Functions are invoked using the *call* operation (for commands) and the *signal* operation (for events). The following code shows a brief excerpt of an implementation of an application that periodically obtains sensor readings (Gay *et al.* 2003).

```
module PeriodicSampling {
    provides interface StdControl;
    uses interface ADC;
    uses interface Timer;
    uses interface Send;
}

implementation {
    uint16_t sensorReading;

    command result_t StdControl.init () {
        return call Timer.start (TIMER_REPEAT, 1000);
    }

    event result_t Timer.fired () {
        call ADC.getData ();
        return SUCCESS;
    }

    event result_t ADC.dataReady (uint16_t data) {
        sensorReading = data;
        ...
        return SUCCESS;
    }
    ....
}
```

In this example, StdControl.init is called at boot time, where it creates a repeat timer that expires every 1000 ms. Upon timer expiration, a new sensor sample is obtained by calling ADC.getData, which triggers the actual sensor data acquisition (ADC.dataReady).

Returning to the TinyOS timer example, the following code sequence shows how the timer service in TinyOS (TimerC) is built by wiring two subcomponents, TimerModule and HWClock (which provides access to the on-chip clock).

```
configuration TimerC {
    provides {
        interface StdControl;
```

```
interface Timer;
    }
}

implementation {
    components TimerModule, HWClock;

    StdControl = TimerModule.StdControl;
    Timer = TimerModule.Timer;

    TimerModule.Clk -> HWClock.Clock;
}
```

In TinyOS, code executes either asynchronously (in response to an interrupt) or synchronously (as a scheduled task). Race conditions can occur when concurrent updates to shared state are performed. In nesC, code that is reachable from at least one interrupt handler is called *asynchronous code* (AC) and code that is only reachable from tasks is called *synchronous code* (SC). Synchronous code is always atomic to other synchronous codes, because tasks are always executed sequentially and without preemption. However, race conditions are possible when shared state is modified from AC or when shared state is modified from SC that is also modified from AC. Therefore, nesC provides programmers with two options to ensure atomicity. The first option is to convert all of the sharing code to tasks (i.e., SC only). The second option is to use *atomic sections* to modify shared state, that is, brief code sequences that nesC will always run atomically. Atomic sections are indicated with the *atomic* keyword, which indicates that a block of statements should be executed atomically, that is, without preemption, as shown in the following code excerpt.

```
...
event result_t Timer.fired () {
    bool localBusy;
    atomic {
        localBusy = busy;
        busy = TRUE;
    }
    ...
}
...
```

Nonpreemption can be obtained by disabling interrupts for the duration of an atomic section. However, to ensure that interrupts are not disabled for too long, no call commands or signal events are allowed within atomic sections.

12.2.2 TinyGALS

TinyGALS (Cheong *et al.* 2003) is a globally asynchronous and locally synchronous (GALS) approach for programming event-driven embedded systems. A TinyGALS program consists of modules, which are composed of components (the most basic elements). A component C has a set of internal variables V_C, a set of external variables X_C, and a set of methods I_C that operate on these variables. Methods are further divided into calls in the ACCEPTS$_C$ set (which can be called by other components) and calls in the USES$_C$ set (which are those needed by C and may belong to other components).

Similar to nesC and TinyOS, TinyGALS defines components using an interface definition and an implementation. For example, a possible interface description of a component DownSample is shown below, where the interface has two methods in the ACCEPTS set and one method in the USES set.

```
COMPONENT DownSample
ACCEPTS {
    void init (void);
    void fire (int in);
};
USES {
    void fireOut (int out);
};
```

The following code sequence shows the corresponding implementation for the Down-Sample component, where _active is an internal boolean variable that ensures that for every other fire() method called, the component will call the fireOut() method with the same integer argument.

```
void init () {
    _active = true;
}
void fire (int in) {
    if (_active) {
        CALL_COMMAND (fireOut) (in);
        _active = false;
    } else {
        _active = true;
    }
}
```

TinyGALS modules consist of one or more components. A module M is a 6-tuple $M = $ (COMPONENTS$_M$, INIT$_M$, INPORTS$_M$, OUTPORTS$_M$, PARAMETERS$_M$, LINKS$_M$), where COMPONENTS$_M$ is the set of components of M, INIT$_M$ is a list of methods of M's components, INPORTS$_M$ and OUTPORTS$_M$ specify the inputs and outputs of the module, PARAMETERS$_M$ is a set of variables external to the components, and LINKS$_M$ specifies the relationships between the method call interfaces and the inputs and outputs of the module. Modules are further connected to each other to form a complete TinyGALS system, where a system is a 5-tuple $S = $ (MODULES$_S$, GLOBALS$_S$, VAR_MAPS$_S$, CONNECTIONS$_S$, START$_S$). The set of modules is described in MODULES$_S$, global variables are described in GLOBAL$_S$, a set of mappings (each of which maps a global variable to a parameter of a module in MODULES$_S$) is contained in VAR_MAPS$_S$, CONNECTIONS$_S$ is a list of the connections between module output ports and input ports, and START$_S$ is the name of an input port of exactly one module, which is used as a starting point for the execution of the system.

The highly structured architecture of TinyGALS can be exploited to automate the generation of scheduling and event handling code, freeing software developers from writing error-prone concurrency control code. Code generation tools can automatically produce all of the necessary code for component links and module connections, system initialization, start of execution, intermodule communication, and global variables reads and writes. Further, through the use of message passing, modules in TinyGALS become decoupled from each other, therefore facilitating their independent development. Each message passed will trigger the scheduler and activate a receiving module. However, this may become quickly inefficient if there is global state that must be updated frequently. Therefore, TinyGALS provides another mechanism, called TinyGUYS (Guarded Yet Synchronous) variables, where modules may read global variables synchronously (without delay), but writes to the variables are asynchronous in the sense that all writes are buffered. The buffer is of size 1, that is, the last module that writes to a variable wins. TinyGUYS variables are updated by the schedule only when it is safe to do so, for example, after one module finishes and before the scheduler triggers the next module.

12.2.3   Sensor Network Application Construction Kit

The Sensor Network Application Construction Kit (SNACK) is a configuration language, component and service library, and compiler for the development of sensor network applications (Greenstein *et al.* 2004). SNACK's goal is to provide *smart libraries* that can be combined to form sensor network applications, while, on one hand, simplifying the development process and, on the other, not losing control over efficiency. For example, to program a sensor node to periodically take temperature and light measurements and forward the sensor data to some sink, it should be possible to write a simple code sequence such as:

```
SenseTemp -> [collect] RoutingTree;
SenseLight -> [collect] RoutingTree;
```

The following examples shows the syntax of SNACK code:

```
service Service {
    src :: MsgSrc;
    src [send:MsgRcv] -> filter :: MsgFilter -> [send] Network;
    in [send:MsgRcv] -> filter;
}
```

Here, $n :: T$ declares an instance named n of a component type T, that is, an instance is effectively an object of the given type. Further, $n[i : \tau]$ indicates an output interface on component n with name i and interface type τ (similarly, $[i : \tau]n$ refers to an input interface). A component *provides* its input interfaces and *uses* its output interfaces.

The SNACK library of components and services contains a variety of components for sensing, aggregation, transmission, routing, and data processing. For example, the messaging architecture of SNACK supports several core components, including Network (which receives messages from and sends messages to the TinyOS radio stack), MsgSink (which ends inbound call chains and destroys buffers it receives), and MsgSrc (which periodically generates empty SNACK messages and passes them on via an outbound interface). The SNACK Timing system has two core components: TimeSrc, which generates a timestamp signal, emitted over its signal interface at a specified minimum rate, and TimeSink, which consumes that signal. `Storage` in SNACK is implemented by components such as Node-Store64M, which implements an associative array of eight-byte values keyed by node ID. Finally, the SNACK `Service` library contains a variety of services, that is, combinations of primitive components. For example, the `RoutingTree` service implements a tree designed to send data up to some root.

12.2.4 Thread-Based Model

The thread-based paradigm is popular in many computing systems and it has recently also found its way into sensor networks. In traditional event-based systems, event handlers are executed in response to events, and these handlers (tasks) run to completion without interruption from other tasks. The main advantage of the thread-based approach is that multiple tasks can make progress in their execution without the concern that a task may block other tasks (or be blocked by other tasks) indefinitely. For example, a task scheduler can execute a task for a certain amount of time, then preempt this task in order to execute another task. This *time-slicing* approach simplifies the programming of sensor systems, but also comes at the cost of increased operating system complexity.

An example of a thread-based operating system for sensor networks is the MANTIS (MultimodAl system for NeTworks of In-situ wireless Sensors) OS, which occupies less than 500 bytes of RAM and about 14 kbytes of flash memory (Bhatti *et al.* 2005). For example, the ATMega128 sensor nodes have 4 kbytes of RAM and 128 kbytes of flash storage, that is, MANTIS OS leaves sufficient space for multiple sensor application threads. Besides memory efficiency, MANTIS OS also aims for energy efficiency by switching the microcontroller to a low-power sleep state after all active threads have called the operating system's *sleep()* function.

The goal of the TinyThread (McCartney and Sridhar 2006) library is to add support for multithreaded programming to sensor networks based on TinyOS and nesC. TinyThread enables procedural programming of sensor nodes and includes a suite of interfaces that provide several blocking I/O operations and synchronization primitives that make multithreaded programming safe and easy.

Protothreads (Dunkels *et al.* 2005) are a very lightweight stackless type of threads. Instead of using a stack for each protothread, all protothreads run on the same stack and context switching is done by stack rewinding. A limitation of protothreads is that contents of variables must be explicitly saved before calling a blocking wait, since variables with function-local scope that are automatically allocated on the stack are not saved across such wait calls.

Finally, Y-Threads (Nitta *et al.* 2006) is another lightweight threading model that provides preemptive multithreading. Application developers identify the preemptable and non-preemptable parts of a program. All threads share a common stack for their nonblocking computations, while each thread has its own stack for blocking calls. The key concept behind this approach is that the blocking portions of a program require only small amounts of stack, therefore achieving better memory utilization compared to other preemptive multithreading approaches.

12.3   Macroprogramming

Macroprogramming refers to a development approach where the focus is not on individual sensor nodes, but on the programming of groups of sensor nodes, including approaches that treat an entire network as a single entity. This section illustrates different approaches to macroprogramming.

12.3.1   Abstract Regions

In-network processing is often performed to address the bandwidth and energy limitations of WSNs. However, decomposing data collection tasks into parallel programs with local communication among sensor nodes can be a challenging problem. Therefore, the goal of *abstract regions* (Welsh and Mainland 2004) is to provide higher-level programming interfaces that hide complex details from the developer, while still being flexible enough to support the implementation of efficient algorithms.

Many sensor applications are often characterized by group-level cooperation, that is, a group of nodes work together to sample, process, and communicate sensor data. Therefore, abstract regions are a communication abstraction intended to simplify the development process by providing a region-based collective communication interface. An abstract region defines the neighborhood relationship between a node and other nodes in the network, for example, as expressed by *"the set of nodes within distance d"*. Specifically, the type of definition of an abstract region will depend on the type of application. Examples of implementations of abstract regions include `N-radio hop` (nodes within N radio hops), `k-nearest neighbor` (k nearest nodes within N radio hops), and `spanning tree` (a spanning tree rooted at a single node, used for aggregating data over the entire network). For example, for regions defined using hop distances, discovery of region members can be achieved using periodic broadcasts (advertisements). Data among region members can be shared using either a "push" (broadcasting updates to neighboring nodes) or "pull" (issue a fetch message to the corresponding node) approach. Reduction is another programming abstraction, which takes a shared variable key and an associative operator (e.g., sum, max, or min) and reduces the shared variable across nodes in the region. In abstract regions based on hop distances, reduction involves collecting shared variable values locally, combining them with the reduction operator, and storing the result in a new shared variable.

12.3.2   EnviroTrack

The EnviroTrack (Abdelzaher *et al.* 2004) object-based middleware library is a programming abstraction geared toward target-tracking sensor applications. Its goal is to free the developer from the details of interobject communication, object mobility, and the maintenance of tracking objects and their state. Similar to abstract regions, EnviroTrack uses the concept of groups. However, instead of concrete descriptions of the shape or size of a group, groups in EnviroTrack are formed by sensors which detect certain user-defined entities in the physical environment, with one group formed around each entity. Groups are identified by *context labels*, which can be thought of as logical addresses that follow the external tracked entity around in the physical environment. Further, objects can be attached to context labels to perform context-specific operations. These *tracking objects* are executed on the sensor group of the context label.

The type of context label depends on the entity being tracked (e.g., a context label of *car* is created wherever a car is tracked). A programmer must provide several pieces of information to declare a context label of some type *e*. First, a function $sense_e()$ describes the sensory signature identifying the tracked environmental target, for example, for a car-tracking application, $sense_e()$ might be a function of magnetometer and motion sensor readings. Whenever the EnviroTrack middleware detects a target, it creates a sensor group around the target. This function is also used to maintain group membership, that is, all nodes that sense the given target (i.e., $sense_e()$ is true) are group members. Next, a programmer declares an environmental state shared by all objects attached to a context label by defining an aggregation function $state_e()$ that acts on the readings of all sensors for which $sense_e()$ is true. Aggregation is performed locally by a sensor node that acts as group leader. The EnviroTrack library contains a variety of distributed aggregation functions such as addition, averaging, and median computation. Finally, the programmer specifies which objects are to be attached to a context label.

12.3.3 Database Approaches

Another commonly used abstraction for sensor network programming is to treat a WSN as a distributed database that can be queried (e.g., using SQL-like queries) to obtain sensor data. A representative example of a distributed query processor for sensor nodes is TinyDB (Madden *et al.* 2005). Here, the network is represented logically as a table (called *sensors*) that has one row per node per instant in time. Each column in this table corresponds to a type of sensor reading such as light, temperature, pressure, etc. A new record in this virtual table (i.e., a new row) is added only when a sensor is queried and this new information is usually stored for a short period of time only. Queries in TinyDB are very much like any other SQL-based database, that is, they use clauses such as SELECT, FROM, WHERE, and GROUP BY to build queries. For example, the following query specifies that each device should report its own identifier (nodeid), light reading, and temperature reading once per second for 10 seconds:

```
SELECT nodeid, light, temp
  FROM sensors
  SAMPLE PERIOD 1s FOR 10s
```

As a result of this query, nodes initiate data collection at the beginning of each epoch (as specified in the SAMPLE PERIOD clause) and the results of such a query are streamed to the root of the network.

TinyDB also supports grouped aggregation queries, that is, as data from an aggregation query flows up the tree, it is aggregated in-network according to an aggregation function and value-based partitioning specified in the query. For example, imagine a user who wishes to use microphone-equipped sensor nodes to monitor the occupancy of a room on a particular floor of a building. Assuming that rooms have multiple sensors, the goal is to look for rooms where the average volume is over a certain threshold. A query for this sensing request could be expressed as:

```
SELECT AVG(volume), room FROM sensors
    WHERE floor = 6
```

```
GROUP BY room
HAVING AVG(volume) > threshold
SAMPLE PERIOD 30s
```

Every 30 s, this query reports all rooms where the volume is above the specified threshold. Each sensor periodically obtains new sensor readings, applies the criteria from the SELECT criteria, and if the criteria are satisfied, the sensor data is forwarded to the sensor's parent node of the tree leading to the root. The parent node listens to the records coming from its children, aggregates its own sensor reading with the records obtained from its children, and forwards the newly obtained aggregate onward to its own parent. This process is continued until the query result has reached the root of the tree.

The main data-processing functions supported by TinyDB are selection and aggregation. A similar approach is taken by Cougar (Bonnet *et al.* 2000), which also represents sensor data as a relational table. Both TinyDB and Cougar focus on resource-efficiency through the use of in-network aggregation. A more sophisticated database approach is taken by SINA (Srisathapornphat *et al.* 2000), which models a sensor network as a collection of distributed objects. SINA supports more complex sensor node collaborations by embedding more powerful SQTL (Sensor Querying and Tasking Language) scripts in an SQL query. The focus of the MiLAN (Heinzelman *et al.* 2004) approach is on Quality-of-Service (QoS), that is, sensor network applications can specify their QoS needs, which the network attempts to meet while maximizing the network lifetime.

A disadvantage of database models for sensor networks is that all nodes are assumed to be homogeneous, for example, the sensors table in TinyDB is structured the same for all sensor nodes. The focus of database systems is on relatively simple data collection applications, targeting resource-constrained environments such as motes.

12.4 Dynamic Reprogramming

It has further become increasingly necessary to support the programming and reprogramming of sensor networks after deployment. It is therefore necessary to provide mechanisms to disseminate code to potentially hundreds or thousands of resource-constrained sensor nodes. One possible approach to address this challenge is to use *virtual machines*. For example, Maté (Levis and Culler 2002) is a small virtual machine implemented on top of TinyOS. A sequence of 24 instructions (each of which is a single byte long) is called a *capsule*, which fits into a single TinyOS packet. Every code capsule also includes type and version information. Maté distinguishes four types of capsules: *message send* capsules, *message receive* capsules, *timer* capsules, and *subroutine* capsules. Programs execute in response to an event, that is, a timer firing, a packet being received, or a packet being sent. Each of these events has a capsule and an execution context. Maté jumps to the first instruction of the capsule and executes until it reaches the **halt** instruction. When a subroutine is called, the return address is pushed onto a return address stack and control jumps to the first instruction of the subroutine. Upon return from the subroutine, an address is taken from the top of the stack and Maté continues at the appropriate instruction.

Trickle (Levis *et al.* 2004) is a controlled flooding protocol for disseminating small pieces of code to all nodes in a sensor network. It uses metadata to describe code, allowing a node to determine if it needs a code update by comparing two different pieces of metadata. A node uses broadcasts to exchange metadata with its neighbors, that is, time is broken into intervals and at a random time during an interval, it broadcasts its metadata if it has not already heard the same metadata from several other nodes. Whenever a node hears another node broadcasting outdated metadata, it broadcasts its own code, thereby giving the outdated node a chance to update its code. Similarly, if a node overhears another node broadcasting newer metadata than its own, it broadcasts its own metadata, thereby triggering its neighbor with the newer code to broadcast it.

Melete (Yu *et al.* 2006) is an extension to Maté in that it supports multiple concurrent applications. It is also an extension to Trickle in that it supports selective dissemination by limiting the dissemination range. That is, code is forwarded only within a *forwarding region*, which covers the desired destination of the code update.

Deluge (Hui and Culler 2004) is another tool to reprogram wireless sensor nodes remotely. Similar to Trickle, Deluge occasionally advertises the most recent code version using broadcast messages. If a node receives an update from a node with an older code version, it responds with its own code version, giving the outdated node an opportunity to detect that its version is old and to request the newer code. To reduce contention, Deluge eliminates redundant advertisements and request messages. It also provides robustness by (i) using a three-phase handshake protocol to ensure that only bidirectional links are used for code updates and (ii) allowing a node to search for a new neighbor to request code if it has not completely received the code after k requests. Finally, Deluge dynamically adjusts the rate of advertisements to allow quick propagation when needed while consuming few resources in the steady state.

The goal of Pump Slowly, Fetch Quickly (PSFQ) (Wan and Campbell 2005) is to distribute data (e.g., pieces of code) from a single source to a number of destinations. The basic concept behind PSFQ is to slowly pace the propagation of packets (pump slowly) and to aggressively fetch lost packets (fetch quickly), where lost packets are detected through out of order packet receptions. Nodes will not relay received packets out of order, that is, a node that detects a lost packet will refrain from forwarding packets received out of order until the lost packet has been recovered. This approach prevents loss events from propagating downstream and allows nodes to recover lost packets from immediate neighbors (because at least one neighbor must have a copy of the missing packets). This *localized recovery* process reduces recovery costs by limiting recovery to single-hop transmissions and by avoiding that a single lost packet causes multiple retransmission requests in the network.

The Push Aggressively with Lazy Error Recovery (PALER) protocol (Miller and Poellabauer 2008) is based on the observation that pushing data downstream and recovering lost packets simultaneously leads to excessive contention and collisions. As a consequence, PALER eliminates the in-order reception requirement and instead pushes all data aggressively to the destinations without delaying data propagations due to lost packets. All nodes in the network keep a list of missing packets and only after the broadcast phase has completed do nodes with missing packets issue retransmission requests to their neighbors. Similar to PSFQ, these retransmission requests do not have to travel multiple hops since a retransmission request will again be handled by an immediate neighbor. If the immediate neighbor cannot retransmit the packet, it must also mean that this neighbor did not receive a copy of the packet and will therefore have issued its own retransmission request to its neighbors. Once this neighbor receives a copy of the missing packet, it can respond to its received retransmission requests. This lazy error recovery approach can significantly reduce collisions, thereby improving both the latency and energy overheads of code distributions in sensor networks.

12.5 Sensor Network Simulators

Many sensor networks consist of hundreds or thousands of nodes distributed over large geographic areas. Further, even with inexpensive hardware components, building large networks of sensors may be prohibitively expensive. Therefore, it is often impractical to implement novel algorithms and protocols on actual networks. As a consequence, simulation tools are particularly important for development and research on new sensor network applications, functionality, and protocols. However, the right choice of simulator is a critical task, since network characteristics can vary widely between different types of sensor networks. Further, the many complex and dynamic relationships and parameters of a WSN make it difficult to obtain realistic models. In general, every simulator typically incorporates the following components: models describing the characteristics of the sensor nodes, a selection of different communication models, models for the physical environment, and tools for collecting and analyzing statistics and for visualization of the collected data and sensor node behavior. This section provides an overview of a few commonly used and representative simulation tools and environments for sensor networks.

12.5.1 Network Simulator Tools and Environments

12.5.1.1 Sensor Network Extensions for ns-2

The *network simulator* (typically called *ns-2*, where the number indicates the current version) is a widely used discrete event simulator targeted at networking research in general. It was written in a combination of C++ and an object-oriented dialect of Tcl, called OTcl. One reason for the popularity of ns-2 is its extensibility. Over time, many enhancements and extensions were developed, for example, to provide support for wireless networks and mobile ad hoc networks. Similarly, a variety of extensions for sensor networks have been created. For example, one such extension adds the notion of a *phenomenon* to a sensor network simulation. A phenomenon describes a physical event such as a chemical cloud or moving vehicle that could be monitored by nearby sensor nodes (Downard 2004). That is, a phenomenon then serves as a trigger for sensor network applications and network activity. The model uses broadcast packets transmitted through a designated channel to represent a phenomenon, that is, the range of phenomena is the set of nodes that can receive these broadcasts. Broadcast packets are generated using the *PHENOM routing protocol*, which emits packets with a certain configurable pulse rate and whose arrival at a sensor node triggers a receive event that is passed to that node's sensor application. Other extensions that have been developed over time include routing protocol implementations, extensions that simulate the type of packets used by sensor applications, and models of multihomed nodes.

12.5.1.2 GloMoSim and QualNet

GloMoSim (Zeng *et al.* 1998) is a simulation tool based on the PARSEC simulation environment (Bagrodia *et al.* 1998). PARallel Simulation Environment for Complex (PARSEC) systems is a C-based simulation language, which is used to represent a set of objects in the physical system as logical processes and interactions among these objects as time-stamped message exchanges. GloMoSim supports a variety of models at different protocol layers, for example, CSMA and MACAW (MAC layer), flooding and DSR (network layer), and TCP and UDP (transport layer). In addition, it supports different node mobility models, for example, the *random waypoint* model (i.e., a node chooses a random destination within the simulated area and moves toward this destination with a specified speed) and the *random drunken* model (i.e., a node periodically moves to a position chosen randomly from its immediate neighboring positions). While GloMoSim is intended for academic use only, a commercialized version of GloMoSim, called QualNet, is produced by Scalable Network Technologies, Inc.

12.5.1.3   JiST/SWANS

A discrete event simulation tool based on Java is JiST (Barr *et al.* 2004), which stands for Java in Simulation Time. The key motivation behind JiST is to create discrete event simulations that can execute efficiently and transparently. Efficiency refers to the ability to execute a given simulation program in parallel, while dynamically optimizing the configuration of the simulation across the available computational resources. Transparency refers to the ability to transform simulation programs automatically to run with simulation time semantics, that is, simulations are instrumented such that no programmer intervention or calls to specialized libraries are needed to support various concurrency, consistency, and reconfiguration protocols.

One of the primary motivations for JiST was to support simulations of ad hoc networks and Scalable Wireless Ad hoc Network Simulator (SWANS) is a simulator built on top of the JiST engine. SWANS is a collection of independent software components that can be aggregated to form complete wireless simulations. The capabilities of JiST/SWANS are comparable to ns-2 and GloMoSim, but it is able to simulate much larger networks (Barr *et al.* 2004).

12.5.1.4   OMNeT++

The Objective Modular Network Testbed (OMNeT++) discrete event simulation environment (Varga and Hornig 2008) is a tool used for the simulations of communication networks, multiprocessors, and various distributed systems. It is an open-source simulator based on C++ that was designed for the simulation of large systems and networks. A model in OMNeT++ consists of modules that communicate with each other using message passing. *Simple modules* can be grouped together to form more complex *compound modules*. A user defines the structure of a module (i.e., the modules and their interconnection) using OMNeT++'s topology description language NED. Further, the OMNeT++ framework includes a graphical editor that can be used to edit network topologies either graphically or in NED source view. Because of its clean design, simulation development is straightforward. However, compared to other tools, its biggest shortcoming is the lack of available protocol models.

12.5.1.5   TOSSIM

A simulator for TinyOS-based wireless sensor networks is TOSSIM (Levis *et al.* 2003). It generates discrete event simulations directly from TinyOS components, therefore running the same code that runs on sensor nodes. TOSSIM replaces low-level components such that hardware interrupts are translated into events in the simulation and the simulator event queue delivers the interrupts that drive the execution of a TinyOS application. Apart from this, TinyOS code runs in the simulator unmodified. TOSSIM works at the bit level, that is, an event is generated for each transmitted or received bit (instead of an entire packet). This allows for experimentations with low-level protocols in addition to higher-level protocols or applications. Similar to most other tools, TOSSIM comes with a visualization tool, which is called TinyViz. TOSSIM scales to thousands of sensor nodes and its advantages include its scalability and extensibility. However, it does not include energy profiling and its use is limited to systems based on TinyOS.

12.5.1.6   EmStar

EmStar (Girod *et al.* 2004) is targeted at high capability nodes called *microservers*, that is, those nodes in a hierarchical sensor network structure that run more complex software than ordinary sensing devices (e.g., motes). EmStar consists of a Linux microkernel extension, libraries, services, and several tools. EmSim operates many virtual nodes in parallel in a simulation that models radio and sensor channels. EmCee runs the EmSim core and is an interface to real low-power radios instead of using a modeled channel. Finally, EmView is a graphical visualizer for EmStar systems.

12.5.1.7   Avrora

Avrora (Titzer and Palsberg 2005) is a flexible simulator framework implemented in Java. Each node is implemented as its own thread and code is executed in an instruction-by-instruction fashion. A key component of Avrora is its implementation of an event queue. Many energy-conscious nodes tend to sleep for large periods of time, for example, using low-power sleep modes where no instructions are executed and the energy consumption is dramatically reduced. The event queue in Avrora takes advantage of this approach to boost the performance of the simulator. That is, when a node sleeps, only a time-triggered event that causes an interrupt can wake up the node. Such an event is inserted into the event queue of the node to be woken up at a certain time in the future. Only such an event, when at the head of the event queue, can affect the simulation when a node sleeps. That is, the simulator can process events in the queue *in order* until one of them triggers a hardware interrupt, which re-awakes the node. In summary, Avrora is a fast and highly scalable simulator that can simulate program executions down to the level of individual clock cycles.

Exercises

12.1 Describe the difference between node-centric and application-centric programming.

12.2 Explain the difference between *provides* and *uses* interfaces in nesC.

12.3 What options does nesC provide to developers to prevent race conditions?

12.4 A common strategy to ensure atomicity is to disable interrupts in an operating system as long as critical operations are being executed. What is the danger of disabling interrupts?

12.5 What are the main advantages and disadvantages of thread-based programming models?

12.6 This chapter introduced several macroprogramming models. Contrast how these different models are able to address multiple (or all) sensor nodes simultaneously.

12.7 Why is it necessary to provide the opportunity to dynamically reprogram a sensor network? What is challenging in distributing a new program to all sensor nodes in the network?

References

Abdelzaher, T., Blum, B., Cao, Q., Chen, Y., Evans D, George, J., George, S., Gu, L., He, T., Krishnamurthy, S., Luo, L., Son, S., Stankovic, J., Stoleru, R., and Wood, A. (2004) EnviroTrack: Towards an environmental computing paradigm for distributed sensor networks. *Proc. of the 24th IEEE International Conference on Distributed Computing Systems (ICDCS), Hachioji, Tokyo, Japan*.

Bagrodia, R., Meyer, R., Takai, M., Chen, Y., Zeng, X., Martin, J., and Song, H.Y. (1998) PARSEC: A parallel simulation environment for complex systems. *IEEE Computer* **31** (10), 77–85.

Barr, R., Haas, Z.J., and van Renesse, R. (2004) JiST: Embedding simulation time into a virtual machine. *Proc. of the 5th EuroSim Congress on Modelling and Simulation, Marne-la-Vallée, France*.

Bhatti, S., Carlson, J., Dai, H., Deng, J., Rose, J., Sheth, A., Shucker, B., Gruenwald, C., Torgerson, A., and Han, R. (2005) MANTIS OS: An embedded multi-threaded operating system for wireless micro sensor platforms. *ACM/Kluwer Mobile Networks and Applications (MONET), Special Issue on Wireless Sensor Networks* **10** (4), 563–579.

Bonnet, P., Gehrke, J., and Seshadri, P. (2000) Querying the physical world. *IEEE Personal Communications* **7** (5), 10–15.

Cheong, E., Liebman, J., Liu, J., and Zhao, F. (2003) TinyGALS: A programming model for event-driven embedded systems. *Proc. of the 18th Annual ACM Symposium on Applied Computing, Melbourne, FL*.

Downard, I. (2004) *Simulating sensor networks in NS2*. Technical Report, NRL/FR/(5522)(0410)073, Naval Research Laboratory, Washington, DC.

Dunkels, A., Schmidt, O., and Voigt, T. (2005) Using prototthreads for sensor node programming. *Proc. of the REALWSN Workshop on Real-World Wireless Sensor Networks, Stockholm, Sweden*.

Gay, D., Levis, P., Behren, R., Welsh, M., Brewer, E., and Culler, D. (2003) The nesC language: A holistic approach to networked embedded systems. *Proc. of the ACM SIGPLAN Conference on Programming Language Design and Implementation (PLDI), San Diego, CA*.

Girod, L., Elson, J., Cerpa, A., Stathopoulos, T., Ramanathan, N., and Estrin, D. (2004) EmStar: A software environment for developing and deploying wireless sensor networks. *Proc. of the USENIX Annual Technical Conference, Boston, MA*.

Greenstein, B., Kohler, E., and Estrin, D. (2004) A sensor network application construction kit (SNACK). *Proc. of the 2nd International Conference on Embedded Networked Sensor Systems (SenSys), Baltimore, MD*.

Heinzelman, W.B., Murphy, A.L., Carvalho, H.S., and Perillo, M.A. (2004) Middleware to support sensor network applications. *IEEE Network* **18** (1), 6–14.

Hui, J.W., and Culler, D. (2004) The dynamic behavior of a data dissemination protocol for network programming at scale. *Proc. of the 2nd ACM Conference on Embedded Networked Sensor Systems (SenSys), Baltimore, MD*.

Levis, P., and Culler, D. (2002) Maté: A tiny virtual machine for sensor networks. *Proc. of the 10th International Conference on Architectural Support for Programming Languages and Operating Systems (ASPLOS), San Jose, CA*.

Levis, P., Lee, N., Welsh, M., and Culler, D. (2003) TOSSIM: Accurate and scalable simulation of entire TinyOS applications. *Proc. of the 1st ACM Conference on Embedded Networked Sensor Systems (SenSys), Los Angeles, CA.*

Levis, P., Patel, N., Culler, D., and Shenker, S. (2004) Trickle: A self-regulating algorithm for code propagation and maintenance in wireless sensor networks. *Proc. of the 1st Symposium on Networked Systems Design and Implementation, San Francisco, CA.*

Madden, S.R., Franklin, M.J., Hellerstein, J.M., and Hong, W. (2005) TinyDB: An acquisitional query processing system for sensor networks. *ACM Transactions on Database Systems* **30** (1), 122–173.

McCartney, W.P., and Sridhar, N. (2006) Abstractions for safe concurrent programming in networked embedded systems. *Proc. of the 4th International Conference on Embedded Networked Sensor Systems (SenSys), Boulder, CO.*

Miller, C., and Poellabauer, C. (2008) PALER: A reliable transport protocol for code distribution in large sensor networks. *Proc. of the 5th IEEE Communications Society Conference on Sensor, Mesh and Ad Hoc Communications and Networks (SECON), San Francisco, CA.*

Nitta, C., Pandey, R., and Ramin, Y. (2006) Y-Threads: Supporting concurrency in wireless sensor networks. *Proc. of the 2nd International Conference on Distributed Computing in Sensor Systems, San Francisco, CA.*

Srisathapornphat, C., Jaikaeo, C., and Shen, C.C. (2000) Sensor information networking architecture. *Proc. of the 3rd International Workshop on Parallel Processing, Toronto, Canada.*

Sugihara, R. and Gupta, R.K. (2008) Programming models for sensor networks: A survey. *ACM Transactions on Sensor Networks* **4** (2), 1–29.

Titzer, B.L., and Palsberg, J. (2005) Nonintrusive precision instrumentation of microcontroller software. *Proc. of the ACM SIGPLAN/SIGBED Conference on Languages, Compilers, and Tools for Embedded Systems, Chicago, IL.*

Varga, A., and Hornig, R. (2008) An overview of the OMNeT++ simulation environment. *Proc. of the 1st International Conference on Simulation Tools and Techniques for Communications, Networks and Systems, Marseilles, France.*

Wan, A.K.L., and Campbell, C.Y. (2005) Pump Slowly, Fetch Quickly (PSFQ): A reliable transport protocol for sensor networks. *IEEE Journal on Selected Areas in Communications* **23** (4), 862–872.

Welsh, M., and Mainland, G. (2004) Programming sensor networks using abstract regions. *Proc. of the 1st Symposium on Networked Systems Design and Implementation, San Francisco, CA.*

Yu, Y., Rittle, L.J., Bhandari, V., and LeBrun, J.B. (2006) Supporting concurrent applications in wireless sensor networks. *Proc. of the 4th International Conference on Embedded Networked Sensor Systems (SenSys), Boulder, CO.*

Zeng, X., Bagrodia, R., and Gerla, M. (1998) GloMoSim: A library for parallel simulation of large-scale wireless networks. *Proc. of the 12th Workshop on Parallel and Distributed Simulation, Banff, Alberta, Canada.*

Index